*Charles
his Book autumn 1836*

*John Garst, his
Book which he Bought
from Charles S Kirkwood
May 28eth 1844*

(These inscriptions were taken from the edition
used in the preparation of this reprint.)

©2004 Algrove Publishing Limited
ALL RIGHTS RESERVED.
No part of this book may be reproduced in any form, including photocopying, without permission in writing from the publishers, except by a reviewer who may quote brief passages in a magazine or newspaper or on radio or television.

Algrove Publishing Limited
36 Mill Street, P.O. Box 1238
Almonte, Ontario, Canada K0A 1A0

Telephone: (613) 256-0350
Fax: (613) 256-0360
Email: sales@algrove.com

Cover photo of the Glade Creek grist mill is by Allen Davis.
Copyright 2004 Allen Davis - Allen Davis Photography.
This mill is located in Babcock State Park near Clifftop, West Virginia. Of special interest, the mill was created by combining parts and pieces from three other mills in the state. The basic structure of the mill came from the Stoney Creek grist mill, which dates back to 1890. The overshot water wheel was salvaged from the Spring Run grist mill (near Petersburg) after a fire destroyed the rest of the mill. Other parts for the mill came from the Onego grist mill near Seneca Rocks. The mill was completed in 1976.

Fully operable, the mill is a re-creation of an earlier Glade Creek mill. In 1900, some 500 such mills were operating in West Virginia.

Text is used with permission: Babcock State Park, West Virginia.

National Library of Canada Cataloguing in Publication

Evans, Oliver, 1755-1819.
    The young mill-wright and miller's guide / Oliver Evans.

(Classic reprint series)
Reprint of 8th ed., originally published: Philadelphia : Carey, Lea &
    Blanchard, 1834.
ISBN 1-894572-90-4

    1. Flour mills. I. Title. II. Series: Classic reprint series (Almonte, Ont.)

TS2145.E83 2004   664'.72   C2003-905706-2

Printed in Canada
#1-7-04

Publisher's Note

This book is the classic work in its field. It has been a standard reference for nearly two centuries now.

Modern readers will find the writing style a bit disconcerting in places because it appears that the writer is debating, not just presenting information. This is a reflection of the fact that the book was written in a period when philosophical societies thrived and theories in all fields were openly debated. Writers often embedded elements of these debates in their written work, continuing oral argument in written form, even down to quoting opposing arguments.

It is all part of the charm of this unique book.

> Leonard G. Lee
> Publisher
> July, 2004
> Almonte, Ontario

How We Make Our Books - *You may not have noticed, but this book is quite different from other softcover books you might own. The vast majority of paperbacks, whether mass-market or the more expensive trade paperbacks, have the pages sheared and notched at the spine so that they may be glued together. The paper itself is often of newsprint quality. Over time, the paper will brown and the spine will crack if flexed. Eventually the pages fall out.*

All of our softcover books, like our hardcover books, have sewn bindings. The pages are sewn in signatures of sixteen or thirty-two pages and these signatures are then sewn to each other. They are also glued at the back but the glue is used primarily to hold the cover on, not to hold the pages together.

We also use only acid-free paper in our books. This paper does not yellow over time. A century from now, this book will have paper of its original color and an intact binding, unless it has been exposed to fire, water, or other catastrophe.

There is one more thing you will note about this book as you read it; it opens easily and does not require constant hand pressure to keep it open. In all but the smallest sizes, all our books will also lie open on a table, something that a book bound only with glue will never do unless you have broken its spine.

The cost of these extras is well below their value and while we do not expect a medal for incorporating them, we did want you to notice them.

A Receipt for tempering mill Stone & Burr picks
Take 2 ozes of Salt peter
" 1 oze of alum
" half pint of Salt
& desolve it in
1½ gallons of rain water

Easter Town
1852

THE

YOUNG MILL=WRIGHT

AND

MILLER'S GUIDE;

ILLUSTRATED BY

TWENTY-EIGHT DESCRIPTIVE PLATES.

BY OLIVER EVANS.

THE EIGHTH EDITION,

WITH ADDITIONS AND CORRECTIONS;

BY

THOMAS P. JONES,

MEMBER OF THE AMERICAN PHILOSOPHICAL SOCIETY, AND CORRESPONDENT OF THE POLYTECHNIC SOCIETY OF PARIS; EDITOR OF THE JOURNAL OF THE FRANKLIN INSTITUTE OF THE STATE OF PENNSYLVANIA, AND LATE

PROFESSOR OF MECHANICS,

IN THAT INSTITUTION:

AND

A DESCRIPTION OF AN IMPROVED MERCHANT FLOUR-MILL,

WITH ENGRAVINGS.

BY C. & O. EVANS, ENGINEERS.

PHILADELPHIA
1834.

Algrove Publishing
Classic Reprint Series

ENTERED, according to the Act of Congress, in the year 1834, by CAREY, LEA & BLANCHARD, in the District Court for the Eastern District of Pennsylvania.

GRIGGS & CO., PRINTERS.

PREFACE.

There are few men whose mechanical inventions have contributed so much to the good of our country as those of Oliver Evans; for my own part, I could name but two, and they are Whitney and Fulton. There have, it is true, within the last thirty years, been a great number of original machines invented, and a great many improvements made on those for which we are indebted to other countries, that do great credit to American genius, and evince a peculiar aptitude to excel in mechanical contrivances: but few, however, of these inventions could be denominated national, although they have been of high importance in the various arts to which they are applied.

The improvements in the flour mill, like the invention of the cotton gin, apply to one of the great staples of our country; and, although nearly forty years have elapsed since Mr. Evans first made his improvements known to the world in the present work, the general superiority of American mills to those even of Great Britain, is still a subject of remark by intelligent travellers. Mr. Evans, however, experienced the fate of most other meritorious inventors; the combined powers of prejudice and of interest deprived him of all benefit from his labours, and, like Whitney, he was compelled to depend upon other pursuits for the means of establishing himself in the world. His reward, as an inventor, was a long continued course of ruinous litigation, and the eventual success of the powerful phalanx which was in league against him.

PREFACE.

It is not the intention of the Editor to pronounce a panegyric on, or to write the history of, Oliver Evans; but his sense of justice, and a confident hope that, in the history of American inventions, posterity may accord to him the place which he really merits, have called forth the preceding remarks.

Mr. Evans made no pretensions to literature; he considered himself, as he really was, a plain, practical man; and the main object of his writing this work was to introduce his inventions to public notice; it has, however, been extensively useful to the mill-wright and the miller as a general treatise, and an edition of it has been published in the French language. The present Editor was employed to revise the work, a few years ago, and a new edition being again called for, the same task has been again assigned to him by the publishers. It has not been thought proper to make any such alterations in it as should destroy its identity; as it would, in that case, be essentially a new work, to which it would not be proper to attach the name of Mr. Evans, as the author; encouraged, however, by the general approval of the alterations and additions formerly made, the Editor has thought himself justified in pursuing, in the present instance, the same course, to a greater extent; and although some theoretical views are interwoven in the general texture of the work, which may be disputable, these can detract but little from its practical utility; and it is hoped that the farther changes which have been made in the phraseology, as well as in some other points, will be found to add to its worth in this respect.

THOMAS P. JONES.

City of Washington, April, 1834.

CONTENTS.

PART I.

Principles of Mechanics and Hydraulics.

ARTICLE	PAGE
1. Preliminary remarks,	9
2. On the essential properties of bodies,	10
3. Axioms, or laws, of motion and rest,	13
4. On absolute and relative motion,	14
5. On momentum,	15
6. On power, or force, and on the motive powers,	16
7. On the effect of collision, or impact,	16
8. On compound motion,	18
9. Of non-elasticity, and of fluidity in impinging bodies,	19
10. Of falling bodies,	21
11. Of bodies descending inclined planes and curved surfaces,	25
12. Of the motion of projectiles,	26
13. Of circular motion and central forces,	27
14. Of the centres of magnitude, motion, and gravity,	30
15. Of the mechanical powers,	31
16. Of the lever,	32
17. General rules for computing the power of an engine,	33
18. Of the different kinds of levers,	34
19. Compound levers,	35
20. Calculating the power of wheel work,	37
21. Power decreases as motion increases,	38
22. No power gained by enlarging undershot water-wheels,	39
23. No power gained by double gearing,	40
24. Of the pulley,	41
25. Of the wheel and axle,	41
26. Of the inclined plane,	42
27. Of the wedge,	42
28. Of the screw,	43
29. Of the fly-wheel and its use,	44
30. On friction,	45
31. On the friction of different substances,	47
32. Mechanical contrivances to reduce friction,	48
33. Of maximums,	50
34. Investigation of old theory,	51
35. New theory doubted,	55
36. True theory attempted,	55
37. Scale of experiments,	57
38. Waring's theory,	59
39. The same continued,	62
40. Doubts respecting it,	63
41. True theory farther sought for,	64
42. " " deduced,	71
43. Finding the velocity of a wheel, theorem for,	73

CONTENTS.

ARTICLE	PAGE
44. Maximum velocity of overshot wheels,	73
Table of velocities of do.	76
Preliminary remarks on hydraulics,	78
45. Of spouting fluids,	79
46. Seventh law of spouting fluids demonstrated,	82
47. Its accordance with practice shown,	83
48. Hydrostatic paradox,	86
49. Practical results of equal pressure,	87
50. To find the pressure on a vessel,	88
51. To find the velocity of spouting fluids,	89
52. Effect of water under a given head,	89
53. Water applied to act by gravity,	91
54. Principles of overshot mills,	93
55. Friction of spouting fluids on apertures,	96
56. Pressure of the air on fluids,	97
57. Of pumps,	98
58. Conveying water under valleys and over hills,	100
59. Definite and indefinite quantities of water striking a wheel,	100
60. Motion of breast and pitch-back wheels,	102
61. Calculating the power of a mill seat,	105
62. Theory and practice compared,	107
63. Observations and experiments on mills in practice,	111
Table of the area and power of mill stones,	116
64. On canals for conveying water to mills,	118
65. On their size and fall,	119
66. Of air pipes, to prevent trunks from bursting,	121
67. Smeaton's experiments concerning undershot wheels,	122
68. " " " overshot wheels,	139
69. " " " wind mills,	147

PART II.
Of the different Kinds of Mills.

70. Of undershot mills, with a table of their proportions and powers, | 153
71. Of tub mills, with a similar table, | 159
72. Of breast and pitch-back wheels, with a table for them, | 164
73. Of overshot wheels, with tables, | 171
74. Rules and calculations in regulating the motion, | 177
75. Rules for finding the pitch circles, | 180
76. The same subject, with a table, | 181
77. Measuring the contents of garners, hoppers, &c. | 185
78. Of the different kinds of gears and forms of cogs, | 187
79. Of spur gears, | 187
80. Of face gears, | 189
81. Of bevel gears, | 191
82. Of matching wheels to make the cogs wear equally, | 193
83. Of rolling screens and fans, | 194
84. Of gudgeons, preventing their heating, &c. | 196
85. Building mill dams, | 199
86. Building mill walls, | 201

PART III.
Description of the Author's Improvements.

87. General account of the author's improvements, | 203
88. Of the elevator, conveyer, hopper-boy, drill, and descender, | 204

CONTENTS.

ARTICLE		PAGE
89. Application of the machines in manufacturing flour,		208
90. Elevating grain from ships,		211
91. A mill for grinding parcels,		213
92. Improved grist mill,		215
93. On elevating from ships, &c., by horse power,		217
94. On the same, by manual power,		218
95. Particular directions for constructing elevators, &c.		221
96. Of the meal elevator,		229
97. " meal conveyer,		231
98. " grain conveyer,		233
99. " hopper boy,		234
100. " drill,		236
101. Utility of these machines,		238
102. Bills of materials for their construction,		240
103. Mill for hulling and cleaning rice,		243

PART IV.

On the Manufacturing of Grain into Flour.

104. Explanation of the principle of grinding,		247
105. Of the draught necessary to be given to the furrows of mill-stones,		250
106. Of facing mill-stones,		254
107. Of hanging do.		256
108. Of regulating the feed and water, in grinding,		258
109. Rule for judging of good grinding,		259
110. Of dressing and sharpening the stones, when dull,		260
111. Of the most proper degree of fineness for flour,		261
112. Directions for grinding wheat mixed with garlic, &c.		262
113. Of grinding middlings over, &c.		264
114. Of the quality of mill-stones to suit that of the wheat,		267
115. Of bolting reels and cloths, with directions for bolting and inspecting flour,		269
116. The duty of the miller,		272
117. Peculiar accidents by which mills are subject to catch fire,		274
118. Observations on improving mills,		275

PART V.

Ellicott's Plans for building Mills.

Prefatory remarks,		277
119. Undershot mills, and laying on the water,		280
120. Directions for making forebays,		281
121. Principle of undershot mills,		282
122. Of breast wheels,		285
123. Of pitch-back wheels,		286
124. Of overshot wheels,		286
125. On the motion of overshot wheels,		287
126. Of gearing,		288
127. Diameter of pitch circles,		289
128. Tables for overshot mills of different falls, &c., &c.		291
129. Constructing undershot wheels,		296
130. Dressing shafts,		298
131. Directions for laying out mortises for arms,		298
132. " for putting in gudgeons,		300

CONTENTS.

ARTICLE		PAGE
133. Directions for constructing cog-wheels,		301
134. " for making sills, spurs, and head blocks,		303
135. Of the best time for cutting, and method of seasoning, cogs,		304
136. Of shanking, putting in, and dressing off cogs,		304
137. Of the little cog-wheel and shaft,		306
138. Directions for making wallowers and trundles,		306
139. " for fixing the head blocks, and hanging the wheels,		307
140. " for sinking the balance ryne,		308
141. " for bridging the spindle,		309
142. " for making the crane and lighter-staff,		309
143. " for making a hoop for the mill-stones,		310
144. " for grinding sand to face the stones,		311
145. " for laying out the furrows in new stones,		312
146. " for making a hopper, shoe, and feeder,		313
147. " for making bolting chests and reels,		314
148. " for setting bolts to go by water,		315
149. " for making bolting wheels,		316
150. Of rolling-screens,		318
151. Of fans,		319
152. Of the shaking sieve,		319
153. Of the use of draughting to build mills by,		321
154. Directions for draughting and planning mills,		322
155. Bills of scantling for a mill,		323
156. Bills of iron work for do.		326
157. Explanation of the plates,		328
158. Of saw-mills, with a table of the dimensions of flutter-wheels,		332
159. Of fulling-mills,		340
160. Management of the saw-mill,		342
161. Mr. French on saw-mills, &c.		344
Rules for discovering new improvements,		347

APPENDIX.

Description of an improved merchant-mill,	357
On the construction of water-wheels, by W. Parkin, &c.	362
Remarks by the Editor,	365
On the distance which bodies fall, and the velocities acquired in consecutive periods of time, with a table,	367
Comparison of different water-wheels, by Mr. Perkins and Mr. Manwaring,	369
Remarks by the Editor,	371

Extracts from Buchanan on Mill-Work.

Strength and durability of the teeth of wheels,	373
Of arranging the numbers of do.	377
Of making patterns for cast-iron,	378
Of malleable or wrought iron gudgeons,	381
Of the bearings of shafts,	382
On the framing of mill-work,	383
On reaction wheels, (extracted from Franklin Journal,)	384
Definition of terms	391

THE YOUNG MILL-WRIGHT'S AND MILLER'S GUIDE.

PART THE FIRST.

CHAPTER I.

MECHANICS.

Of the general Properties of Bodies, and the first Principles of Mechanics.

ARTICLE 1.

PRELIMINARY REMARKS.

ALTHOUGH there are many good, practical workmen who are entirely ignorant of the theory of mechanics as a science, it will be universally acknowledged that an acquaintance with the general properties of matter, and the laws of motion would not only be gratifying to every intelligent mind, but would introduce a certainty into many mechanical operations which would ensure their success: and this is a truth, with the importance of which the author of this work was so fully impressed, that he devoted a whole chapter to its consideration. The present editor has thought it best to alter and modify the original work, but he has been careful not only to retain all that appeared to him important in it, but to make such additions, and give such an arrangement to the whole, as have appeared to him calculated to place the subjects of which it treats in a more familiar light.

It is only, however, those properties of bodies, and those laws of motion, which most intimately concern the practical mechanician, that it is thought proper, here, to treat at any length, as any thing farther would be entirely foreign to the object of this work.

ARTICLE 2.

ON THE ESSENTIAL PROPERTIES OF BODIES.

There are certain properties of bodies which belong to matter in all its forms; these are called its essential properties, as they are those without which it cannot exist: these are *Extension, Figure, Impenetrability, Divisibility, Mobility, Inertia,* and *Attraction.*

Extension. We become acquainted with the existence of matter only by the space which it occupies. We cannot conceive of a body without length, breadth, and thickness, which are the three dimensions of extension. These vary greatly from each other in different bodies; and in some they are all equal to each other, as in the sphere and the cube.

Figure, or shape, is the necessary result of extension, and constitutes its limits. The business of the machinist is to give to various substances those figures, or shapes, which shall adapt them to his purpose.

Impenetrability is that property, by which a body occupies a certain space, which cannot, at the same time, be occupied by another body. If a nail be driven into a piece of wood, it removes a portion of the latter out of its way. Water and other fluids may be made to enter the pores of wood, but it is manifest that two distinct particles of matter cannot exist in the same space with each other.

Divisibility is the susceptibility of matter to be divided into any number of parts. If, in conceiving of the minuteness of the particles of matter, we carry the imagination to its utmost limits, we must confess that a single particle must contain as many halves, quarters, and eighths, as the largest masses. We are not to conclude from this, however, that matter is actually infinitely divisible, although it is mathematically so. It is probable that the Creator has formed masses of matter of certain minute particles, which are infinitely hard, and incapable, from their nature, of mechanical division.

Mobility is one of those essential properties of matter, which form the very foundation of operative mechanics, as it is the capability of matter to be moved from the place, or space, which it now occupies. No mechanical operation, indeed, or any other change, can be effected in matter without motion.

Inertia, or inactivity, is that negative property of matter by which it resists every change of state, whether of rest or of motion. By this term we mean to express the fact that matter is powerless; that if at rest, it has nothing within itself tending to put it into motion; and if in motion, its own tendency is to continue to move, which it would consequently do perpetually, but for those extraneous resistances to which every thing upon the surface of the earth is subjected. The term *vis inertia*, or the power of inertia, is altogether objectionable, although it is very frequently employed. If inertia were a power existing in a body, it must be in some definite quantity, capable of being expressed in numbers, and of resisting a force less than itself; but it is a fact, that any force impressed, however small, will move any body, however great.

Attraction is that power which exists in particles or in masses, of matter, by which they tend to approach each other. It has been divided into five kinds: the attraction of *Gravitation*, of *Cohesion* or aggregation, of *Magnetism*, of *Electricity*, and *Chemical attraction*. It is the two former only of these attractions which claim particular attention in their relationship to mechanics.

The attraction of *cohesion* is that power by which particles of matter become united together and form masses. We could conceive of the existence of matter without attraction, but it must be in its original constituent particles only, unformed into masses; all matter, however, is manifestly endowed with this property, and its particles are, therefore, capable of being united together. In order that the attraction of cohesion may be exerted, it is necessary that the particles of matter be in contact with each other, as it does not take place at sensible distances. By sawing, filing, grinding, and many other mechanical operations, we destroy the attraction of cohesion; and this, indeed, is the great object of these processes. In those bodies which are capable of undergoing fusion, as the metals, we can readily restore this attraction, by subjecting the disentegrated particles to this process.

The attraction of *Gravitation* is manifested in masses as well as in particles of matter; by it all the bodies in nature tend to approach each other. The sun, the earth, the moon, and all the planets, notwithstanding their immense distances, are subjected to this universal law. A stone, or other substance, if unsupported, falls to the earth, in consequence of the attraction existing between it and the earth. What we call weight, results from this attraction, and is the measure of its force or power, in different bodies. The weight of a body is the sum of the attractive force exerted upon its individual particles. A piece of lead, weighing two pounds, contains twice as many particles as another weighing but one pound, and it is therefore drawn to the earth with double the force. It might be supposed that, in consequence of this double quantity of attraction, the piece of two pounds would fall with double the velocity of that of one pound; but, upon making the experiment, the time of their fall will be precisely the same in each. This arises from the inertia of matter, by which, when at rest, it tends to remain so; and, therefore, to move a double quantity with the same velocity, must require a double force. Gravitation must be considered as acting equally on each particle, and consequently, there exists no reason why a piece weighing two pounds should fall with any greater rapidity than would its two halves, were it divided. Light bodies, which expose a large surface to the air, are retarded in their fall by the resistance which it presents; were that removed, a feather would fall with the same velocity as a piece of lead.

This fact is of high importance in practical mechanics, as, in the greater number of instances, gravitation is the active agent in moving machines; and in the construction of all, it is an element which must enter into the calculation of their power.

ARTICLE 3.

AXIOMS, OR LAWS, OF MOTION AND REST.

1. Every body in a state of rest, will remain so; and every body in motion, will continue to move in a right line, until a change is effected by the agency of some mechanical force.
2. The change from rest to motion, and from motion to rest, is always proportional to the force producing these changes.
3. Action and reaction are always equal, and in directions contrary to each other; or, when two bodies act upon each other, the forces are always equal, and directed towards contrary parts.

The first of these laws results, necessarily, from the inertia of matter. The assertion, however, that a body in motion would continue to move in a right line, may require some illustration. That motion when once communicated would never cease, is fairly inferred from the fact, that the motion is continued in the exact proportion in which the obstruction is diminished. A pendulum will vibrate longer in air than in water, and longer still in an exhausted receiver, and stops at last in consequence of the friction on its points of suspension, and the imperfection of the vacuum.

When a stone is thrown in a horizontal direction, as motion is constantly retarded, it also moves in a curve, and eventually falls to the ground. The retardation, in this case, is exactly proportioned to the density of the air, and the curve in which it moves is the consequence of the force of gravity, which is always drawing it towards the earth: the curve in which it moves is determined by this known force, and is precisely proportionate to it. It necessarily follows, that, if the cause of retardation, and of deflexion were both removed, that the body would continue its course in a right line. The preceding remarks may serve to illustrate the second, as well as the first law.

The third law is confirmed by all our observations on the motions of the heavenly bodies, and by all our experiments. If a glass bottle be struck by a hammer, or a hammer by a glass bottle, the bottle will in either case be broken by the same degree of moving power: were the hammer equally fragile with the bottle, both would be broken. If a stone be thrown against a pane of glass, the glass will be broken, and the stone be retarded, in exact proportion to the resistance offered by the glass.

To assert the contrary of this law, would be to maintain an absurdity; for if action and reaction be not equal, one must be greater than the other, which would be to say that the effect was greater than, or not equal to, the cause.

ARTICLE 4.

ON ABSOLUTE AND RELATIVE MOTION.

The idea intended to be conveyed by the term *motion* is too familiar to require a definition.

Motion is either absolute, or relative.

Absolute motion is the removal of a body from one part of space to another, as the motion of the earth in its orbit.

Relative motion is the change of place which one body undergoes in relationship to another: such, for example, as the difference of motion in the flight of two birds, or the sailing of two ships.

Were all the articles upon the surface of the earth to retain their respective situations, they would still be in absolute motion with the earth in space, but they would experience no relative motion, and would appear to us to be at rest.

In the theory of mechanics, much information is derived from our knowledge of the laws observed by the heavenly bodies in their absolute motions; but, in practical mechanics, we have to do with relative motion, only.

On equable, accelerated, and retarded Motion.

Time must, of necessity, enter into the idea of motion, as it is the measure of its *velocity*. Thus, a body which passes the distance of two miles in an hour, moves with twice the velocity of another, which, in the same time, travels but one mile.

A body in motion may continue to move with the same velocity throughout its whole course; its motion is then said to be *equable:* or,

Its motion may be perpetually increasing, as is the case with falling bodies. This is denominated *accelerated motion.*

Retarded motion, is that which is continually decreasing; such is the motion of a stone, or of a cannon ball, projected perpendicularly upwards.

The cause of the equable acceleration of falling bodies, and the retardation of such as are projected upwards from the earth, will be rendered clear, by attending to the article on falling bodies.

ARTICLE 5.

OF MOMENTUM.

It is known to every one that if the velocity of a moving body be increased, the force with which it will strike against another body will be increased also: the fact is equally familiar, that if the weight of a body in motion be increased, the result will be similar. It is evident, therefore, that the force with which a body in motion strikes against another body, must be in the compound ratio of its velocity, and its mass, or quantity of matter. This force is called its *momentum,* which is the product of its *quantity of matter* multiplied by its *quantity of motion;* or, in other words, its *weight* multiplied by its *velocity.*

The effects produced by the collision of bodies against each other differ greatly in those which are elastic, from those that are nonelastic, which will be more particularly noticed presently.

ARTICLE 6.

ON POWER, OR FORCE, AND ON THE MOTIVE POWERS.

Force, or *Power,* in a mechanical sense, is that which causes a change in the state of a body, from motion to rest, or from rest, to motion.

When two or more forces act upon a body, in such a way as to destroy the operation of each other, there is then said to be an *equilibrium of forces.*

The Motive Powers, are those which we employ to produce motion in machines: these are, the strength of men, and of other animals; the descent of weights; the force of water in motion; wind, or the motion of the air; the elasticity of springs, and the elastic force of steam. The whole of these are included in the two principles of *Gravitation* and *Elasticity.*

Attempts have been made to employ other agents as motive powers, but these have either failed altogether, or have not been attended with that success which justifies the giving to them a place in a practical work. Among these may be mentioned magnetism; electricity; condensed air; air rendered more elastic by heating it; explosive gases and fulminating compounds.

ARTICLE 7.

ON THE EFFECT OF COLLISION, OR IMPACT.

The striking of bodies against each other is denominated collision, or impact.

Bodies are divided into *elastic,* and *nonelastic.* By elastic bodies are intended those which resume their dimensions and form, when the force which changed them is removed. Nonelastic bodies are those which not only change their forms, when struck, but remain permanently altered in this particular. Although there are no solid bodies which possess either of these properties in perfection, yet the difference between those which are most, and those which are least elastic, is sufficiently great to justify the division.

Ivory, and hardened steel are eminently elastic. Such bodies, when struck together, become flattened at the point of contact; but immediately resuming their form, they react upon each other, and rebound. Lead and soft clay are nonelastic: if two balls of either of these substances be struck together, a permanent flattening is produced at their points of contact, and they do not rebound.

If two nonelastic bodies, A and B, fig. 1, each having the same quantity of matter, move towards each other with equal velocities, they will come into contact, as at A B, in the centre, where they will remain at rest after the stroke, because their momentums were equal, and in opposite directions. That is, if each have two pounds of matter, and a velocity which we may call ten, the momentum of each is twenty; and just sufficient, therefore, to destroy each other.

If, on the contrary, the bodies be perfectly elastic, they will recede from each other with the same velocity with which they met. In the former case, a permanent indentation was produced on the bodies; in the present, the flattening is instantaneous only, and the particles resuming their former position and arrangement, react upon each other with a force equal to their action, and, after the stroke, recede with undiminished velocity.

If two nonelastic bodies, A and B fig. 2 moving in the same direction with different velocities, impinge upon each other, they will move on together after the stroke with such velocity, as, being multiplied into the sum of their weights, will produce the sum of the momentums which they had before the stroke: that is, if each weigh one pound, and A has 3, and B 4 degrees of velocity, the sum of their momentums is 12; $1 \times 8 + 1 \times 4 = 12$: then after the stroke their velocity will be 6; which multiplied into their quantity of matter 2, produces 12. The quantity of motion before and after the stroke, or, which is the same thing, their momentums, will be unchanged.

If, on the contrary, they had both been elastic, and movingas before, then, after the stroke, A would have moved with four, and B with eight degrees of velocity: they would consequently have interchanged velocities, but the quantity of motion would remain unchanged.

If A and B be nonelastic bodies, equal in quantity of matter, and A moving with a velocity 10, come into contact with B at rest, they will move on together with the velocity 5. The quantity of motion will therefore remain unchanged, a double mass moving with one half the velocity. If the bodies A and B be both elastic, B, after the stroke, will fly off with the velocity 10, and A will remain at rest. The quantity of motion will, as before, remain unchanged. To understand this difference between elastic and nonelastic bodies, we may suppose that when the two elastic bodies come into contact with each other, they tend to move on together, like the nonelastic, with one half the velocity of the body A; that is, A gives half its motion to B; but, being elastic, the impinging parts, which give way, instantaneously resume their form, and react upon each other with a force equal to their first action, which drives A back with a velocity 5, and B forward with an equal velocity: the effect of which must be to leave A at rest, and to accumulate the whole motion in B.

ARTICLE 8.

ON COMPOUND MOTION.

If a body be struck by two equal forces, in contrary directions, it will remain unmoved; but if the forces, instead of acting on the body in directions exactly opposite, strike it in two directions inclined to each other, motion will be communicated to the body so struck; but its direction will not be that of either of the striking bodies, but somewhere between them, dependent upon the power of the blows respectively. The motion in this case is manifestly compounded of the two possessed by the striking bodies, and is therefore called a *compound motion*.

If a body A, fig. 4, receive two strokes, or impulses at the same time, in different directions, one which would propel it from A to B in a given time, and another which would propel it from A to D in an equal time, then this compound force will propel it from A to C, in the same time in which it would have arrived at B or D by one impulse only. If lines be drawn from C, to join B, and D, the parallelogram A B C D, will be formed, in the diagonal of which the compound motion was performed. If the two impulses had been equal, then A D would be equal to A B, and the parallelogram would become a square.

ARTICLE 9.

OF NON-ELASTICITY, AND OF FLUIDITY, IN IMPINGING BODIES.

If A and B, fig. 3, be two columns of matter in motion, meeting each other, and equal in non-elasticity, quantity, and velocity, they will meet at the dotted line e e, destroy each other's motion, and remain at rest, provided none of their parts separate.

But if A be elastic, and B non-elastic, when they meet at e e, B will give way by battering up, and both will move a little farther; that is, half the distance that B shortens.

But if B be a column of fluid, and when it strikes A, flies of in a lateral direction perpendicular to A, then whatever is the sum total of the momentums of these particles laterally, has not been communicated to A.

But with what proportion of the striking velocity the fluid, after the stroke, will move in the lateral direction, I do not find determined; but from some experiments I have made, I suppose it to be more than one half; because water falling four feet, and striking a horizontal plane with 16,2 feet velocity will cast some few drops to the distance of 9 feet (say 10 feet, allowing one foot to be lost by friction, &c.) which we must suppose take their direction at an angle of 45 degrees; because a body projected at an angle of 45 degrees, will describe the greatest possible horizontal range. It is known also, that a body falling 4 feet, and reflected with its acquired velocity 16,2 feet, at 45 degrees, will reach 16 feet horizontal range, or 4 times the distance of the fall. Therefore, by this rule, $\frac{1}{4}$ of 10 feet, equal to 2,5 feet, is the fall that will produce the velocity necessary to this effect, viz. velocity 12,64 feet per second, about $\frac{3}{4}$ of the striking velocity.

This side force cannot be applied to produce any farther forward force, after it has struck the first obstacle, because its action and reaction then balance each other: which I demonstrated by fig. 27.

Let A be an obstacle, against which the column of water G A, of quantity 16 with velocity per second 16, strikes; as it strikes A, suppose it to change its direction, at right angles with $\frac{3}{4}$ velocity and to strike B B, then to change again and strike forward against C C, and backwards against D D: then again in the side direction E E; and again in the forward and backward directions, all of which forces counteract, and balance, each other.

Therefore, if we suppose the obstacle A to be the float of an undershot water wheel, the water can be of no farther service in propelling it, after the first impulse, but rather a disadvantage; because the elasticity of the float will cause it to rebound in a certain degree, and, instead of keeping fully up with the float it struck, to react back against the next float. It will be better, therefore, to let it escape freely as soon as it has fully made the stroke; not sooner, however, as it will require a certain space to act in, which will be in direct proportion to the distance between the floats.

From these considerations, we may conclude, that the greatest effect to be obtained from striking fluids, will not amount to more than half the power which gives them motion, and much less, if they be not applied to the best advantage: and also, that the effect produced by the collision of non-elastic bodies, will be in proportion to their non-elasticity.

ARTICLE 10.

OF FALLING BODIES.

Bodies descending freely by their gravity, in vacuo, or in a non-resisting medium, are subject to the following laws:—

1st. They are equally accelerated.

It is evident, that, in every equal part of time, the body must receive an equal impulse from gravity, which will propel it an equal distance, and give it an equal additional velocity; it will, therefore, produce equal effects in equal times; and the velocity will be proportioned to the time.

2d. Their velocity is always in proportion to the time of their fall, and the time is as the square root of the distance fallen.

If the velocity, at the end of one second, be 32,4 feet at the end of two seconds, it will be 64,8, at the end of three seconds, 97,2 feet per second, and so on.

3d. The spaces through which they pass are as the squares of the times or velocities.

That is, as the square of one second is to the space passed through, 16,2, so is the square of two seconds, which is 4, to 64,8 feet, passed through at the end of 2 seconds; and so on, for any number of seconds. Therefore the spaces passed through at the end of every second, will be as the square numbers 1, 4, 9, 16, 25, 36, &c. and the spaces passed through, in each second separately, will be as the odd numbers 1, 3, 5, 7, 9, 11, 13, 15, &c.

4th. Their velocities are as the square root of the space descended through, and their force, to produce effect, as their distances fallen, directly.

That is, as the square root of 4, which is 2, is to 16,2, the velocity acquired in falling four feet; so is the square root of any other distance, to the velocity acquired in falling that distance.

5th. The space passed through the first second, is very nearly 16,2 feet, and the velocity acquired, at the lowest point, is 32,4 feet, per second.

6th. A body will pass through twice the space, in a horizontal direction, with the last acquired velocity of the descending body, in the same time that its fall required.

That is, suppose the body as it arrives at the lowest point of its fall, and has acquired its greatest velocity, was to be turned in a horizontal direction, or that the acceleration from gravity was at that moment to cease, and the velocity to continue uniform, it would then pass over double the distance that it had descended through, in the same time.

7th. The total sum of the effective impulse acting on falling bodies to give them velocity, is in direct proportion to the space descended through;* and their velocity being as the square root of the space descended through, or, which is the same as the square root of the total impulse. Therefore,

8th. Their momentums, or force to produce effects, are as the squares of their velocities,† or directly as their distances fallen through; and the times expended in producing the effects, are as the square root of the distance fallen through.

That is, if a body fall 16 feet, and strike a non-elastic body, such as soft lead, clay, &c., it will strike with velocity 32, and produce a certain effect in a certain time. Again, if it fall 64 feet, it will strike with velocity 64, and produce a quadruple effect, in a double time; because if a perfectly elastic body fall 16 feet (in one second of time, and strike a perfectly elastic plane, with velocity 32 feet,) it will rise 16 feet in one second of time. Again, if the body fall two seconds of time, it will fall 64 feet, and strike with velocity 64, and rise 64 feet in two seconds of time. Now, if we call the rising of the body the effect of the striking velocity (which it really is) then all will appear clear. I am aware that any thing here advanced, which is contrary to the opinion of learned and ingenious authors, ought to be doubted, unless found to agree with practice.

* This is evident from the consideration, that in every equal part of distance it descends through, it receives an equally effective impulse from gravity. Therefore, 4 times the distance gives 4 times the effective impulse.

† This is evident, when we consider, that a quadruple distance, or impulse, produces only double velocity, and that a quadruple resistance will be required to stop double velocity; consequently, their force is as the squares of their velocities, which brings them to be directly as their distances descended through: and this agrees with the second law of spouting fluids, Art. 45.

A TABLE

OF THE

MOTION OF FALLING BODIES.

SUPPOSED IN VACUO.

Distance passed through in feet.	The velocity acquired by the fall, in feet and parts, counted per second.	Seconds of time that a body is supposed to be falling.	Distance passed through in said time, in feet and parts.	Velocity per second acquired at the end of every second, in feet and parts.
1	8.1	.125	.25	4.
2	11.4	.25	1.01	8.1
3	14.	.5	4.05	16.2
4	16.2	.75	9.11	24.3
5	18.	1	16.2	32.4
6	19.84	2	64.8	64.8
7	21.43	3	145.8	97.2
8	22.8	4	259.2	129.6
9	24.3	5	305.	162.
10	25.54	6	583.2	194.4
11	26.73	7	793.8	226.8
12	28.	8	1036.8	259.2
13	29.16	9	1312.2	291.6
14	30.2	10	1620.	324.
15	31.34	30	14580.	972.
16	32.4	60	58320.	1944.
17	33.32			
18	34.34			
19	35.18			
20	36.2			
21	37.11			
36	48.6			
49	56.7			
64	64.8			
100	81.			
144	97.2			

MECHANICS.

A SCALE OF THE MOTION OF FALLING BODIES.

Time of fall in seconds.	Spaces fallen, each being equal to 16.2 feet.	In this table the time is divided into seconds, and the absolute distances are proportioned to this division; but the ratios are the same, whether minutes, hours, or any other period, be taken as the unit of time.	Space fallen through in each second, in the ratio of the odd numbers.	Ratio of velocity acquired; as the sq. root of the distance fallen.	Absolute distance fallen at the end of each second.
	0				feet.
1″	1	........a................................ Velocity, in feet, acquired at the end of 1″=32.4 feet.	1	1	16.2
	2				
	3				
2″	4	............b................ Velocity acquired at the end of 2″=64.8 feet.	3	2	64.8
	5				
	6				
	7				
	8				
3″	9	....................c...... Velocity acquired at the end of 3″=97.2 feet.	5	3	145.8
	10				
	11				
	12				
	13				
	14				
	15				
4″	16	..............................d. Velocity acquired at the end of 4″=129.6 feet.	7	4	259.2

This scale shows at one view, all the laws observed by falling bodies. The body O would fall from O to 1, equal to 16,2 feet, in the first second, and acquire a velocity that would carry it 32,4 feet from I to *a*, horizontally, in the next second, by laws 5 and 6; this velocity would also carry it down to three in the same time; but its gravity, producing equal effects, in equal times, will accelerate it so much as to take it to 4 in the same time, by law 1. It will now have a velocity of 64,8 feet per second, that will take it to *b* horizontally, or down to 8, but gravity will help it on to 9 in the same time. Its velocity will now be 97,2 feet; which will take it horizontally to *c*, or down to 15, but gravity will help it on to 16; and its last acquired velocity will be 129,6 feet, per second, which would carry it to *d* horizontally.

If either of these horizontal velocities be continued, the body will pass over double the distance it fell, in the same time, by law 6.

Again, if O be perfectly elastic, and falling, strikes a perfectly elastic plane, either at 1, 3, 5, or 7, the effective force of its stroke will cause it to rise again to O in the same space of time it took to fall.

This shows, that in every equal part of distance, it received an equally effective impulse from gravity, and that the total sum of the effective impulse is as the distance fallen directly—and the effective force of the stroke will be as the squares of the velocities, by laws 7 and 8.

ARTICLE 11.

OF BODIES DESCENDING INCLINED PLANES AND CURVED SURFACES.

Bodies descending inclined planes and curved surfaces, are subject to the following laws:—

1. They are equably accelerated, because their motion is the effect of gravity.
2. The force of gravity propelling the body A, fig. 5, to descend an inclined plane A D, is to the absolute gra-

vity of the body, as the height of the plane A C is to its length A D.

3. The spaces descended through are as the squares of the times.

4. The times in which the different planes A D, A H, and A I, or the altitude A C, are passed over, are as their lengths respectively.

5. The velocities acquired in descending such planes, in the lowest points, D, H, I, or C, are all equal.

6. The times and velocities of bodies descending through planes alike inclined to the horizon, are as the square roots of their lengths.

7. Their velocities, in all cases, are as the square roots of their perpendicular descent.

From these laws or properties of bodies descending inclined planes, are deduced the following corollaries; namely:—

1. That the times in which a body descends through the diameter A C, or any chord A a, A e, or A i, are equal: hence,

2. All the chords of a circle are described in equal times.

3. The velocity acquired in descending through any arch, or chord of an arch, of a circle, as at C, in the lowest point C, is equal to the velocity that would be acquired in falling through the perpendicular height F C.

Pendulums in motion have the same properties, the rod or string acting as the smooth, curved surface.

For illustrations of these properties, see Kater and Lardner's Mechanics, p. 79, or any general treatise on that subject.

ARTICLE 12.

ON THE MOTION OF PROJECTILES.

A projectile is a body thrown, or projected, in any direction; such as a stone from the hand, water spouting from any vessel, a ball from a cannon, &c., fig. 6.

Every projectile is acted upon by two forces at the same time; namely, Impulse and Gravity.

By impulse, or the projectile force, the body will pass over equal distances, A B, B C, &c., in equal times, by 1st general law of motion, Art. 7, and by gravity, it descends through the spaces A G, G H, &c., which are as the squares of the times, by 3d law of falling bodies, Art. 9. Therefore, by these forces compounded, the body will describe the curve A Q, called a parabola; and this will be the case in all, except perpendicular directions: the curve will vary with the elevation, yet it will still be what is called a parabola.

If the body be projected at an angle of 45 degrees elevation, it will be thrown to the greatest horizontal distance possible; and, if projected with double velocity, it will describe a quadruple range.

ARTICLE 13.

OF CIRCULAR MOTION AND CENTRAL FORCES.

If a body A, fig. 7, be suspended by a string A C, and caused to move round the centre C, that tendency which it has to fly from the centre, is called the centrifugal force; and the action upon a body, which constantly solicits it towards a centre, is called the centripetal force. This is represented by the string, which keeps the body A, in the circle A M. Speaking of these two forces indifferently, they are called central forces.*

The particular laws of this species of motion, are,

1. Equal bodies describing equal circles in equal times, have equal central forces.

2. Unequal bodies describing equal circles in unequal times, their central forces are as their quantities of matter multiplied into their velocities.

---

* It may be well to observe here, that this central force is no real power, but only an effect of the power that gives motion to the body. Its inertia causes it to recede from the centre, and fly off in a direct tangent, with the circle it moves in: therefore this central force can neither add to, nor diminish the power of any mechanical or hydraulic engine, unless it be by friction and inertia, where water is the moving power and the machine changes its direction.

3. Equal bodies describing unequal circles in equal times, their velocities and central forces are as their distances from their centres of motion, or as the radii of their circles.*

4. Unequal bodies describing unequal circles in equal times, their central forces are as their quantities of matter multiplied into their distances from their centres, or the radii of their circles.

5. Equal bodies describing equal circles in unequal times, their central forces are as the squares of their velocities; or, in other words, a double velocity generates a quadruple central force.† Therefore,

6. Unequal bodies describing equal circles in unequal times, their central forces are as their quantities multiplied into their velocities.

7. Equal bodies describing unequal circles with equal celerities, their central forces are inversely as their distances from their centres of motion, or the radii of their circles.‡

* This shows, that when mill-stones are of unequal diameters, and revolve in equal times, the largest should have the draught of their furrows less, in proportion as their central force is more; which is in inverse proportion; also, that the draught of a stone should vary, and be in inverse proportion to the distance from the centre. That is, the greater the distance, the less the draught.

Hence we conclude, that if stones revolve in equal times, their draught must be equal near the centre: that is, so much of the large stones, as is equal to the size of the small ones, must be of equal draught. But that part which is greater, must have less draught in inverse proportion; as the distance from the centre is greater, the furrows must cross at so much less angle; which will be nearly the case (if their furrows lead to an equal distance from their centres) at any considerable distance from the centre of the stone; but near the centre the angles become greater than the proportion: if the furrows be straight, as appears by the lines, g 1, h 1, g 2, h 2, g 3, h 3, in fig. 1, Pl. XI. the angles near the centre are too great, which seems to indicate that the furrows of mill-stones should not be straight, but a little curved; but what this curve should be is very difficult to lay down exactly in practice. By theory it should be such as to cause the angle of furrows crossing, to change in inverse proportion with the distance from the centre, which will require the furrows to curve more, as they approach the centre.

† This shows that mill stones of equal diameters, having their velocities unequal, should have the draught of their furrows, as the square roots of their number of revolutions per minute. Thus, suppose the revolutions of one stone, the furrows of which are correctly made, to be 81 per minute, and the mean draught of the furrows 5 inches, and found to be right, the revolutions of the other stone to be 100; then, to find the draught, say, as the square root of 81, which is 9, is to the 5 inches draught, so is the square root of 100, which is 10, to 4,5 inches, the draught required (by inverse proportion,) because the draught must decrease as the central force increases.

‡ That is, the greater the distance, the less the central force. This shows that mill-stones of different diameters, having their peripheries revolving with equal

8. Equal bodies describing unequal circles, having their central forces equal; their periodical times are as the square roots of their distances.

9. Therefore the squares of the periodical times are proportional to the cubes of their distances, when neither the periodical times nor the celerities are given. In that case,

10. The central forces are as the squares of the distances inversely.*

velocities, should have the angle of draught, with which their furrows cross each other, in inverse proportion to their diameters, because their central forces are as their diameters, by inverse proportion, directly; and the angle of draught should increase, as the central force decreases, and decrease, as it increases.

But here we must consider, that, to give stones of different diameters equal draughts, the distance of their furrows from the centre, must be in direct proportion to their diameters. Thus, as 4 feet diameter is to 4 inches draught, so is 5 feet diameter to 5 inches draught. To make the furrows of each pair of stones cross each other at equal angles, in all proportional distances from the centre, see fig. 1, Plate XI. where g b, g d, g f, h a, h c, and h e, show the direction of the furrows of the 4, 5, and 6 feet stones, with their proportional draughts; now, it is obvious that they cross each other at equal angles, because the respective lines are parallel, and cross in each stone, near the middle of the radius; which shows that in all proportional distances, they cross at equal angles, consequently their draughts are equal.

But the draught must be farther increased, with the diameter of the stone, in order to increase the angle of draught in the inverse ratio, as the central force decreases.

To do which, say,—If the 4 feet stone has central force equal 1, what central force will the 5 feet stone have? Answer: ,8 by the 7th law.

Then say,—If central force 1 require 5 inches draught, for a 5 feet stone, what will central force 8 require? Answer: 6,25 inches draught. This is, supposing the verge of each stone to move with equal velocity. This rule may bring out the draught nearly true, provided there be not much difference between the diameter of the stones. But it appears to me, that neither the angle with which the furrows cross, nor the distance of the point from the centre, to which they direct, is a true measure of the draught.

* These are the laws of circular motion and central forces. For experimental demonstrations of them, see Furguson's Lectures on Mechanics, page 27 to 47.

I may here observe that the whole planetary system is governed by these laws of circular motion and central forces. Gravity acting as the string, is the centripetal force; and as the power of gravity decreases, as the square of the distance increases, and as the centripetal and centrifugal forces must always be equal, in order to keep the body in a circle: hence appears the reason why the planets most remote from the sun have their motion so slow, while those near him have their motion swift; because their celerities must be such as to create a centrifugal force equal to the attraction of gravity

ARTICLE 14.

OF THE CENTRES OF MAGNITUDE, MOTION, AND GRAVITY.

The centre of magnitude is that point which is equally distant from all the external parts of a body.

2. The centre of motion is that point which remains at rest, while all other parts of the body move round it.

3. The centre of gravity of bodies, is of great consequence to be well understood, it being the principle of much mechanical motion; it possesses the following particular properties.

1. If a body be suspended on this point, as its centre of motion, it will remain at rest in any position.

2. If a body be suspended on any other point than its centre of gravity, it can rest only in such position, that a right line drawn from the centre of the earth through the centre of gravity, will intersect the point of suspension.

3. When this point is supported, the whole body is kept from falling.

4. When this point is at liberty to descend in a right line, the whole body will fall.

5. The centre of gravity of all homogeneal bodies, as squares, circles, spheres, &c., is the middle point in a line connecting any two opposite points or angles.

6. In a triangle, it is in a right line drawn from any angle to bisect the opposite side, and at the distance of one-third of its length from the side bisected.

7. In a hollow cone, it is in a right line passing from the apex to the centre of the base, and at the distance of one third of the side from the base.

8. In a solid cone, it is one-fourth of the side from the base, in a line drawn from the apex to the centre of the base.

The solution of many curious phenomena, as, why many bodies stand more firmly on their bases than others; and why some bodies lean considerably over without falling, depends upon a knowledge of the position of the centre of gravity.

Hence appears the reason, why wheel-carriages, loaded with stones, iron, or any heavy matter, will not overturn so easily, as when loaded with wood, hay, or any light article; for when the load is not higher than a b, fig. 22, a line from the centre of gravity will fall within the centre of the base at c; but if the load be as high as d, it will then fall outside the base of the wheels at e, consequently it will overturn. From this appears the error of those, who hastily rise in a coach or boat, when it is likely to overset, thereby throwing the centre of gravity more out of the base, and increasing their danger.

CHAPTER II.

ARTICLE 15.

OF THE MECHANICAL POWERS.

Having premised and considered all that is necessary for the better understanding those machines called mechanical powers, we now proceed to treat of them. They are six in number; namely:

The Lever, the Pulley, the Wheel and Axle, the Inclined Plane, and the Screw.

These are called Mechanical Powers, because they increase our power of raising or moving heavy bodies. Although they are six in number, yet they are all governed by one simple principle, which I shall call the First General Law of Mechanical Powers; it is this, *the momentums of the power and weight are always equal, when the engine is in equilibrio.*

Momentum, here means the product of the weight of the body multiplied into the distance it moves; that is, the power multiplied into its distance moved, or into its distance from the centre of motion, or into its velocity, is equal to the weight multiplied into its distance moved, or into its distance from the centre of motion, or into its velocity; or, the power multiplied into its perpendicular descent, is equal to the weight multiplied into its perpendicular ascent.

The Second General Law of Mechanical Powers, is,

The power of the engine, and velocity of the weight moved, are always in the inverse proportion to each other; that is, the greater the velocity of the weight moved, the less it must be; and the less the velocity, the greater the weight may be: and that universally in all cases.

The Third General Law, is,

Part of the original power is always lost in overcoming friction, inertia, &c., but no power can be gained by engines, when time is considered in the calculation.

In the theory of this science, we suppose all planes to be perfectly smooth and even, levers to have no weight, cords to be perfectly pliable, and machines to have no friction: in short, all imperfections are to be laid aside, until the theory is established, and then proper allowances are to be made for them.

ARTICLE 16.

OF THE LEVER.

A bar of iron, of wood, or of any other inflexible material, one part of which is supported by a fulcrum or prop, and all other parts turn or move on that prop, as their centre of motion, is called a lever; when the lever is extended on each side of the prop, these extensions are called its arms; the velocity or motion of every part of these arms, is directly as its distance from the centre of motion, by the third law of circular motion.

With respect to the lever, when in equilibrium,— Observe the following laws:—

1. The power and weight are to each other, inversely as their distances from the prop, or centre of motion.

That is, the power P, fig. 8, Plate I. which is one multiplied into its distance B C, from the centre 12, is equal to the weight 12 multiplied into its distance A B 1; each product being 12.

2. The power is to the weight, as the distance the weight moves, is to the distance the power moves, respectively.

That is, the power multiplied into its distance moved, is equal to the weight multiplied into its distance moved.

3. The power is to the weight, as the perpendicular ascent of the weight, is to the perpendicular descent of the power.

That is, the power multiplied into its perpendicular descent, is equal to the weight multiplied into its perpendicular ascent.

4. Their velocities are as their distances from their centre of motion, by the 3d law of circular motion, p. 28.

These simple laws hold universally true, in all mechanical powers or engines; therefore it is easy (from these simple principles) to compute the power of any engine, either simple or compound; for it is only to find how much swifter the power moves than the weight, or how much farther it moves in the same time; and so much is the power (and time of producing it) increased, by the help of the engine.

ARTICLE 17.

GENERAL RULES FOR COMPUTING THE POWER OF ANY ENGINE.

1. Divide either the distance of the power from its centre of motion, by the distance of the weight from its centre of motion. Or,

2. Divide the space passed through by the power, by the space passed through by the weight, (this space may be counted either on the arch, or on the perpendicular described by each,) and the quotient will show how much the power is increased by the help of the engine; then multiply the power applied to the engine, by that quotient, and the product will be the power of the engine, whether simple or compound.

EXAMPLES.

Let A B C, Plate I. fig. 8, represent a lever; then, to compute its power, divide the distance of the power P from its centre of motion B C 12, by the distance A B 1, of the weight W, and the quotient is 12: the power is increased 12 times by the engine; which, multiply by the power applied 1, produces 12, the power of the engine at A, or the weight W, that will balance P, and hold the engine in equilibrio. But suppose the arm A B to be continued to E, then, to find the power of the engine, divide the distance B C 12, by B E 6, and the quotient is two; which multiplied by 1, the power applied, produces 2, the power of the engine, or weight w to balance P.

Or divide the perpendicular descent C D of the power equal 6, by the perpendicular ascent E F equal 3; and the quotient 2, multiplied by the power P equal 1, produces 2, the power of the engine at E.

Or divide the velocity of the power P equal 6, by the velocity of the weight w equal 3; and the quotient 2, multiplied by the power 1, produces 2, the power of the engine at E. If the power P had been applied at 8, then it would have required to have been $1\frac{1}{2}$ to balance W, or w: because $1\frac{1}{2}$ times 8 is 12, which is the momentum of both weights W and w. If it had been applied at 6, it must have been 2; if at 4, it must have been 3; and so on for any other distance from the prop or centre of motion.

ARTICLE 18.

OF THE DIFFERENT KINDS OF LEVERS.

There are four kinds of Levers.

1. The most common kind, where the prop is placed between the weight and power, but generally nearest the weight, as otherwise, there would be no gain of power.

2. When the prop is at one end, the power at the other, and the weight between them.

3. When the prop is at one end, the weight at the other, and the power applied between them.

4. The bended lever, which differs only in form, but not in properties, from the others.

Those of the first and second kind, have the same properties and powers, and produce real mechanical advantage, because they increase the power; but the third kind produces a decrease of power, and is only used to increase velocity, as in clocks, watches, and mills, where the first mover is slow, and the velocity is increased by the gearing of the wheels.

The levers which nature employs in the machinery of the human frame, are of the third kind; for when we lift a weight by the hand, the muscle that exerts the force to raise the weight, is fastened at about one-tenth of the distance from the elbow to the hand, and must exert a force ten times as great as the weight raised; therefore, he that can lift 56 lbs. with his arm at a right angle at the elbow, exerts a force equal to 560 lbs. by the muscles of his arm.

ARTICLE 19.

OF COMPOUND LEVERS.

Several levers may be applied to act one upon another, as 2 1 3 in fig. 9, Plate I. where No. 1 is of the first kind, No. 2 of the second, and No. 3 of the third. The power of these levers, united to act on the weight W, is found by the following rule, which will hold universally true in any number of levers united, or wheels (which operate on the same principle) acting upon one another.

RULE.

1st. Multiply the power P, into the length of all the driving levers successively, and note the product.

2d. Then multiply all the leading levers into one another successively, and note the product.

3d. Divide the first product by the last, and the quotient will be the weight W, that will hold the machine in equilibrio.

This rule is founded on the first law of the lever, Art. 16, and on this principle; namely:

Let the weight W, and power P, be such, that when suspended on any compound machine, whether of levers united, or of wheels and axles, they hold the machine in equilibrio: then if the power P be multiplied into the radius of all the driving wheels, or lengths of the driving levers, and the product noted, and the weight W multiplied into the radius of all the leading wheels, or length of the leading levers, and the product noted, these products will be equal. If we had taken the velocities, or the circumferences of the wheels, instead of their radii, they would have been equal also.

On this principle is founded all rules for calculating the power and motion of wheels in mills, &c. See Art. 20.

EXAMPLES.

Given, the power P equal to 4, on lever 2, at 8 distance from the centre of motion. Required, with what force lever 1, fastened at 2 from the centre of motion of lever 2, must act, to hold the lever 2 in equilibrio.*

By the rule 4×8 the length of the long arm is 32, and this divided by 2, the length of the short arm, gives 16, the force required.

Then 16 on the long arm, lever 1, at 6 from the centre of motion. Required, the weight on the short arm, at 2, to balance it.

By the rule, 16×6=96, which divided by 2, the short arm, gives 48, for the weight required.

Then 48 is on the lever 3, at 2 from the centre. Required, the weight at 8 to balance it.

Then 48×2=96, which divided by 8, the length of the long arm, gives 12, the weight required.

Given, the power P=4, on one end of the combination

---

* In order to abbreviate the work, I shall hereafter use the following Algebraic signs; namely:

of levers. Required, the weight W, on the other end, to hold the whole in equilibrio.

Then by the rule, 4×8×6×2=384 the product of the power multiplied into the length of all the driving levers, and 2×2×8=32 the product of all the leading levers, and 384÷32=12 the weight W required.

ARTICLE 20.

CALCULATING THE POWER OF WHEEL WORK.

The same rule holds good in calculating the power of machines consisting of wheels, whether simple or compound, by counting the radii of the wheels as the levers; and because the diameters and circumferences of circles are proportional, we may take the circumferences instead of the radii, and it will be the same result. Then again, because the number of cogs in the wheels constitute the circle, we may take the number of cogs and rounds instead of the circle or radii, and the result will still be the same.

Let fig. 11, Plate II. represent a water-mill (for grinding grain) double geared.

Number 8 The water-wheel,
        4 The great cog-wheel,
        2 The wallower,
        3 The counter cog-wheel,
        1 The trundle,
        2 The mill-stones,

And let the above numbers also represent the radius of each wheel in feet.

Now suppose there be a power of 500 lbs. on the water-wheel, required what will be the force exerted on the mill-stone, 2 feet from the centre.

The sign + plus, or more, for addition.
        — minus, or less, for substraction.
        × multiplied, for multiplication.
        ÷ divided, for division.
        = equal, for equality.

Then, instead of 8 more 4 equal 12, I shall write 8+4=12. Instead of 12 less 4 equal 8, 12—4=8. Instead of 6 multiplied by 4 equal 24, 6×4=24. And instead of 24 divided by 3 equal 8, 24÷3=8.

Then by the rule, 500×8×2×1=8000, and 4×3×2= 24, by which divide 8000, and it *quotes* 333,33 lbs. the power or force required, exerted on the mill-stone two feet from its centre, which is the mean circle of a 6 feet stone.—And as the velocities are as the distance from the centre of motion, by the third law of circular motion, Art. 13, therefore, to find the velocity of the mean circle of the stone 2, apply the following rule; namely:

1st. Multiply the velocity of the water wheel into the radii or circumferences of all the driving wheels, successively, and note the product.

2d. Multiply the radii or circumferences of all the leading wheels, successively, and note the product; divide the first by the last product, and the quotient will be the answer.

But observe here, that the driving wheels in this rule, are the leading levers in the last rule.

EXAMPLES.

Suppose the velocity of the water-wheel to be 12 feet per second; then by the rule 12×4×3×2=288 and 8×2 ×1=16, by which divide the first product 288, and this gives 18 feet per second, the velocity of the stone 2 feet from its centre.

ARTICLE 21.

POWER DECREASES AS MOTION INCREASES.

It may be proper to observe here, that as the velocity of the stone is increased, the power to move it is decreased, and as its velocity is decreased, the power on it to move it is increased, by the second general law of mechanical powers. This holds universally true in all engines that can possibly be contrived; which is evident from the first law of the lever, when in equilibrium, namely, the power multiplied into its velocity or distance moved, is equal to the weight multiplied into its velocity or distance moved.

Hence the general rule to compute the power of any engine, simple or compound, Art. 17. If you have the moving power, and its velocity or distance moved, given, and the velocity or distance of the weight, then, to find the weight, (which, in mills, is the force to move the stone, &c.) divide that product by the velocity of the weight, or mill-stone, &c. and this gives the weight or force exerted on the stone to move it. But a certain qauntity or proportion of this force is lost from friction in order to obtain a velocity to the stone; which is shown in Art. 31.

ARTICLE 22.

NO POWER GAINED BY ENLARGING UNDERSHOT WATER-WHEELS.

This seems a proper time to show the absurdity of the idea of increasing the power of the mill, by enlarging the diameter of the water-wheel, on the principle of lengthening the lever; or by double gearing mills where single gears will do; because the power can neither be increased nor diminished by the help of engines, while the velocity of the body moved is to remain the same.

EXAMPLE.

Suppose we enlarge the diameter of the water-wheel from 8 to 16 feet radius, fig. 11, Plate II. and leave the other wheels unaltered; then, to find the velocity of the stone, allowing the velocity of the periphery of the water-wheel to be the same (12 feet per second;) by the rule $12 \times 4 \times 3 \times 2 = 288$, and $16 \times 2 \times 1 = 32$, by which divide 288, which gives 9 feet in a second, for the velocity of the stone.

Then, to find the power by the rule for that purpose, Art. 20, $500 \times 16 \times 2 \times 1 = 16000$, and $4 \times 3 \times 2 = 24$, by which divide 16000, it gives 666,66 lbs. the power. But as velocity as well as power, is necessary in mills, we shall be obliged, in order to restore the velocity, to enlarge the great cog-wheel from 4 to 8 radius.

Then, to find the velocity, 12×8×3×2=576, and 16×2×1=32, by which divide 576, it gives 18, the velocity as before.

Then, to find the power by the rule, Art. 20, it will be 333,33 as before.

Therefore no power can be gained, upon the principle of lengthening the lever, by enlarging the water-wheel.

The true advantages that large wheels have over small ones, arise from the width of the buckets bearing but a small proportion to the radius of the wheel; because if the radius of the wheel be 8 feet, and the width of the bucket or float-board but 1 foot, the float takes up 1-8 of the arm, and the water may be said to act fairly upon the end of the arm, and to advantage. But if the radius of the wheel be but 2 feet, and the width of the float 1 foot, part of the water will act on the middle of the arm, and of course, to disadvantage, as the float takes up half the arm. The large wheel also serves the purpose of a fly-wheel (Art. 30;) it likewise keeps a more regular motion, and casts off back water better. (See Art. 70.)

But the expense of these large wheels is to be taken into consideration, and then the builder will find that there is a maximum size, (see Art. 44,) or a size that will yield him the greatest profit.

ARTICLE 23.

NO POWER GAINED, BUT SOME LOST, BY DOUBLE GEARING MILLS.

I might go on to show that no power or advantage is to be gained by double gearing mills, upon any other principles than the following; namely:

1. When the motion necessary for the stone cannot be obtained without having the trundle too small, we are obliged to have the pitch of the cogs and rounds, and the size of the spindle, large enough to bear the stress of the

power; and this pitch of gear, and size of spindle, may bear too great a proportion to the radius of the trundle, (as does the size of the float to the radius of the water-wheel, Art. 22,) and may work hard. There therefore may be a loss of power on that account, greater than that resulting from friction in double gearing.

2. By double gearing, the mill may be made more convenient for two pair of stones to one water-wheel.

Many and great have been the losses sustained by mill-builders on account of their not properly understanding these principles. I have often met with water wheels of large diameter, where those of half the size and expense would answer better; and double gears, where single would be preferable.

ARTICLE 24.

OF THE PULLEY.

2. The pulley is a mechanical power well known. One pulley, if it be moveable with the weight, doubles the power, because each rope sustains half the weight.

If two or more pulleys be joined together in the common way, then the easiest mode of computing their power is, to count the number of ropes that join to the lower or moveable block, and so many times is the power increased; because all these ropes have to be shortened, and all run into one rope (called the fall) to which the moving power is applied. If there be 4 ropes, the power is increased fourfold. See Plate I. fig. 10.

The objection to this engine is, that there is great loss of power, by the friction of the pulleys, and in the bending of the ropes.

ARTICLE 25.

OF THE WHEEL AND AXLE.

3. The wheel and axle, fig. 17, is a mechanical power, similar to the lever of the first kind; therefore, when

the power is to the weight, as the diameter of the axle is to the diameter of the wheel; or when the power multiplied into the radius of the wheel is equal to the weight multiplied into the radius of the axle, this engine is in equilibrium.

The loss of power is but small in this instrument, because it has but little friction.

ARTICLE 26.

OF THE INCLINED PLANE.

4. The inclined plane is the fourth mechanical power; and in this the power is to the weight, as the perpendicular height of the plane is to its length. This is of use in rolling heavy bodies, such as barrels, hogsheads, &c. into wheel carriages &c., and for letting them down again. See Plate I. fig. 5. If the height of the plane be half its length, then half the force will roll the body up the plane, that would lift it perpendicularly to the same height, but it has to travel double the distance.

ARTICLE 27.

OF THE WEDGE.

5. The wedge is only an inclined plane. Whence, in the common form of it, the power applied will be to the resistance to be overcome, as the thickness of the wedge is to the length thereof. This is a very useful mechanical power, and, for some purposes, excels all the rest; because with it we can effect what we cannot with any other in the same time; and its power, I think, may be computed in the following manner.

If the wedge be 12 inches long and 2 inches thick, then the power to hold it in equilibrio is as 1 to balance 12 resistance; that is, 12 resistance pressing on each side of the wedge; and when struck with a mallet, the whole force of the weight of the mallet, added to the whole force of the power exerted in the stroke, is communicated to the wedge in the time it continues to move: and this force, to produce effect, is as the square of the velocity with which the mallet strikes, multiplied into its weight; therefore, the mallet should not be too large, because it may be too heavy for the workman's strength, and will meet too much resistance from the air, so that it will lose more by lessening the velocity, than it will gain by its weight. Suppose a mallet of 10 lbs. strike with 5 velocity, its effective momentum is 250; but if it strike with 10 velocity, then its effective momentum is 1000. The effects produced by the strokes will be as 250 to 1000; and all the force of each stroke, except what may be destroyed by the friction of the wedge, is added in the wedge, until the sum of these forces amounts to more than the resistance of the body to be split, which therefore, must give way; but when the wedge does not move, the whole force is destroyed by the friction; therefore, the less the inclination of the sides of the wedge, the greater the resistance we can overcome by it, because it will be easier moved by the stroke.

ARTICLE 28.

OF THE SCREW.

6. The screw is the last mentioned mechanical power, and may be denominated, a circular inclined plane, (as will appear by wrapping a paper, cut in form of an inclined plane, round a cylinder.) It is used in combination with a lever of the first kind, (the lever being applied to force the weight upon the inclined plane :) this compound instrument is a mechanical power, of extensive use, both for pressure, and raising great weights. The power applied is to the weight it will raise, as the distance through which the weight moves, is to the distance through which the power moves; that is, as the distance of two contiguous threads of the screw is to the circle the power describes, so is the power to the weight it will raise. If the distance of the thread be half an inch,

and the lever be fifteen inches radius, and the power applied be 10 lbs. then the power will describe a circle of 94 inches, while the weight rises half an inch; then, as half an inch is to 94 inches, so is 10 lbs. to 1880 lbs. the weight the engine would raise with 10 lbs. power. But this is supposing the screw to have no friction, of which it has a great deal.

ARTICLE 29.

OF THE FLY WHEEL, AND ITS USE.

Before I dismiss the subject of mechanical powers, I shall take some notice of the fly-wheel, the use of which is to regulate the motion of engines; it is best made of cast iron, and should be of a circular form, that it may not meet with much resistance from the air.

Many have supposed this wheel to be an increaser of power, whereas it is, in reality, a considerable destroyer of it: which appears evident, when we consider that it has no motion of its own, but receives all its motion from the first mover; and as the friction of the gudgeons, and the resistance of the air are to be overcome, this cannot be done without the loss of some power: yet this wheel is of great use in many cases; namely:

1st. For regulating the power, where it is irregularly applied, such as the treadle and crank moved by the foot or hand; as in spinning-wheels, turning-lathes, flax-mills, or where steam is applied, by a crank, to produce a circular motion.

2d. Where the resistance is irregular, or by jerks, as in saw-mills, forges, slitting-mills, powder-mills, &c. the fly-wheel, by its inertia, regulates the motion; because, if it be very heavy, it will require a great many little shocks or impulses of power to give it a considerable velocity; and it will, of course, require as many equal shocks to resist or destroy the velocity it has acquired.

While a rolling or slitting-mill is running empty, the

force of the water is employed in generating momentum in the fly-wheel; which force accumulated in the fly, will be sufficient to continue the motion without much abatement, while the sheet of metal is running between the rollers; whereas, had the force of the water been lost while the mill was empty, its motion might be destroyed before the metal passed through the rollers. Where water is scarce, its effect may be so far aided by a fly-wheel, as to overcome a resistance to which the direct force of the water is unequal, that is, where the power is required at intervals only.

A heavy water-wheel frequently produces all the effect of a fly-wheel, in addition to its direct office.

ARTICLE 30.

ON FRICTION.

We have hitherto considered the action and effect of the mechanical powers, as they would answer to the strictness of mathematical theory, were there no such thing as friction or rubbing of parts upon each other; but it is generally allowed, that one-fourth of the effect of a machine is, at a medium, destroyed by it: it will be proper to treat of it next in course.

From what I can gather from different authors, and by my own experiments, it appears that the doctrine of friction is as follows, and we may say it is subject to the following laws; namely:

Laws of Friction.

1. Friction is greatly influenced by the smoothness or roughness, hardness or softness, of the surfaces rubbing against each other.
2. It is in proportion to the pressure, or load; that is, a double pressure will produce a double amount of friction, a triple pressure a triple amount of friction, and so of any other proportionate increase of the load.
3. The friction does not depend upon the extent of surface, the weight of the body remaining the same.

Thus, if a parallelopiped, say of four inches in width and one in thickness, as F. plate II. fig. 13, be made smooth, and laid upon a smooth plane A. B. C. D. and the weight P. hung over a pulley, it will require the weight P. to draw the body F along, to be equal, whether it be laid on its side or on its edge.

The experiments of Vince led him to conclude that the law, as thus laid down, was not correct; but those more recently performed justify the conclusion, that it is so, the deviations being so trifling, as not to affect the general result.

4. The friction is greater after the bodies have been allowed to remain for some time at rest, in contact with each other, than when they are first so placed; as for example, a wheel turning upon gudgeons will require a greater weight to start it after remaining for some hours at rest, than it would at first.

The cause of this appears to be, that the minute asperities which exist even upon the smoothest bodies, gradually sink into the opposite spaces, and thus hold upon each other.

It is for the same reason, that a greater force is required to set a body in motion, than to keep it in motion. If about $\frac{1}{3}$ the amount of a weight be required to move that weight along in the first instance, $\frac{1}{4}$ will suffice to keep it in motion.

5. The friction of axles does not at all depend upon their velocity; thus, a rail-road car travelling at the rate of twenty miles an hour, will not have been retarded by friction, more than another which travels only ten miles in that time.

It appears, therefore, from the three last laws, that the amount of friction is as the pressure directly, without regard to surface, time, or velocity.

6. Friction is greatly diminished by unguents, and this diminution is as the nature of the unguents, without reference to the substances moving over them. The kind of unguent which ought to be employed depends principally upon the load; it ought to suffice just to prevent the bodies from coming into contact with each other. The lighter the weight, therefore, the finer and more fluid should be the unguent, and vice versa.

ARTICLE. 31.

ON THE FRICTION OF DIFFERENT SUBSTANCES.

It is well known that in general the friction of two dissimilar substances is less than that of similar substances, although alike in hardness. The most recent experiments upon this subject are those of Mr. Rennie, of England, performed in the year 1825, and published in the philosophical transactions. Many of the experiments were performed upon substances which do not concern the present work; those with the metals, and other hard substances, were tried both with and without unguents. The following facts were deduced from those in which unguents were not employed:

Table showing the amount of friction (without unguents) of different substances, the insistent weight being 36 lbs., and within the liits of abrasion of the softer substances.

	Parts of the whole weights.
Brass on wrought iron	7.38
Brass on cast iron	7.11
Brass on steel	7.20
Soft steel on soft steel	6.85
Cast iron on steel	6.62
Wrought iron on wrought iron	6.26
Cast iron on cast iron	6.12
Hard brass on cast iron	6.00
Cast iron on wrought iron	5.87
Brass on brass	5.70
Tin on cast iron	5.59
Tin on wrought iron	5.53
Soft steel on wrought iron	5-28

With unguents it was found that, with gun metal on cast iron, with oil intervening, the insistent weight being 10 cwt. the friction amounted to $\frac{1}{3\cdot13}$ of the pressure; that by a diminution of weight, the friction was rapidly diminished.

That cast iron on cast iron, under similar circumstances, showed less friction; and that this was still farther diminished by hog's lard.

That yellow brass, on cast iron, with anti-attrition composition of black lead and hog's lard, increased friction with light weights, and greatly diminished it with heavy weights, showing extremely irregular results.

That yellow brass, on cast iron, with tallow, gave the least friction, and may therefore be considered the best substance under the circumstances tried.

That yellow brass, on cast iron, with soft soap, gave the second best result, being superior to oil.

ARTICLE 32.

OF MECHANICAL CONTRIVANCES, TO REDUCE FRICTION.

Friction is considered as of two kinds, the first is occasioned by the rubbing of the surfaces of bodies against each other, the second by the rolling of a circular body, as that of a carriage wheel upon the ground, or rollers placed under a heavy load. In the preceding articles first kind of friction has been considered; it is that which we most frequently have to encounter, and which produces the greatest expenditure of power. When the parts can be made to roll over each other, the resistance is greatly diminished. To change one into the other has been the object of those mechanical contrivances denominated friction wheels, and friction rollers.

A, in plate II. fig. 14, may represent the gudgeon of a wheel set to run upon the peripheries of two wheels C. C, which pass each other; these are called friction wheels. This gudgeon, instead of grinding, or rubbing its surface or the surface on which it presses, carries that surface with it, causing the wheels C, C, to revolve. A gudgeon B, is sometimes set upon a single wheel, with supporters to keep it on, which produces an analogous effect.

Less advantage, however, has been derived from friction wheels in heavy machinery, than had been anticipated; and it has been found, in many cases, that they do not compensate for the expense of construction, and their liability to get out of order. The rubbing friction still exists on their gudgeons, and it has frequently happened that instead of turning them, the gudgeon resting upon them, has rolled round, whilst they have remained at rest.

The principle of the roller has already been noticed, and its mode of action is shown in fig. 15. plate II, where A B may represent a body of a 100 tons weight, with the under side perfectly smooth and even, set on rollers perfectly hard and smooth, rolling on a horizontal plane, C D, perfectly hard, smooth, and horizontal. If these rollers stand precisely parallel to each other, the least imaginable force would move the load; even a spiders web would be sufficient, were time allowed to overcome the inertia. These suppositions, however, can never be realized, and although in this mode of action there will be the least possible rubbing friction, there will be enough to produce considerable resistance.

It has been attempted to apply this principle to wheel carriages, to the sheaves of blocks on ship board, and to the axles of other machinery, by an ingenious contrivance called Garnett's friction rollers, for which a patent was obtained in England about fifty years ago, by an American gentleman from New Jersey. This contrivance is shown at fig. 16, Plate II. The outside ring B. C. D. may represent the box of a carriage wheel; the inside circle A the axle; the circles a a a a a a the rollers round the axle, and between it and the box; the inner ring is a thin plate for the pivots of the rollers to run in to keep them at a proper distance from each other. When the wheel turns, the rollers pass round on the axle, and on the inside of the box, and that almost without friction, because there is no rubbing of the parts in passing one another.

Such friction rollers, from the use of which so much was expected, have not been found to answer in practice. If not made with the most perfect accuracy, they *gather* as they roll, and thus increase the friction. In carriages, and, indeed, in every kind of machine, subject to an irregular jolting motion, the rollers, and the cylinder within which they revolve, soon become indented, and are then worse than useless.

ARTICLE 33.

OF MAXIMUMS, OR THE GREATEST EFFECTS OF ANY MACHINE.

The effect of a machine is the distance to which it moves a body of a given weight, in a given time; or, in other words, the resistance which it overcomes. The weight of the body, multiplied into its velocity, is the measure of this effect.

The theory published by philosophers, and received and taught as true, for several centuries past, is, that any machine will work with its greatest perfection when it is charged with just 4-9ths of the power that would hold it in equilibrio, and then its velocity will be just ⅓ of the greatest velocity, of the moving power.

To explain this, we may suppose the water-wheel, Plate II. fig. 17, to be of the undershot kind, 16 feet diameter, turned by water issuing from under a 4 feet head, with a gate drawn 1 foot wide, and 1 foot high, then the force will be 250 lbs., because that is the weight of the column of water above the gate, and its velocity will be 16,2 feet per second, as shall be shown under the head of Hydraulics; the wheel will then be moved by a power of 250 lbs., and if let run empty, will move with a velocity of 16 feet per second; but if the weight W be hung by a rope to the axle of two feet diameter, and we continue to add to it until it stops the wheel, and holds it in equilibrio, the weight will be found to be 2000 lbs. by the rule, Art. 19; and then the effect of the machine is nothing, because the velocity is nothing: but as we decrease the weight W, the wheel begins to move, and its velocity increases accordingly; and then the product of the weight multiplied into its velocity, will increase until the weight is decreased to 4-9ths of 2000=888,7, which multiplied into its velocity or distance moved, will produce the greatest effect, and the velocity of the wheel will then be ⅓ of 16 feet, or 5,33 feet per second. So say those who have treated of it.

This will probably appear plainer to a beginner, if he conceives this wheel to be applied to work an elevator, as E, Plate II, fig. 17, to hoist wheat, and suppose that the buckets, when all full, contain 9 pecks, and will hold the wheel in equilibrio, it is evident it will then hoist none, because it has no motion; and, in order to obtain motion, we must lessen the quantity in the buckets, when the wheel will begin to move, and hoist faster and faster until the quantity is decreased to 4-9ths or 4 pecks, and then, by the theory, the velocity of the machine will be ⅓ of the greatest velocity, when it will hoist the greatest quantity possible in a given time: for if we lessen the quantity in the buckets below 4 pecks, the quantity hoisted in any given time will be lessened; this is the established theory.

ARTICLE 34.

OLD THEORY INVESTIGATED.

In order to investigate this theory, and the better to understand what has been said, let us consider as follows; namely:

1. That the velocity of spouting water, under 4 feet head, is 16 feet per second, nearly.

2. The section or area of the gate drawn, in feet, multiplied by the height of the head in feet, gives the cubic feet in the whole column, which multiplied by 62,5 (the weight of a cubic foot of water) gives the weight or force of the whole column pressing on the wheel.

3. that the radius of the wheel, multiplied by the force, and that product divided by the radius of the axle, gives the weight that will hold the wheel in equilibrio.

4. That the absolute velocity of the wheel, subtracted from the absolute velocity of the water, leaves the re-

lative velocity with which the water strikes the wheel when in motion.

5. That as the radius of the wheel, is to the radius of the axle, so is the velocity of the wheel, to the velocity of the weight hoisted on the axle.

6. That the effects of spouting fluids, are as the squares of their velocities (see Art. 45, law 6,) but the instant force of striking fluids is as their velocities simply. See Art. 8.

7. That the weight hoisted, multiplied into its perpendicular ascent gives the effect.

8. That the weight of water expended, multiplied into its perpendicular descent, gives the power used per second.

On these principles I have calculated the following scale; first supposing the force of striking fluids, to be as the square of their striking or relative velocity, which brings out the maximum agreeably to the old theory, namely:

When the load at equilibrio is 2000, then the maximum load is 888,7=4-9ths of 2000, the effect being then greatest, namely, 591,98, as appears in the 6th column; and then the velocity of the wheel is 5,333 feet per second, equal to $\frac{1}{3}$ of 16, the velocity of the water, as appears in the 5th line of the scale: but there is an evident error in the first principle of this theory, by counting the instant force of the water on the wheel to be as the square of its striking velocity, it cannot, therefore, be true. See Art. 41.

I then calculate upon this principle, namely: that if the instant force of striking fluids, is as their velocity simply, then the load that the machine will carry, with its different velocities, will also be as the velocity simply, as appears in the 7th column; and the load, at a maximum, is 1000 lbs.=$\frac{1}{2}$ of 2000, the load at equilibrio, when the velocity of the wheel is 8 feet=$\frac{1}{2}$ of 16, the velocity of the water per second; and then the effect is at its greatest, as shown in the 8th column, namely, 1000, as appears in the 4th line of the scale.

This I call the new theory, (because I found that

William Waring had also, about the same time, established it, see Art. 37,) namely, that when any machine is charged with just one-half of the load that will hold it in equilibrio, its velocity will be just one-half of the natural velocity of the moving power, and then its effect will be at a maximum, or the greatest possible.

It thus appears that a great error has been long overlooked by philosophers, and that this has rendered the theory of no use in practice, but led many into expensive failures.

MECHANICS. [Chap. 2.

A Scale for determining the Maximum Charge, and Velocity, of Undershot Mills.

Ratio of the power and effect at a maximum, the power being 4000 in each case.			Maximum by new theory 4 to 1 10 to 1.47 Maximum by old theory.
Effect, by the new theory.	feet.		750 937 1000 937 878 859 750 375
Weight hoisted, according to the new theory.	lbs.		500 750 1000 1250 1332 1375 1500 1750 2000
Effect by the old theory.	feet.		0 187.5 351 500 585.9 591.98 590.6 562.5 382.7
Weight hoisted, according to the old theory.	lbs.		0 125 281 500 781.2 888.7 945 1125 1531 2000
Velocity of the weight ascending.	feet.		2 1.5 1.25 1 .75 .666 .625 .5 .25 0
Velocity with which the water strikes the wheel in motion, or relative velocity.	feet.		0 4 6 8 10 10.666 11 12 14 16
Velocity of the wheel per second, by supposition.	feet.		16 12 10 8 6 5.333 5 4 2 0

Radius of the wheel - - - feet.	8
Radius of the axle - - -	1
Section of the gate in square feet,	1
Height of the head of water -	4
Velocity of the water per second -	16
Weight of the column of water pressing on the wheel	lbs. 250
The weight that holds the wheel in equilibrio	2000

Chap. 2.] MECHANICS. 55

ARTICLE 35.

NEW THEORY DOUBTED.

Although I know that the velocity of the wheel, by this new theory, is (though rather slow,) much nearer to general practice than by the old, yet I am led to doubt its correctness, for the following reasons; namely:

There are 16 cubic feet of water, equal to 1000 lbs. expended in a second; which multiplied by its perpendicular descent, 4 feet, produces the power 4000. The ratio of the power and effect by the old theory, is as 10 to 1,47, and by the new, as 4 to 1, as appears in the 9th column of the scale; this is a proof that the old theory is incorrect, and sufficient to make us suspect that there is some error in the new. And as the subject is of the greatest consequence in practical mechanics, I therefore have endeavoured to discover a true theory, and will show my work, in order that if I establish a theory, it may be the easier understood, if right, or detected, if wrong.

ARTICLE 36.

ATTEMPT TO DEDUCE A TRUE THEORY.

I constructed the apparatus fig. 18, Plate II, which represents a simple wheel with a rope passing over it, and the weight P, 100 lbs. at one end to act by its gravity, as a power to produce effects, by hoisting the weight w at the other end.

This, seems to be on the principles of the lever, and overshot wheel; but with this exception, that the quantity of descending matter, acting as power, will still be the same, although the velocity will be accelerated, whereas, in overshot wheels, the power on the wheel is inversely, as the velocity of the wheel.

Here we must consider,

1. That the perpendicular descent of the body P, per second, multiplied into its weight, shows the power.
2. That the weight w, when multiplied into its perpendicular ascent, gives the effect.
3. That the natural velocity of the falling body P, is 16

feet the first second, and the distance it has to fall 16 feet.

4. That we suppose that the weight w, or resistance, will occupy its proportional part of the velocity; that is, if w be=½ P, the velocity with which P will then descend, will be ½ 16=8 feet per second.

5. If w be=P, there can be no velocity, consequently no effect; and if w=o, then P will descend 16 feet in a second, but produce no effect, because the power, although 1600 per second, is applied to hoist nothing.

Upon these principles I have calculated the following scale.

A SCALE

FOR DETERMINING THE

Maximum Charge and Velocity of 100 *lbs.*

DESCENDING BY ITS GRAVITY.

Power applied on the wheel.	Natural velocity in feet per second, of the power falling freely.	Weight w hoisted, or the resistance in lbs.	Proportion of the velocity occupied by the resistance, or weight w, hoisted.	Proportion of the velocity left in motion, which is the velocity of both power and weight.	Effect, which is the weight w, multiplied into its ascent per second.	Power, which is the power P, multiplied into its descent per second.	Ratio of the power and effect.	
lbs.	feet.	lbs.	feet.	feet.	0	1600	10..0	
100	16							
		1	.16	15.84	15.84	1584	10 : 0	
		10	1.6	14.4	144	1440	10 : 1	
		20	3.2	12.8	256	1280	10 : 2	
		30	4.8	11.2	336	1120	10 : 3	
		40	6.4	9.6	384	960	10 : 4	Maximum,
		50	8.	8	400	800	10 : 5	by new theory.
		60	9.6	6.4	384	640	10 : 6	
		70	11.2	4.8	336	480	10 : 7	
		80	12.8	3.2	256	320	10 : 8	
		90	14.4	1.6	144	160	10 : 9	
		99	15.84	.16	15.8	16	10 : 99	
		100	16.	0.	0.	0		

By this scale it appears, that when the weight w is =50=½ P the power, the effect is at a maximum, namely, 400, as appears in the 6th column, when the velocity is half the natural velocity, namely, 8 feet per second; and then the ratio of the power to the effect is as 10 to 5, as appears in the 8th line.

By this scale it appears, that all engines that are moved by one constant power, which is equally accelerated in its velocity, must be charged with weight or resistance equal to half the moving power, in order to produce the greatest effect in a given time; but if time be not regarded, then the greater the charge, so as to leave any velocity, the greater the effect, as appears by the 8th column. So that it appears that an overshot wheel, if it be made immensely capacious, and to move very slowly, may produce effects in the ratio of 9,9 to 10 of the power.

ARTICLE 37.

SCALE OF EXPERIMENTS.

The following is a scale of actual experiments made to prove whether the resistance occupies its proportion of the velocity, in order that I might judge whether the foregoing scale was founded on true principles; the experiments were not very accurately performed, but were often repeated, and the results were always nearly the same. See Plate II. fig. 18.

A SCALE
OF
EXPERIMENTS.

Power applied on the wheel, in pounds.	Distance it had to descend, in feet.	Weight, in pounds, hoisted the whole distance.	Equal parts of time [each being two beats of a watch] in which the weight was hoisted the whole distance.	Distance in feet, that the weight moved in one of the equal parts of time, found by dividing 40, the whole distance, by the number of equal parts of time taken up in the ascent.	Effect, found by multiplying the weight w into the velocity, or distance ascended in one of those parts of time.	Power, found by multiplying the weight of P into its descent, in one of those parts of time.	Ratio of the power and effect.	Effect, supposing it to be as the square of the velocity of the weight; found by multiplying the weight into the square of its velocity.
7	40	7		0	0			
		6	20	2×6	12	14	10:8.5	24
		5	15.5	2.6×5	13	18.2	10:7.1	33.8
		4	12	3.33×4	13.32	23.31	10:5.7	44.35
		3.5	10	4×3.5	14	28	10:5.	maximum new theory.
		3	9	4.44×3	13.32	31.08	10:4.2	59.14
		2	6.5	6×2	12	42	10:2.8	72 maximum
		1	6	6.6×1	6.6	46.2	10:1.4	33.56
		0	5	8	0	56		

By this scale it appears, that when the power P falls freely without any load, it descends 40 feet in five equal parts of time; but, when charged with 3,5 lbs. $=\frac{1}{2}$ P, which was 7 lbs., it then takes up 10 of those parts of time to descend the same distance; which seems to show, that the charge occupies its proportional part of the whole velocity, which was wanted to be known, and the maximum appears as in the last scale. It also shows that the effect is not as the weight multiplied into the square of its ascending velocity, this being the measure of the effect that would be produced by the stroke on a non-elastic body.

Atwood, in his Treatise on Motion, gives a set of accurate experiments, to prove (beyond doubt) that the conclusion I have drawn is right; namely:—That the charge occupies its proportional part of the whole velocity.

These experiments partly confirmed me in what I have called the New Theory; but still doubting, and after I had formed the foregoing tables, I called, for his assistance, on the late ingenious and worthy friend, William Waring, teacher in the Friend's Academy, Philadelphia, who informed me that he had discovered the error in the old theory, and corrected it in a paper which he had laid before the Philosophical Society of Philadelphia, wherein he had shown that the velocity of the undershot water-wheel, to produce a maximum effect, must be just one half the velocity of the water.

ARTICLE 38.

WILLIAM WARING'S THEORY.

The following are extracts from the above mentioned paper, published in the third volume of the transactions of the American Philosophical Society, held at Philadelphia, p. 144.

After his learned and modest introduction, in which he shows the necessity of correcting so great an **error as** the old theory, he begins with these words; namely:

"But, to come to the point, I would just premise these

DEFINITIONS.

If a stream of water impinge against a wheel in motion, there are three different velocities to be considered appertaining thereto; namely:

1st. The absolute velocity of the water.

2d. The absolute velocity of the wheel.

3d. The relative velocity of the water to that of the wheel; that is, the difference of the absolute velocities, or the velocity with which the water overtakes or strikes the wheel.

Now the mistake consists in supposing the momentum, or force of the water against the wheel, to be in the duplicate ratio of the relative velocity; whereas,

Prop. 1.

The force of an invariable stream impinging against a mill-wheel in motion, is in the simple proportion of the relative velocity.

For, if the relative velocity of a fluid against a single plane be varied either by the motion of the plane or of the fluid from a given aperture, or both, then the number of particles acting on the plane, in a given time, and likewise the momentum of each particle being respectively as the relative velocity, the force, on both these accounts, must be in the duplicate ratio of the relative velocity, agreeably to the common theory, with respect to this single plane; but the number of these planes, or parts of the wheel acted on in a given time, will be as the velocity of the wheel, or inversely as the relative velocity; therefore, the moving force of the wheel must be as the simple ratio of the relative velocity. Q. E. D.

Or the proposition is manifest from this consideration, that while the stream is invariable, whatever be the velocity of the wheel, the same number of particles, or quantity of fluid, must strike it somewhere or other in a given time; consequently, the variation of the force is only on account of the varied impingent velocity of the same body, occasioned by a change of motion in the wheel; that is, the momentum is as the relative velocity.

Now, this true principle, substituted for the erroneous one in use, will bring the theory to agree remarkably with the notable experiments of the ingenious Smeaton, published in the Philosophical Transactions of the Royal Society of London, for the year 1751, vol. 51; for which the honorary annual medal was adjudged by the society, and presented to the author by their president.

An instance or two of the importance of this correction may be adduced, as follows:

Prop. II.

The velocity of a wheel, moved by the impact of a stream, must be half the velocity of the fluid, to produce the greatest effect possible.

V=the velocity, M=the momentum, of the fluid.
v=the velocity, P=the power, of the wheel.

Then $V-v$=their relative velocity, by definition 3d.

And, as $V:V-v::M:\frac{M}{V}\times\overline{V-v}=P$, (Prop. 1,) which $\times v=P$, $v=\frac{M}{V}\times \overline{Vv-v^2}$=a maximum; hence $Vv-v^2$= a maximum and its fluxion (v being a variable quantity) $=Vv-2vv=0$; therefore $=\frac{1}{2}V$; that is, the velocity of the wheel=half that of the fluid, at the place of impact when the effect is a maximum. Q. E. D.

The usual theory gives $v=\frac{1}{3}V$, where the error is not less than one-sixth of the true velocity.

Wm. Waring."

Philadelphia, 7th
  9th mo. 1790.

I here omit quoting Proposition III. as it is altogether algebraical, and refers to a figure; I am not writing for men of science, but for practical mechanics.

ARTICLE 39.

Extract from a farther paper, read in the Philosophical Society, April 5th, 1793.

"Since the Philosophical Society were pleased to favour my crude observations on the theory of mills with a publication in their transactions, I am apprehensive some part thereof may be misapplied; it being therein demonstrated, that 'the force of an invariable stream, impinging against a mill-wheel in motion, is in the simple direct ratio of the relative velocity.' Some may suppose that the effect produced should be in the same proportion, and either fall into an error, or finding by experiment, the effect to be as the square of the velocity, conclude the new theory to be not well founded; I therefore wish there had been a little added, to prevent such misapplication, before the Society had been troubled with the reading of my paper on that subject, perhaps something like the following.

The maximum effect of an undershot wheel, produced by a given quantity of water, in a given time, is in the duplicate ratio of the velocity of the water; for the effect must be as the impetus acting on the wheel, multiplied into the velocity thereof: but this impetus is demonstrated to be simply as the relative velocity, Proposition I., and the velocity of the wheel, producing a maximum, being half of the water by Proposition II., is likewise as the velocity of the water; hence the power acting on the wheel, multiplied into the velocity of the wheel, or the effect produced, must be in the duplicate ratio of the velocity of the water. Q. E. D.

COROLLARY. Hence the effect of a given quantity of water, in a given time, will be as the height of the head, because this height is as the square of the velocity. This also agrees with experiment.

If the force, acting on the wheel, were in duplicate ratio of the water's velocity, as is usually asserted, then the effect would be as the cube thereof, when the quantity of water and time are given, which is contrary to the result of experiment."

ARTICLE 40.

WARING'S THEORY DOUBTED.

From the time I first called on William Waring, until I read his publication on the subject (after his death,) I had rested partly satisfied with the new theory, as I have called it, with respect to the velocity of the wheel, at least; but finding that he had not determined the charge, as well as the velocity, by which we might have compared the ratio of the power and the effect produced, and that he had assigned somewhat different reasons for the error, and having found the motion to be rather too slow to agree with practice, I began to suspect the whole, and resumed the search for a true theory, thinking that perhaps no person had ever yet considered every thing that affects the calculation; I therefore premised the following

POSTULATES.

1. A given quantity of perfectly elastic, or solid matter, impinging on a fixed obstacle, its effective force is as the squares of its different velocities, although its instant force may be as its velocities simply, because the distance it will recede after the stroke through any reristing medium, will be as the squares of its impinging velocities.
2. An equal quantity of elastic matter impinging on a fixed obstacle with a double velocity, produces a quadruple effect, their effects are as the squares of their velocities. Consequently—
3. A double quantity of said matter, impinging with a

double velocity, produces an octuple effect, or their effects are as the cubes of their velocities, Art. 47 and 67.

4. If the impinging matter be non-elastic, such as fluids, then the instant force will be but half, but the ratio will be the same in each case.

5. A double velocity, through a given aperture, gives a double quantity to strike the obstacle or wheel; therefore the effects will be as the cubes of the velocity. See Art. 47.

6. But a double relative velocity cannot increase the quantity that is to act on the wheel; therefore, the effect can only be as the square of the velocity, by postulate 2.

7. Although the instant force and effects of fluids striking on fixed obstacles, are only as their simple velocities, yet their effects, on moving wheels, are as the squares of their velocities; because, 1st, a double striking velocity gives a double instant force, which bears a double load on the wheel; and, 2d. a double velocity moves the load a double distance in an equal time, and a double load moved a double distance, is a quadruple effect.

ARTICLE 41.

SEARCH FOR A TRUE THEORY, COMMENCED ON A NEW PLAN.

It appears that we have applied wrong principles in our search after the true theory of the maximum velocity, and load of undershot water-wheels, or other engines moved by a constant power, that does not increase or decrease in quantity on the engine, as on an overshot water-wheel, as the velocity varies.

Let us suppose water to issue from under a head of 16 feet, on an undershot water-wheel; then, if the wheel move freely with the water, its velocity will be 32,4 feet per second, but will bear no load.

Again; suppose we load it, so as to make its motion equal only to the velocity of water spouting from under a head of 15 feet; it appears evident that the load will then be just equal to the 1 foot of the head, the **velocity** of which is checked; and this load multiplied into the velocity of the wheel; namely: 31,34×1=31,34, for the effect.

This appears to be the true principle, from which we must seek the maximum velocity and load, for such engines as are moved by one constant power; and on this principle I have calculated the following scale.

A SCALE
FOR DETERMINING THE
TRUE MAXIMUM VELOCITY AND LOAD
FOR
UNDERSHOT WHEELS.

Total head of water in action.	Head of water left unbalanced to give motion to the wheel.	Velocity of the wheel in feet per second, being equal the velocity of the water from under the head left unbalanced.	Load of the wheel, being equal that part of the total head, the motion of which is checked.	Effect per second, being the velocity of the wheel, multiplied by the load.	
feet.	feet.	feet.			
16	16	32.4	0	0	
	15	31.34	1	31.34	
	14	30.2	2	60.4	
	12	28	4	112	
	10	25.54	6	153.24	
	8	22.8	8	182.4	
	7	21.43	9	192.87	
	6	19.84	10	198.4	
	5·66	19.27	10.33	198.95	
	5·35	18.71	10.66	199.44	Maximum motion and load.
	5	18	11	198	
	4	16.2	12	194.4	
	3	14	13	172	
	2	11.4	14	159.6	
	1	8.1	15	120	
	0	0	16	0	

In this scale let us suppose the aperture of the gate to be a square foot; then the greatest load that will balance the head will be 16 cubic feet of water and the different loads will be shown in cubic feet of water.

It appears, by this scale, that when the wheel is loaded with 10,66 cubic feet of water, just ⅔ of the greatest load, its velocity will be 18,71 feet per second, just ,577 parts of the velocity of the water, and the effect produced is then at a maximum, or the greatest possible, namely: 199,44.

To make this more plain, let us suppose A B, Plate II. fig. 19, to be a fall of water of 16 feet, which we wish to apply to produce the greatest effect possible, by hoisting water on its side, opposite to the power applied. First, on the undershot principle, where the water acts by its impulse only. Let us suppose the water to strike the wheel at I, then, if we let the wheel move freely without any load, it will move with the velocity of the water, namely, 32,4 feet per second, but will produce no effect, if the water issue at C; although there be 32,4 cubic feet of water expended, under 16 feet perpendicular descent. Let the weight of a cubic foot of water be represented by unity or 1, for ease in counting; then 32,4×16 will show the power expended, per second, namely, 518,4; and the water it hoists multiplied into its perpendicular ascent, or height hoisted, will show the effect. Then, in order to obtain effect from the power, we load the wheel; the simplest way of doing which is, to cause the tube of water C D, to act on the back of the bucket at I; then, if C D be equal to A B, the wheel will be held in equilibrio; this is the greatest load, and the whole of the fall A B is balanced, and no part left to give the wheel velocity; therefore the effect=o. But if we make C D=12 feet of A B, then from 4 to A,=4 feet, is left unbalanced, to give velocity to the wheel, which being loaded with 12 feet, would be exactly balanced by 12 on the other side, and left perfectly free to move either way by the least force applied beyond this balance. Therefore, it is evident, that the whole pressure or force of 4 feet of A B, will act to give velocity to the wheel, and, as there is no resistance to oppose the pressure of these 4 feet, the velocity will be that of water spouting from under a 4 feet head, namely, 16,2 feet per second, which is shown by the horizontal line, 4=16,2; and the perpendicular line, 12=12, represents the load of the wheel; the rectangle or product of these two lines forms a parallelogram, the area of which is a true representation of the effect, namely, the load 12 multiplied into 16,2, the distance it moves per second=194,4, the effect. In like manner we may try the effect of different loads; the less the load, the greater will be the velocity. The horizontal lines all show the velocity of the wheel, produced by the respective heads left unbalanced, and the perpendicular lines show the load on the wheel; and we find, that when the load is 10,66=$\frac{2}{3}$ 16, the load at equilibrio, the velocity of the wheel will be 18,71 feet per second, which is $\frac{577}{1000}$ parts, or a little less than 6 tenths, or $\frac{2}{3}$ the velocity of the water, and the effect is 199,44 the maximum or greatest possible; and if the aperture of the gate be 1 foot, the quantity will be 18, 71 cubic feet per second. The power being 18,71 cubic feet expended per second, multiplied by 16 feet, the perpendicular descent, produces 299,36, the ratio of the power and effect, being as 10 to $6\frac{8}{10}$, or nearly as 3 : 2; but this is supposing none of the force lost by non-elasticity.

This may appear plainer, if we suppose the water to descend in the tube A B, and, by its pressure, to raise the water in the tube C D; for it is evident, that if we raise the water to D, we have no velocity, therefore, effect=0. Then again, if we open the gate at C, we have 32,4 feet per second velocity; but because we do not hoist the water to any height, effect is=0. Therefore, the maximum is somewhere between C and D. Then supppose we open gates of 1 foot area, at different heights, the velocity will show the quantity of cubic feet raised; which multiplied by the perpendicular height of the gate from C, or height raised, gives the effect, and the maximum as before. But here we must consider, that in both these cases the water acts as a perfectly definite quantity, which will produce effects equal to elastic bodies, or equal to its gravity (See Art. 59,) which is unattainable in practice: whereas, when it acts by percussion only, it communicates only half of its original force, on account of its non-elasticity, the other half being spent in splashing about; therefore the true effect will be $\frac{38}{100}$ (a little more than $\frac{1}{3}$) of the moving power; because nearly $\frac{1}{3}$ is lost to obtain velocity, and half of the remaining $\frac{2}{3}$ is lost by non-elasticity. These are the reasons why the effect produced by an undershot wheel is only half of that produced by an overshot wheel, the perpendicular descent and quantity of water being equal. And this agrees with Smeaton's experiments (see Art. 68;) but if we suppose the velocity of the wheel to be one-third that of the water=10,8, and the load to be $\frac{4}{9}$ of 16, the greatest load at equilibrio, which is=7,111, as by old theory, then the effect will be 10,8×4,9 of 16 =76,79 for the effect, which is quite too little, the moving power being 32,4 cubic feet of water, multiplied by 16 feet descent=518,4; the effect by this theory being less than $\frac{15}{100}$ of the power, about half equal to the effect by experiment, which effect is set on the outside of the dotted circle in fig. 19. The dotted lines join the corner of the parallelograms, formed by the lines that represent the loads and velocities, in each experiment or supposition, the areas of which truly represent the effect, and the dotted line A a d x, meeting the perpendicular line x E in the point x, forming the parallelogram ABCx, truly represents the power =518,4.

Again, if we suppose the wheel to move with half the velocity of the water, namely, 16,2 feet per second; and to be loaded with half the greatest load=8, according to Waring's theory; then the effect will be 16.2×8=129,6 for the effect, about $\frac{23}{100}$ of the power, which is still less than by experiment. All this seems to confirm the maximum brought out on the new principles.

But, if we suppose, according to the new principle, that, when the wheel moves with the velocity of 16,2 feet per second, which is the velocity of a 4 feet head, it will then bear as a load the remaining 12 feet, then the effect will be 16,2×12=194,4, which nearly agrees

Chap. 2.]   MECHANICS.   69

with practice: but as most mills in practice move faster, and few slower, than what I call the true maximum, this shows it to be nearest the truth; the true maximum velocity being ,577 of the velocity of the water, and the mills in practice moving with ⅔, and generally quicker.*

This scale also establishes a true maximum charge for an overshot wheel; that is, if we suppose the power, or quantity of water on the wheel at once, to be always the same, even although the velocity vary, which would be the case, if the buckets were kept always full: for, suppose the water to be shot into the wheel at a, and by its gravity to raise the whole water again on the opposite side; then as soon as the water rises in the wheel to d, it is evident that the wheel will stop, and the effect bb =o; therefore, we must let the water out of the wheel, before it rises to d, which will be, in effect, to lose part of the power to obtain velocity. If the buckets both descending and ascending, carry a column of water 1 foot square, then the velocity of the wheel will show the quantity hoisted as before, which multiplied by the perpendicular ascent, shows the effect; and the quantity expended multiplied by the perpendicular descent shows the power; and we find, that when the wheel is loaded with ⅔ of the power, the effect will be at a maximum; that is, the whole of the water is hoisted, ⅔ of its whole

---

* The reason why the wheel bears so great a load at a maximum, appears to be as follows; namely:—

A 16 feet head of water over a gate of 1 foot, issues 32,4 cubic feet of water in a second, to strike the wheel in the same time, that a heavy body will take up in falling through the height of the head. Now, if 16 cubic feet of elastic matter were to fall 16 feet, and strike an elastic plane, it would rise by the force of the stroke to the height from whence it fell; or, in other words, it will have force sufficient to bear a load of 16 cubic feet.

Again, if 32 cubic feet of non-elastic matter, moving with the same velocity (with which the 16 feet of elastic matter struck the plane,) strike a wheel in the same time, although it communicate only half the force that gave it motion; yet, because there is a double quantity striking in the same time, the effects will be equal; that is, it will bear a load of 16 cubic feet, or the whole column to hold it in equilibrio.

Again, to check the whole velocity, requires the whole column that produces the velocity; consequently, to check any part of the velocity, will require such a part of the column, as is equal to the part checked; and we find by Art. 41, that, to check the velocity of the wheel, so as to be ,577 of the velocity of the water, it requires 2-3ds of the whole column, and this is the maximum load. When the velocity of the wheel is multiplied by 2-3ds of the column, it produces the effect, which will be to the power, as 38 to 100; or, as 3,8 to 10, somewhat more than 1-3d, and the friction and resistance of the air may reduce it to 1-3d.

descent; or ⅔ of the water, the whole of the descent; therefore, the ratio of the power to the effect is as 3 to 2; or double the effect of an undershot wheel: but this is supposing the quantity in the buckets to be always the same, whereas, in overshot wheels, the quantity in the buckets is inversely as the velocity of the wheel; that is, the slower the motion of the wheel, the greater the quantity in the buckets, and the greater the velocity, the less the quantity: but, again, as we are obliged to let the overshot wheel move with a considerable velocity, in order to obtain a steady, regular motion to the mill, we shall find this charge to be always nearly right; hence, I deduce the following theory.

ARTICLE 42.

THEORY.

A TRUE THEORY DEDUCED.

This scale seems to have shown,

1. That when an undershot mill moves with ,577 or nearly ,6 of the velocity of the water, it will then bear a charge, equal to ⅔ of the load that will hold the wheel in equilibrio, and then the effect will be at a maximum. The ratio of the power to the effect will be as 3 to 1, nearly.

2. That when an overshot wheel is charged with ⅔ of the weight of the water acting upon the wheel, then the effect will be at a maximum; that is, the greatest effect that can be produced by said power in a given time, and the ratio of the power to the effect will be as 3 to 2, nearly.

3. That ⅓ of the power is necessarily lost, to obtain velocity, or to overcome the inertia of the matter; and this will hold true with all machinery that requires velocity as well as power. This I believe to be the true theory of water-mills, for the following reasons; namely:

1. The theory is deduced from original reasoning, without depending much on calculation.

2. It agrees better than any other theory, with the ingenious Smeaton's experiments.

3. It agrees best with real practice, according to the best of my information

Yet I do not wish any person to receive it implicitly, without first informing himself whether it be well founded, and in accordance with actual experience: for this reason I have quoted the experiments of Smeaton at full length, in this work, that the reader may compare them with the theory.

TRUE THEOREM FOR FINDING THE MAXIMUM CHARGE FOR UNDERSHOT WHEELS.

As the square of the velocity of the water, or wheel empty, is to the height of the head, or pressure, which produced that velocity, so is the square of the velocity of the wheel loaded, to the head, pressure, or force, which will produce that velocity; and this pressure, deducted from the whole pressure or force, will leave the load moved by the wheel, on its periphery or verge, which load, multiplied by the velocity of the wheel, shows the effect.

PROBLEM.

Let $V=32,4$, the velocity of the water or wheel,
$P=16$, the pressure, force, or load, at equilibrio,
$v=$the velocity of the wheel, supposed to be $16,2$ feet, per second,
$p=$the pressure, force, or head, to produce said velocity,
$l=$the load on the wheel,
Then to find l, the load, we must first find p;
Then, by
Theorem $VV : P :: vv : p,$
And $P-p=l$
$$VVp=vvP$$
$$p=\frac{vvP}{VV}=4$$
$l=P-p=12,$ the load.

Which, in words at length, is, the square of the velocity of the wheel, multiplied by the whole force, pressure, or head of the water, and divided by the square of the velocity of the water, quotes the pressure, force, or head of water, that is left unbalanced by the load to produce the velocity of the wheel; which pressure, force, or head, subtracted from the whole pressure, force, or head, leaves the load that is on the wheel.

ARTICLE 43.

Theorem for finding the velocity of the wheel, when we have the velocity of the water, load at equilibrio, and load on the wheel given.

As the square root of the whole pressure, force or load at equilibrio, is to the velocity of the water, so is the square root of the difference, between the load on the wheel, and the load at equilibrio, to the velocity of the wheel.

PROBLEM.

Let V=velocity of the water=32,4,
  P=pressure, force, head, or load at equilibrio=16,
  l=the load on the wheel, suppose 12,
  v=velocity of the wheel,
Then by the
Theorem $\sqrt{P}:V::\sqrt{P-l}:v$
And $\sqrt{P}\times v = V\sqrt{P-l}$

$$v = \frac{V\sqrt{P-l}}{\sqrt{P}} = 16,2 \quad \left\{ \begin{array}{c} \text{The velocity of the} \\ \text{wheel.} \end{array} \right.$$

That is, in words at length, the velocity of the water 32,4, multiplied by the square root of the difference, between the load on the wheel, 12, and the load at equilibrio 16=2=64,8, divided by the square root of the load at equilibrio, quotes 16,2, the velocity of the wheel.

Now, if we seek for the maximum, by either of these theorems, it will be found as in the scale, fig. 19.

Perhaps here may now appear the true cause of the error in the old theory, Art. 34, by supposing the load on the wheel, to be as the square of the relative velocity of the water and wheel.

And of the error in what I have called the new theory, by supposing the load to be in the simple ratio of the relative or striking velocity of the water, Art. 38; whereas it is to be found by neither of these proportions.

Neither the old nor the new theory agrees with practice; therefore, we may suspect they are both founded in error.

But if what I call the true theory should be found to accord with experience, the practitioner need not be much concerned on what it is founded.

ARTICLE 44.

Of the Maximum Velocity for Overshot Wheels, or those that are moved by the weight of the Water.

Before I dismiss the subject of maximums, I think it best to consider, whether this doctrine will apply to the motion of the overshot wheel. It seems to be the general opinion of those who consider the matter, that it will not; but that the slower the wheel moves, provided it be capacious enough to hold all the water, without losing any until it be delivered at the bottom of the wheel, the greater will be the effect, which appears to be the case in theory (see Art. 36;) but how far this theory will hold good in practice is to be considered. Having met with the ingenious James Smeaton's experiments, where he shows, that when the circumference of his little wheel, of 24 inches diameter, (head 6 inches) moved with about 3,1 feet per second (although the greatest effect was diminished about $\frac{1}{20}$ of the whole) he obtained the best effect, with a steady, regular motion. Hence he concludes about three feet to be the best velocity for the circumference of overshot mills. See Art. 68. I undertook to compare this theory of his with the best mills in practice, and, finding that those of about 17 feet diameter generally moved about 9 feet per second, being treble the velocity assigned by Smeaton, I began to doubt the theory, which led me to inquire into the principle that moves an overshot wheel; and this I found to be that of a body descending by its gravity, and subject to all the laws of falling bodies (Art. 10,) or of bodies descending inclined planes, and curved surfaces (Art. 11;) the motion being equably accelerated in the whole of its descent, its velocity being as the square root of the distance descended through; and, that the diameter of the wheel was the distance through which the water descended. From thence I concluded, that the velocity of the circumference of overshot wheels was as the square root of their diameters, and of the distance the water has to descend, if it be a breast or a pitch-back wheel: then, taking Smeaton's experiments, with his wheel of two feet diameter, for a foundation, I say, As the square root of the diameter of Smeaton's wheel is to its maximum velocity, so is the square root of the diameter of any other wheel, to its maximum velocity. Upon these principles I have calculated the following table; and, having compared it with at least 50 mills in practice, found it to agree so nearly with all those best constructed, that I have reason to believe it is founded on true principles.

If an overshot wheel move freely, without resistance, it will require a mean velocity between that of the water coming on the wheel, and the greatest velocity it would acquire, by falling freely through its whole descent: therefore, this mean velocity will be greater than the velocity of the water coming on the wheel; consequently, the backs of the buckets will overtake the water, and drive a great part of it out of the wheel. But, the velocity of the water being accelerated by its gravity, overtakes the wheel, perhaps half way down, and presses on the buckets, until it leaves the wheel: therefore the water presses harder upon the buckets in the lower than in the upper quarter of the wheel. Hence appears the reason why some wheels cast their water; which is always the case, when the head is not sufficient to give it velocity enough to enter the buckets. But this depends also much on the position of the buckets, and the direction of the shute into them. It, however, appears evident that the head of water above the wheel, should be nicely adjusted to suit the velocity of the wheel. Here we may consider, that the head above the wheel acts by percussion, or on the same principles with the undershot wheel; and, as we have shown (Art. 40) that the undershot wheel should move with nearly 2-3ds of the velocity of the water, it appears, that we should allow a head over the wheel, that will give such velocity to the water, as will be to that of the wheel as 3 to 2. Thus, the whole descent of the water of a millseat should be nicely divided between head and fall, to suit each other, in order to obtain the best effect, and a steady-moving mill. First, find the velocity with which the wheel will move, by the weight of the water, for any diameter you may suppose you will take for the wheel, and divide said velocity into two parts; then try if your head be such as will cause the water to come on with a velocity of 3 such parts, making due allowances for the friction of the water, according to the aperture. See Art. 55. Then, if the buckets and the direction of the shute be right, the wheel will receive the water well, and move to the best advantage, keeping a steady, regular motion when at work, loaded or charged with a resistance equal to 2-3ds of its power. (Art. 41, 42.)

A TABLE
OF
VELOCITIES OF THE CIRCUMFERENCE
OF
OVERSHOT WHEELS

Suitable to their Diameters, or rather to the Fall, after the Water strikes the wheel; and of the head of Water above the Wheel, suitable to said Velocities; also of the Number of Revolutions the Wheel will perform in a Minute, when rightly charged.

Diameter of the wheel in feet.	Velocity of its circumference, in feet and parts, per second.	Head of water above the wheel to give velocity as 3 to 2 of the wheel, in feet and parts.	Additional head to overcome the friction of the aperture, by conjecture only.	Total head of water.	Number of revolutions of the wheel per minute.
2	3.1				
3	3.78				
4	4.38				
5	4.88				
6	5.36				
7	5.8				
8	6.19				
9	6.57	1.41	.1	1.51	14.3
10	6.92	1.64	.1	1.74	13.
11	7.24	1.84	.1	1.94	12.6
12	7.57	2.	.2	2.2	12.
13	7.86	2.17	.3	2.47	11.54
14	8.19	2.34	.4	2.74	11.17
15	8.47	2.49	.5	2.99	10.78
16	8.76	2.68	.6	3.28	10.4
17	9.	2.8	.7	3.5	10.1
18	9.28	3.	.8	3.8	9.8
19	9.5	3.13	.9	4.03	9.54
20	9.78	3.34	1.	4.34	9.3
21	10.	3.49	1.05	4.54	9.1
22	10.28	3.76	1.1	4.86	8.9
23	10.5	3.84	1.15	4.99	8.7
24	10.7	3.97	1.2	5.27	8.5
25	10.95	4.2	1.25	5.45	8.3
26	11.16	4.27	1.3	5.57	8.19
27	11.36	4.42	1.35	5.77	8.03
28	11.54	4.56	1.4	5.96	7.93
29	11.78	4.7	1.45	6.15	7.75
30	11.99	4.9	1.5	6.4	7.63

This doctrine of maximums is very interesting, and is to be met with, in many occurrences through life.

1. It has been shown, that there is a maximum load and velocity for all engines, to suit the power and velocity of the moving power.

2. There is also a maximum size, velocity, and feed for mill-stones, to suit the power; a maximum velocity for rolling screens, and bolting-reels, by which the greatest work can be done in the best manner, in a given time.

3. A maximum degree of perfection and closeness, with which grain is to be manufactured into flour, so as to yield the greatest profit by the mill in a day or week, and this maximum is continually changing with the prices in the market, so that what would be the greatest profit at one time, will sink money at another. See Art. 113.

4. A maximum weight for mallets, axes, sledges, &c., according to the strength of those that use them.

A true attention to the principles of maximums, will prevent us from running into many errors.

CHAPTER III.

HYDRAULICS.

PRELIMINARY REMARKS.

The science which treats upon the mechanical properties and effects of water and other fluids, has most commonly been divided into two branches, HYDROSTATICS and HYDRAULICS. *Hydrostatics* treats of the weight, pressure, and equilibrium of fluids, when in a state of rest. *Hydraulics* treats of water in motion, and the means of raising, conducting, and using it for moving machinery, or for other purposes. These two divisions are so intimately connected with each other, that the latter could not be at all understood without an acquaintance with the former; and it is not necessary, in a work like the present, to treat of them separately. Considered abstractedly, the same laws obtain in the pressure and motion of water, as those which belong to solid bodies; and in the last chapter, on Mechanics, this similarity has led to some notice of the effects produced by water, which, strictly speaking, would belong to the present. In doing this, utility has been preferred to a strict adherence to system.

In treating of the elementary principles of Hydraulics, it is necessary to proceed upon theoretical principles; but let it always be recollected that from various causes resulting from the constitution of fluids, and particularly from that essential property in them, the perfect mobility of their particles among each other, the phenomena actually exhibited in nature, or in the processes of art, in which the motion of water is concerned, deviate so very considerably from the deductions of theory, that the latter must be considered as a very imperfect guide to the practical mill-wright and engineer. It is not to be inferred from this circumstance, that such theoretical investigations are false and useless; they are still approximations, which serve as guides to a certain extent. Their defectiveness arises from our inability to form an estimate of the many disturbing causes which influence the motion of fluids; whilst in the mechanics of solids we have, in many cases, no other correction to make in our theoretical deductions, than to allow for the effect of friction.

"The only really useful method of treating a branch of knowledge so circumstanced, is to accompany a very concise account of such general principles as are least inapplicable to practice, by proportionately copious details of the most accurate experiments which have been instituted, with a view to ascertain the actual circumstances of the various phenomena." (*Lardner's Hydrostatics.* Such has been the course pursued, to a considerable extent, by the author of this work, and in pursuing this subject, under the present head of Hydraulics, we shall consider only such parts of the science as immediately relate to our purpose: namely, such as may lead to the better understanding of the principles and powers of water, acting on mill-wheels, and conveying water to them.

ARTICLE 45.

OF SPOUTING FLUIDS.

Spouting fluids observe the following laws:—
1. Their velocities and powers, under equal pressures, or equal perpendicular heights, and equal apertures, are equal in all cases.*
2. Their velocities, under different pressures or perpendicular heights, are as the square roots of those pres-

---

* It makes no difference whether the water stands perpendicularly, or inclined, above the aperture, [see Plate III. fig. 22.] provided the perpendicular height be the same; or whether the quantity be great or small, provided it be sufficient to keep the fluid up to the same height.

sures or heights, and their perpendicular heights or pressures, are as the squares of their velocities.*

3. Their quantities expended through equal apertures, in equal times, under unequal pressures, are as their velocities simply.†

4. Their pressures or heights being the same, their effects are as their quantities expended.‡

5. Their quantities expended being the same, their effects are as their pressure, or height of their head directly.§

6. Their instant forces with equal apertures, are as the squares of their velocities, or as the height of their heads directly.

7. Their effects are as their quantities, multiplied into the squares of their velocities.‖ See Art. 46.

8. Therefore, their effects or powers with equal apertures are as the cubes of their velocities.¶

---

* This law is similar to the 4th law of falling bodies, their velocities being as the square root of their spaces passed through; and by experiment it is known, that water will spout from under a 4 feet head, with a velocity of 16,2 feet, per second, and from under a 16 feet head, 32,4 feet per second, which is only double to that of a 4 feet head, although there be a quadruple pressure. Therefore, by this law, we can find the velocity of water spouting from under any given head: for as the square root of 4 equal 2, is to 16,2 its velocity, so is the square root of 16 equal 4, to 32,4 squared, to 16, its head: by which ratio we can find the head that will produce any velocity.

† It is evident, that a double velocity will vent a double quantity.

‡ If the pressure be equal, the velocity must be equal; and it is evident, that double quantity, with equal velocity, will produce a double effect.

§ That is, if we suppose 16 cubic feet of water to issue from under a 4 feet head in a second, and an equal quantity to issue in the same time from under a 16 feet head, then their effects will be as 4 to 16. But we must note, that the aperture in the last case, must be only half of that in the first, as the velocity will be double.

‖ This is evident, from this consideration; namely: that a quadruple impulse is required to produce a double velocity, by law 2d, where the velocities are as the square roots of their heads: therefore, their effects must be as the squares of their velocities.

¶ The effects of striking fluids with equal apertures are as the cubes of their velocities, for the following reasons; namely: 1st. If an equal quantity strike with double velocity, the effect is quadruple on that account by the 7th law; and a double velocity expends a double quantity by 3d law; therefore, the effect is augmented to the cube of the velocity.—The theory for undershot wheels agrees with this law also.

9. Their velocity, under any head, is equal to the velocity that a heavy body would acquire in falling from the same height.*

10. Their velocity is such, under any head or height, as will pass over a distance equal to twice the height of the head, in a horizontal direction, in the time that a heavy body falls the distance of the height of the head.

11. Their action and re-action are equal.†

12. They being non-elastic, communicate only half their real force by impulse, in striking obstacles; but by their gravity produce effects, equal to elastic or solid bodies.‡

A SCALE

Founded on the 3d, 6th, and 7th laws, showing the effects of striking Fluids with different Velocities.

Aperture.	Multiplied by the	Velocity.	Is equal the	Quantity expended.	Which multiplied by the	Square of the velocity.	Is equal the	Effect.	Which is as the	Cubes of the velocity.
1	×	1	=	1	×	1	=	1	as	1
1	×	2	=	2	×	4	=	8	as	8
1	×	3	=	3	×	9	=	27	as	27
1	×	4	=	4	×	16	=	64	as	64

* The falling body is acted on by the whole force of its own gravity, in the whole of its descent through any space; and the whole sum of this action that is acquired as it arrives at the lowest point of its fall is equal to the pressure of the whole head or perpendicular height above the issue; therefore their velocities are equal.

† That is, a fluid re-acts back against the penstock with the same force that it issues against the obstacle it strikes; this is the principle by which Barker's mill, and all those that are denominated improvements thereon, move.

‡ When non-elastic bodies strike an obstacle, one half of their force is spent in a lateral direction, in changing their figure, or in splashing about. See Art. 9.

For want of due consideration or knowledge of this principle, many have been the errors committed by applying water to act by impulse, when it would have produced a double effect by its gravity.

ARTICLE 46.

DEMONSTRATION OF THE 7TH LAW OF SPOUTING FLUIDS.

Let A F, (plate III. fig. 26,) represent a head of water 16 feet high, and suppose it divided into 4 different heads of 4 feet each, as B C D E; then suppose we draw a gate of 1 foot square at each head successively, always sinking the water in the head, so that it will be but 4 feet above the centre of the gate in each case.

Now it is known that the velocity under a 4 feet head, is 16,2 feet per second; to avoid fractions say 16 feet, which will issue 16 cubic feet of water per second, and for sake of round numbers, let unity or 1 represent the quantity of a cubic foot of water; then, by the 7th law the effect will be as the quantity multiplied by the square of the velocity; that is, 16 multiplied by 16 is equal to 256, which, multiplied by 16, the quantity, is equal to 4096, the effect of each 4 feet head; and 4096 multiplied by 4 is equal to 16384, for the sum of effects of all the 4 feet heads.

Then as the velocity under a 16 feet head is 32,4 feet, to avoid fractions say 32, the gate must be draw to only half the size, to vent the 16 cubic feet of water per second as before (because the velocity is double;) then to find the effect, 32 multiplied by 32 is equal to 1024; which multiplied by 16, the quantity, gives the effect 16384. equal the sum of all the 4 feet heads, which agrees with the practice and experience of the best teachers. But if their effects were as their velocities simply, then the effect of each 4 feet head would be, 16 multiplied by 16, equal to 256; which, multiplied by 4, is equal to 1024, for the sum of the effects of all the 4 feet heads; and 16 multiplied by 32 equal to 512, for the effect of the 16 feet head, which is only half of the effect of the same head when divided into 4 parts; which is contrary to both experiment and reason.

Again, let us suppose the body A of quantity 16, to be perfectly elastic, to fall 16 feet and strike F, a perfectly elastic plane, it will (by laws of falling bodies) strike with a velocity of 32 feet per second, and rise 16 feet to A again.

But if it fall only to B, 4 feet, it will strike with a velocity of 16 feet per second, and rise 4 feet to A again. Here the effect of the 16 feet fall is 4 times the effect of the 4 feet fall, because the body rises 4 times the height.

But if we count the effective momentum of their strokes to be as their velocities simply, then 16 multiplied by 32 is equal to 512, the momentum of the 16 feet fall; and 16 multiplied by 16 is equal to 256; which, multiplied by 4, is equal to 1024, for the sum of the momentums of the strokes of 16 feet divided into 4 equal falls, which is absurd. But if we count their momentums to be as the squares of their velocities, the effects will be equal.

Again, it is evident that whatever impulse or force is required to give a body velocity, the same force or resistance will be required to stop it; therefore, if the impulse be as the square of the velocity produced, the force or resistance will be as the squares of the velocity also. But the impulse is as the squares of the velocity produced, which is evident from this consideration: Suppose we place a light body at the gate B, of 4 feet head, and pressed with 4 feet of water; when the gate is drawn it will fly off with a velocity of 16 feet per second; and if we increase the head to 16 feet, it will fly off with 32 feet per second. Then, as the square of 16 equal to 256 is to the square of 32 equal to 1024, so is 4 to 16. Q. E. D.

ARTICLE 47.

THE 7TH LAW IS IN ACCORDANCE WITH PRACTICE.

Let us compare this 7th law with the theory of undershot mills, established Art. 41, where it is shown that the power is to the effect as 3 to 1. By the 7th law, the quantity shown by the scale, Plate II. to be 32,4 multiplied by 1049,76 the square of the velocity, which is equal to 3401,2124, the effect of the 16 feet head; then, for the effect of a 4 feet head, with equal apertures, quantity by scale 16,2, multiplied by 262,44, the velocity squared, is equal to 425,15 28, the effect of a 4 feet head; here the ratio of the effect is as 8 to 1.

Then, by the theory, which shows that an undershot wheel will raise 1-3d of the water that turns it, to the whole height from which it descended, the 1-3d of 32,4 the quantity, being equal to 10,8, multiplied by 16, perpendicular ascent, which is equal to 172,8, effect of a 16 feet head: and 1-3d of 16,2 quantity, which is equal to 5,4 multiplied by 4, perpendicular ascent, is equal to 21,6 effect of a 4 feet head, by the theory: and here again the ratio of the effects is as 8 to 1; and,

as 3401,2124, the effect of 16 feet head, } by 7th law,
is to 425,1825, the effect of a 4 feet head, }

so is 172,8, the effect of 6 feet head, } by the theory
to 21,6, the effect of 4 feet head, }

The quantities being equal, their effects are as the height of their heads directly, as by 5th law, and as the squares of their velocities, as by the 7th law. Hence it appears, that the theory agrees with the established laws.

Application of the Laws of Motion to Undershot Wheels.

To give a short and comprehensive detail of the ideas I have collected from different authors, and from the result of my own reasoning on the laws of motion and of spouting fluids, as they apply to move undershot mills, I refer to fig. 44, Plate V.

Let us suppose two large wheels, one of 12 feet, and

the other of 24 feet radius, the circumference of the largest will then be double that of the smallest: and let A 16, and C 16, be two penstocks of water, of 16 feet head each, then,—

1. If we open a gate of 1 square foot at 4, to admit water from the penstock A 16, to impinge on the small wheel at I, the water being pressed by 4 feet head, will move 16 feet per second (we omit fractions.) The instant pressure or force on that gate, being four cubic feet of water, it will require a resistance of 4 cubic feet of water from the head C 16 to stop it, and hold it in equilibrio, (but we suppose the water cannot escape, unless the wheel moves, so that no force be lost by non-elasticity.) Here equal quantities of matter, with equal velocities, have their momentums equal.

2. Again, suppose we open a gate of 1 square foot at A 16 under 16 feet head, it will strike the large wheel at k, with velocity 32, its instant force or pressure being 16 cubic feet of water, it will require 16 cubic feet resistance, from the head C 16. to stop or balance it. In this case, the pressure, or instant force, is quadruple to the first, and so is the resistance, but the velocity only double. In these two cases the forces and resistances being equal quantities, with equal velocities, their momentums are equal.

3. Again, suppose the head C 16 to be raised to E, 16 feet above 4, and a gate drawn 1-4th of a square foot, then the instant pressure on the float I of the small wheel, will be 4 cubic feet, pressing on 1-4th of a square foot, and will exactly balance 4 cubic feet, pressing on 1 square foot from the head A 16; and the wheel will be in equilibrio, (supposing the water cannot escape until the wheel moves as before,) although the one has power of velocity 32, and the other only 16, feet per second; their loads at equilibrio are equal, consequently, their loads at a maximum velocity and charge will be equal, but their velocities different.

Then, to try their effects, suppose, first, the wheel to move by the 4 feet head, its maximum velocity to be half the velocity of the water, which is 16, and its maximum load to be half its greatest load, which is 4, by Waring's theory; then the velocity 16÷2 multiplied by the load 4÷2=16, the effect of the 4 feet head, with 16 cubic feet expended; because the velocity of the water is 16, and the gate 1 foot.

Again, suppose it to move by the 16 feet head and gate of 1-4th of a foot; then the velocity 32÷2 multiplied by the load 4÷2 = 32, the effect, with but 8 cubic feet expended, because, the velocity of the water is 32, and the gate but 1-4th of a foot.

In this case the instant forces are equal, each being 4; but the one moving a body only 1-4th as heavy as the other, moves with velocity 32, and produces effect 32, while the other, moving with velocity 16, produces effect 16. A double velocity, with equal instant pressure, produces a double effect, which seems to be according to the Newtonian theory. And in this sense the momentums of bodies in motion are as their quantities, multiplied into their simple velocities, and this is what I call the instant momentums.

But when we consider, that in the above case it was the quantity of matter put in motion, or water expended, that produced the effect, we find that the quantity 16, with velocity 16, produced effect 16; while quantity 8, with velocity 32, produced effect 32. Here the effects are as their quantities, multiplied into the squares of their velocities; and this I call the effective momentums.

Again, if the quantity expended under each head had been equal, their effects would have been 16 and 64, which is as the squares of their velocities, 16 and 32.

4. Again, suppose both wheels to be on one shaft, and let a gate of 1-8th of a square foot be drawn at 16 C, to strike the wheel at K, the head being 16 feet, the instant pressure on the gate will be 2 cubic feet of water, which is half of the 4 feet head with 1 foot gate, from A 4 striking at I; but the 16 feet head, with instant pressure 2, acting on the great wheel, will balance 4 feet on the small one, because the lever is of double length, and the wheels will be in equilibrio. Then, by Waring's theory, the greatest load of the 16 feet head being 2, its load at a maximum will be 1, and the velocity of the water being 32, the maximum velocity of the wheel will be

16. Now the velocity 16×1=16, the effect of the 16 feet head; and gate of 1-8th of a foot, the greatest load of the 4 feet head being 4, its maximum load 2, the velocity of the water 16, and the velocity of the wheel 8: now 8×2=16, the effect. Here the effects are equal, and here, again, the effects are as the instant pressures, multiplied into their simple velocities: and the resistances that would instantly stop them must be equal thereto, in the same ratio.

But when we consider, that in this case the 4 feet head expended 16 cubic feet of water, with velocity 16, and produced effect 16; while the 15 feet head expended only 4 cubic feet of water, with velocity 32, and produced effect 16, we find that the effects are as their quantities, multiplied into the squares of their velocities.

And when we consider, that the gate of 1-8th of a square foot, with velocity 32, produced effects equal to the gate of 1 square foot, with velocity 16, it is evident, that if we make the gates equal, the effects will be as 8 to 1; that is, the effects of spouting fluids, with equal apertures, are as the cubes of their velocities; because, their instant forces are as the squares of their velocities, by 6th law, although the instant forces of solids are as their velocities simply, and their effects as the squares of their velocities, a double velocity does not double the quantity of a solid body to strike in the same time.

ARTICLE 48.

THE HYDROSTATIC PARADOX.

The pressure of fluids is as their perpendicular heights, without any regard to their quantity: and their pressure upwards is equal to their pressure downwards. In short, their pressure is every way equal, at any equal distance from their surface.

In a vessel of cubic form, whose sides and bottom are equal, the pressure on each side is just half the pressure on the bottom; therefore, the pressure on the bottom and sides is equal to three times the pressure on the bottom.

And, in this sense, fluids may be said to act with three times the force of solids. Solids act by gravity only, but fluids by gravity and pressure jointly. Solids act with a force proportional to their quantity of matter; but fluids act with a pressure proportional to their altitude only.

To explain the law, that the pressure of fluids is as their perpendicular heights, let A B C D, Plate III. fig. 22, be a vessel of water of a cubical form, with a small tube, as H, fixed therein; let a hole of the same size with the tube be made at o. and covered with a piece of pliant leather, nailed thereon, so as to hold the water, Then fill the vessel with water by the tube H, and it will press upwards against the leather, and raise it in a convex form, requiring just as much weight to press it down, as will be equal to the weight of water in the tube H. Or if we set a glass tube over the hole at o, and pour water therein, we shall find that the water in the tube o, must be of the same height of that in the tube H, before the leather will subside, even if the tube o be much larger than H; which shows, that the pressure upwards is equal to the pressure downwards; because the water pressed up against the leather with the whole weight of the water in the tube H. Again, if we fill the vessel by the tube I, it will rise to the same height in H that it is in I; the pressure being the same in every part of the vessel as if it had been filled by H; and the pressure on the bottom of the vessel will be the same, whether the tube H be of the whole size of the vessel, or only one quarter of an inch diameter. For suppose H to be 1-4th of an inch diameter, and the whole top of the vessel of leather, as at o, and we pour water down H, it will press the leather up with such force, that it will require a column of water of the whole size of the vessel, and height of H, to cause the leather to subside. Q. E. D.

ARTICLE 49.

PRACTICAL RESULTS OF THIS EQUAL PRESSURE.

And again, suppose we make two holes in the vessel, one close to the bottom, and the other in the bottom, both of one size, the water will issue with equal velocity out of each; this may be proved by holding equal vessels under each, which will be filled in equal time; this shows, that the pressure on the sides and bottom is equal under equal distances from the surface. And this velocity will be the same whether the tube be filled by pipe I, or H, or by a tube the whole size of the vessels, provided the perpendicular height be equal in all cases.

From what has been said, it appears, that it makes no difference in the power of water in mill-wheels, whether it be brought on in an open forebay and perpendicular penstock, or down an inclining one, as I C; or under ground in a close trunk, in any form that may best suit the situation and circumstances, provided that the trunk be sufficiently large to supply the water fast enough to keep the head from sinking.

This principle of the Hydrostatic Paradox has sometimes operated in undershot mills, by pressing up against the bottom of the buckets, thereby destroying or counteracting, in great part, the force of impulse. See Art. 59.

ARTICLE 50.

The weight of a cubic foot of water is found by experience, to be 1000 ounces avoirdupois, or 62,5 lbs. On the principles explained in Art. 48 and 49, is founded the following

THEOREM.

The area of the base or bottom, or any part of a vessel, of whatever form, multiplied by the greatest perpendicular height of any part of the fluid, above the centre of the base or bottom, whatever be its position with the horizon, produces the pressure on the bottom of said vessel.

PROBLEM I.

Given, the length of the sides of the cubic vessel (fig. 22, Pl. III.) 6 feet, required the pressure on the bottom when full of water.

Then 6×6=36 feet the area, multiplied by 6, the altitude, =216, the quantity or cubic feet of water, pressing on the bottom; which multiplied by 62,5=13500 lbs. the whole pressure on the bottom.

PROBLEM II.

Given, the height of a penstock of water 31,5 feet, and its dimensions at bottom 3 by 3 feet, inside, required the pressure on three feet high of one of its sides, measuring from the bottom.

Then, 3×3=9 the area, multiplied by 30 feet, the perpendicular height or head above the centre of the 3 feet on the side=270 cubic feet of water pressing, which ×62,5=16875 lbs. the pressure on one yard square, which shows what great strength is required to hold the water under such great heads.

ARTICLE 51.

RULE FOR FINDING THE VELOCITY OF SPOUTING WATER.

It has been found by experiment, that water will spout from under a 4 feet head, with a velocity equal to 16,2 feet per second, and from under a 16 feet head, with a velocity equal to 32,4 feet per second.

On these experiments, and the 2d law of spouting fluids, is founded the following theorem, or general rule, for finding the velocity of water under any given head.

THEOREM.

As the square root of a 4 feet head (=2) is to 16,2 feet, the velocity of the water spouting under it, so is the square root of any other head, to the velocity of the water spouting under it.

PROBLEM I.

Given, the head of water 16 feet, required the velocity of water spouting under it.

Then, as the square root of 4 (=2) is to 16,2, so is the square root of 16, (=4) to 32,4, the velocity of the water under the 16 feet head.

PROBLEM II.

Given, a head of water of 11 feet, required the velocity of water spouting under it.

Then, as 2:16,2::3,316:26,73 feet per second, the velocity required.

ARTICLE 52.

EFFECT OF WATER UNDER A GIVEN HEAD.

From the 1st and 2d laws of spouting fluids, (Art. 45.) the theory for finding the maximum charge and velocity of undershot wheels, (Art. 41,) and from the principle of non-elasticity, the following theorem is deduced for finding the effect of any gate drawn under any given head, upon an undershot water-wheel.

THEOREM.

Find by the theorem (Art. 50,) the instantaneous pressure of the water, which is the load at equilibrio, and 2-3ds thereof is the maximum load, which, multiplied by ,577 of the velocity of the water, under the given head, (found by the theorem, Art. 51,) produces the effect.

PROBLEM.

Given, the head 16 feet, gate 4 feet wide, ,25 of a foot drawn, required the effect of an undershot wheel, per second. The measure of the effect to be the quantity, multiplied into its distance moved (velocity,) or into its perpendicular ascent.

Then, by the theorem (Art. 50) $4 \times ,25 = 1$ square foot (the area of the gate) $\times 16 = 16$ the cubic feet pressing; but, for the sake of round numbers, we call each cubic foot 1, and although 32,4 cubic feet strike the wheel per second, yet, on account of non-elasticity, only 16 cubic feet is the load at equilibrio, and 2-6ths of 16 is 10,666, the maximum load.

Then, by theorem (Art. 51) the velocity is 32,4 ,577 of which is $=18,71$, the maximum velocity of the wheel $\times 10,66$, the load $=199,4$, the effect.

This agrees with Smeaton's observations, where he says, (Art. 67,) "It is somewhat remarkable, that though the velocity of the wheel in relation to the velocity of the water, turns out to be more than 1-3d, yet the impulse of the water, in case of the maximum, is more than double of what is assigned by theory; that is, instead of 4-9ths of the column, it is nearly equal to the whole column." Hence I conclude, that non-elasticity does not operate so much against this application, as to reduce the load to be less than 2-3ds. And when we consider, that 32,4 cubic feet of water, or a column 32,4 feet long, strikes the wheel, while it moves only 18,71 feet, the velocity of the wheel being to the velocity of the water as 577 to 1000, may not this be the reason why the load is just 2-3ds of the head, which brings the effect to be just ,38 (a little more than 1-3d of the power?) This I admit, because it agrees with experiment, although it be difficult to assign the true reason thereof. See Annotation, Art. 42.

Therefore, ,577 the velocity of the water =18,71, multiplied by 2-3ds of 16, the whole column, or instantaneous pressure, pressing on the wheel—Art. 50—which is 10,66, produces 199,4, the effect. This appears to be the true effect, and if so, the true theorem will be as follows; namely:

THEOREM.

Find by the theorem Art. 50, the instantaneous pressure of the water, and take 2-3ds for the maximum load; multiply by ,577 of the velocity of the water—which is the velocity of the wheel—and the product will be the effect.

Then 16 cubic feet, the column, multiplied by 2-3ds =10,66, the load, which multiplied by 18,71, the velocity of the wheel, produces 199,4, for the effect; and if we try different heads and different apertures, we find the effects to bear the ratio to each other, that is agreeable to the laws of spouting fluids.

ARTICLE 53.

WATER APPLIED ON WHEELS TO ACT BY GRAVITY.

When fluids are applied to act on wheels to produce effects by their gravity, they act on very different principles from the foregoing, producing double effects to what they do by percussion, and then their powers are directly as their quantity, or weight, multiplied into their perpendicular descent.

DEMONSTRATION.

Let D. B. fig. 19. Plate III. be a lever, turning on its centre or fulcrum A. Let the long arm A B represent the perpendicular descent, 16 feet, the short arm A D a descent of 4 feet, and suppose water to issue from the trunk F, at the rate of 50 lbs. in a second, falling into the buckets fastened to the lever at B. Now, from the principles of the lever, Art. 16, it is evident, that 50 lbs. in a second at D, will balance 200 lbs. in a second, at D, issuing from the trunk G, on the short arm; because $50 \times 16 = 800$, and $4 \times 200 = 800$. Perhaps it may appear plainer, if we suppose the perpendicular line or diameter F C, to represent the descent of 16 feet, and the diameter G I a descent of 4 feet. By the laws of the lever—Art. 16—it is shown, that, to multiply 50 into its perpendicular descent 16 feet or distance moved, is $= 200$ multiplied into its perpendicular descent 4 feet, or distance moved; that is, $53 \times 16 = 200 \times 4 = 800$; that is, their power is as their quantity, multiplied into their perpendicular descent; or, in other words, a fall of 4 feet will require 4 times as much water, as a fall of 16 feet, to produce equal power and effects. Q. E. D.

Upon these principles is founded the following simple theorem, for measuring the power of an overshot mill, or of a quantity of water, acting upon any mill-wheel by its gravity.

THEOREM.

Cause the water to pass along a regular canal, and multiply its depth in feet and parts, by its width in feet and parts, for the area of its section, which product multiply by its velocity per second in feet and parts, and the product is the cubic feet used per second, which multiplied by 62,5 lbs. the weight of one cubic foot, produces the weight of water per second, that falls on the wheel, which, multiplied by its whole perpendicular descent, gives a true measure of its power.

PROBLEM I.

Given, a mill-seat with 16 feet fall, width of the canal 5,333 feet, depth 3 feet, velocity of the water passing along it 2,03 feet per second, required the power per second.

Then, 5,333×3=15,999 feet, the area of the section of the stream, multiplied by 2,03 feet, the velocity, is equal 32,4 cubic feet, the quantity per second, multiplied by 62,5, is equal 2025 lbs. the weight of the water per second, multiplied by 16, the perpendicular descent, is equal 32400, for the power of the seat per second.

PROBLEM II.

Given, the perpendicular descent 18,3, width of the gate 2,66 feet, height ,145 of a foot, velocity of the water per second issuing on the wheel, 15,76 feet, required the power.

Then, 2,66×,145=,3857 the area of the gate, ×15,76 the velocity=6,178 cubic feet expended per second, ×62,5=375,8 lbs. per second, ×18,3 feet perpendicular descent=6877 for the measure of the power per second; which has ground 3,75 lbs. per minute, equal ,375 bushels in an hour, with a five feet pair of burr stones.

ARTICLE 54.

INVESTIGATION OF THE PRINCIPLES OF OVERSHOT MILLS.

Some have asserted, and many believed, that water is applied to great disadvantage on the principle of an overshot mill; because, say they, there are never more than two buckets, at once, that can be said to act fairly on the end of the lever, (as the arms of the wheel are called in these arguments.) But we must examine well the laws of bodies descending inclined planes, and curved surfaces. See Art. 11. This matter will be cleared up, if we consider the circumference of the wheel to be the curved surface: for the fact is, that the water acts to the best advantage, and produces effects equal to what it would, in case the whole of it acted upon the very end of the lever, in the whole of its perpendicular descent. The want of a knowledge of this fact has led to many fatal errors in the application of water.

DEMONSTRATION.

Let A B C, Plate III. fig. 20, represent a water-wheel, and F H a trunk, bringing water to it from a 16 feet head. Now, suppose F G and 16 H to be two penstocks under equal heads, down which the water descends, to act on the wheel at C, on the principle of an undershot, on opposite sides of the float C with equal apertures: it will be evident from the principles of hydrostatics, shown by the paradox, (Art. 48, and the first law of spouting fluids, Art. 45,) that the impulse, and pressure, will be equal from each penstock respectively. Although the one be an inclined plane, and the other a perpendicular, their forces are equal, because their perpendicular heights are so; (Art. 48;) therefore, the wheel will remain at rest, because each side of the float is pressed on by a column of water of equal size and height, as represented by the lines on each side of the float. Then, suppose we shut the penstock F G, and let the water down the circular one r x, which is close to the point of the buckets; this makes it obvious, from the same principles, that the wheel will be held in equilibrio, if the columns of each side be equal. For, although the column in the circular penstock is longer than the perpendicular one, yet, because part of its weight presses on the lower side of the penstock, its pressure on the float is due only to its perpendicular height.

Then, again, suppose the column of water in the circular penstock to be instantly thrown into the buckets, it is evident, that the wheel will be still held in equilibrio, and each bucket will then bear a proportional part of the column that the bucket C bore before; and that part of the weight of the circular column, which rested on the under side of the circular penstock, is now on the gudgeons of the wheel. This shows that the effect of a stream, applied on an overshot wheel, is equal to the effect of the same stream, applied on the end of the lever, in its whole perpendicular descent, as in fig. 21, where the water is shot into the buckets fastened to a strap or chain, revolving over two wheels; and here the whole force of the gravity of the column acts on the very end of the lever, in the whole of the descent. Although the length of the column in action, in this case, is only 16 feet, whereas, on a 16 feet wheel, the length of the column in action is 25,15, yet their powers are equal.

Again, if we divide the half circle into three arches Ab, be, eC, the centre of gravity of the upper and lower arches will fall near the point a, 3,9 feet from the centre of motion, and the centre of gravity of the middle arch, near the point o, 7,6 feet from the centre of motion. Now, each of these arches is 8,38 feet, and 8,38 ×2×3,9=65,36, and 8,38×7,6 feet=63,07, which two products added=128,43, for the momentum of the circular column, by the laws of the lever, and for the perpendicular column 16×8 the radius of the wheel=128, for the momentum; by which it appears, that if we could determine the exact points on which the arches act, the momentums would be equal: all which shows, that the power of water on overshot wheels, is equal to the whole power it can any way produce, through the whole of its perpendicular descent, except what may be lost to obtain velocity (Art. 41,) overcome friction, or by spilling a part of the water before it gets to the bottom of the wheel. Q. E. D.

I may add, that I have made the following experiment; namely: I fixed a truly circular wheel on nice pivots, to avoid friction, and took a cylindrical rod of thick wire, cutting one piece exactly the length of half the circumference of the wheel, and fastening it to one side, close to the rim of the wheel its whole length, as at G x r a. I then took another piece of the same wire, of a length equal to the diameter of the wheel, and hung it on the opposite side, on the end of the lever or arm, as at B, and the wheel was in equilibrio. Q. E. D.

ARTICLE 55.

OF THE FRICTION OF THE APERTURES OF SPOUTING FLUIDS.

The doctrine of this species of friction appears to be as follows:—

1. The ratio of the friction of round apertures, is as their diameters, nearly; while the quantity expended, is as the squares of their diameters.

2. The friction of an aperture of any regular or irregular figure, is as the length of the sum of the circumscribing lines, nearly; the quantities being as the areas of the aperture.* Therefore,

3. The less the head or pressure, and the larger the aperture, the less the ratio of the friction; therefore,

4. This friction need not be much regarded, in the large openings or apertures of undershot mills, where the gates are from 2 to 15 inches in their shortest sides; but it very sensibly affects the small apertures of high overshot or undershot mills, with great heads, where their shortest sides are from five-tenths of an inch to two inches.†

* This will plainly appear, if we consider that the friction does sensibly retard the velocity of the fluid to a certain distance; say half an inch from the side or edge of the aperture, towards its centre; and we may reasonably conclude, that this distance will be nearly the same in a 2 and 12 inch aperture; so that in the 2 inch aperture, a ring on the outside, half an inch wide, is sensibly retarded, which is about 3-4ths of the whole; while, in the 12 inch aperture, there is a ring on the outside half an inch wide, retarded about one-sixth of its whole area.

† This seems to be proved by Smeaton, in his experiments; (see table, Art. 67;) where, when the head was 33 inches, the sluice small, drawn only to the 1st hole, the velocity was only such as is assigned by theory, to a head of 15,85 inches, which he calls virtual head. But when the sluice was larger, drawn to the 6th hole, and head 6 inches, the virtual head was 5,33 inches. But seeing there is no theorem, yet discovered, by which we can truly determine the quantity or effect of the friction, according to the size of the aperture, and height of the head; we, cannot, therefore, by the established laws of hydrostatics, determine exactly the velocity or quantity expended through any small aperture; which renders the theory in these cases but little better than conjecture.

ARTICLE 56.

OF THE PRESSURE OF THE AIR ON FLUIDS.

Under certain circumstances, the rise of water is caused by the pressure of the air on the surface of its reservoir, or source; and this pressure is equal to that of a head of water of about $33\frac{1}{3}$ feet perpendicular height; under which pressure, or height of head, the velocity of spouting water is 46,73 feet per second.

If, therefore, we could by any means take off the pressure of the atmosphere, from any one part of the surface of a fluid, that part would spout up with a velocity of 46,73 feet per second, and rise to the height of $33\frac{1}{3}$ feet nearly.

All syphons, or cranes, and all pumps for raising water by suction, as it is called, act on this principle.—Let fig. 23, Pl. III. represent a cask of water, with a syphon therein, to extend $33\frac{1}{3}$ feet above the surface of the water in the cask. Now, if the bung be made perfectly air-tight round the syphon, so that no air can get into the cask, and the cask be full, and if all the air be then drawn out of the syphon, the fluid will not rise in the syphon, because the air cannot get to it to press it up; but take out the plug P, and let the air into the cask, to press on the surface of the water, and it will spout up the short leg of the syphon B A, with the same force and velocity, as if it had been pressed with a head of water $33\frac{1}{3}$ feet high, and will run into the long leg and fill it. If we then turn the cock c, and let the water run out, its weight in the long leg will overbalance the weight in the short one, drawing the water out of the cask until it sinks so low, that the leg B A will be $33\frac{1}{3}$ feet high, above the surface of the water in the cask; it will then stop, because the weight of water in the legs, in which it rises, will be equal to the weight of a column of the air of equal size, and of the whole height of the atmosphere. The water will not run out of the leg A C, but will stand $33\frac{1}{3}$ feet above its mouth, because the air will press up the mouth C, with a force that will balance 33⅓ feet of water in the leg C A. This will be the case let the upper part of the leg be of any size whatever—and there will be a vacuum at the upper end of the syphon.

It must not, however, be supposed that if the mouth C be left open, after the water has ceased running, that the portion of it which is in the leg A C, will remain there, as air will be gradually admitted, and will press upon the upper end of the column A B, which will then descend in both legs.

ARTICLE 57.

OF PUMPS.

Let fig. 24, Pl. III. represent a pump of the common kind used for drawing water out of wells. The moveable valve or bucket A, is cased with leather, which springs outwards, and fits the tube so nicely, that neither air nor water can pass freely by it. When the lever L is worked, the valve A opens as it descends, letting the air or water pass through it. As it ascends again the valve shuts, the water which is above the bucket A is raised, and there would be a vacuum between the valves, but the weight of the air presses on the surface of the water in the well, at W, forcing it up through the valve B, to fill the space between the buckets; and as the valve A descends, B shuts, and prevents the water from descending again. But if the upper valve A be set more than 33⅓ feet above the surface of the water in the well, the pump cannot be made to draw, because the pressure of the atmosphere will not cause the water to rise more than 33⅓ feet. Although in theory the water would rise to the height stated, yet, in point of fact, the distance between the valve in the piston, and the surface of the water in the well, ought never to exceed 24 or 25 feet, or, from the imperfection of workmanship, and other causes, the pump will *lose water;* and will cease to act.

A TABLE FOR PUMPMAKERS.

Height of the pump, in feet, above the surface of the well.	Diameter of the bore. inches.	100 parts of an inch	Water discharged in a minute, in wine measure. Galls.	Pints.
10	6	93	81	6
15	5	66	54	4
20	4	90	40	7
25	4	38	32	6
30	4	00	27	2
35	3	70	23	3
40	3	46	20	3
45	3	27	18	1
50	3	10	16	3
55	2	95	14	7
60	2	84	13	5
65	2	72	12	4
70	2	62	11	5
75	2	53	10	7
80	2	45	10	2
85	2	38	9	5
90	2	31	9	1
95	2	25	8	5
100	2	18	8	1

The preceding table is extracted from Ferguson's Lectures, and its use is pointed out by him in the subjoined quotation: before giving which, however, it will be proper to remark, that it is a common practice to make the bore in the lower part of the pump-tree smaller than the chamber, under the erroneous supposition that there will be a less weight of water to lift in this than in a larger bore. The consequence of this is, that the water has to rush with greater velocity in order to fill the capacity of the chamber, by which much friction is caused, and much power wasted.

"All pumps should be so constructed as to work with equal ease in raising the water to any given height above the surface of the well: and this may be done by observing a due proportion between the diameter of that part of the pump bore in which the piston or bucket works, and the height to which the water must be raised.

"For this purpose I have calculated the above table; in which the handle of the pump is supposed to be a lever, increasing the power five times: that is, the distance or length of that part of the handle that lies between the pin on which it moves, and the top of the pump-rod to which it is fixed, to be only one-fifth part of the length of the handle, from the said pin to the part where the man who works the pump applies his force or power.

"In the first column of the table, find the height at which the pump must discharge the water above the surface of the well; then in the second column you have the diameter of that part of the bore in which the piston or bucket works, in inches and hundredth parts of an inch; in the third column is the quantity of water (in wine measure) that a man of common strength can raise in a minute.— And by constructing according to this method, pumps of all heights may be wrought by a man of ordinary strength so as to be able to hold out for an hour."

ARTICLE 58.

OF CONVEYING WATER UNDER VALLEYS AND OVER HILLS.

Water, by its own pressure, and the pressure of the atmosphere, may be conveyed under valleys and over hills, to supply a family, a mill, or a town. In fig. 20, Pl. III. F H is a canal for conveying water to a mill-wheel: now let us suppose F G 16 H to be a tight tube or trunk, —the water being let in at F, it will descend from F to G, and its pressure at F will cause it to rise to H, which shows how it may be conveyed under a valley; and it may be conveyed over a hill by a tube, acting on the principle of the syphon. (Art. 56.) But some who have had occasion to convey water, under any obstacle, for the convenience of a mill, have gone into the following expensive error: they have made the tube at G 16, smaller than they would if it had been on a level; because, say they, a greater quantity will pass through a tube, pressed by the head G F, than on a level: but, it should be considered that the head G F, is balanced by the head H 16, and the velocity through the tube G 16, will be such only, as a head equal to the difference between the perpendicular height of G F, and H 16, would give it (see Art. 41, fig. 19;) therefore, it should be as large at G 16, as if on a level.

ARTICLE 59.

OF THE DIFFERENCE IN THE FORCE OF INDEFINITE AND DEFINITE QUANTITIES OF WATER STRIKING A WHEEL.

DEFINITIONS.

1. By an indefinite quantity of water we here mean a river, or quantity, much larger than the float of the wheel; so that, when it strikes the float, it has liberty to move or escape from it in every lateral direction.

2. By a definite quantity of water we mean a quantity

passing through a given aperture, along a shute, to strike a wheel; but as it strikes the float, it has liberty to escape in every lateral direction.

3. By a perfectly definite quantity, we mean a quantity passing along a close tube, so confined that when it strikes the float, it has not liberty to escape in any lateral direction.

First, When a float of a wheel is struck by an indefinite quantity, the float is struck by a column of water, the section of which is equal to the area of the float; and as this column is confined on every side by the surrounding water, which has equal motion, it cannot escape sideways without some resistance; more of its force, therefore, is communicated to the float, than would be, if it had free liberty to escape in every direction.

Secondly, The float being struck by a definite quantity, with liberty to escape freely in every lateral direction, it acts as the most perfectly non-elastic body; therefore (by Art. 9) it communicates only a part of its force, the other part being spent in the lateral direction. Hence it appears, that in the application of water to act by impulse, we should draw the gate as near as possible, to the float-board, and confine it, as much as possible from escaping sideways as it strikes the float; but taking care, at the same time, that we do not bring the principle of the Hydrostatic Paradox into action. (Art. 48.)

What proportion of the force of the water is spent in a lateral direction is not determined.

4. A perfectly definite quantity striking a plane, communicates its whole force, because no part can escape sideways; and is equal in power to an elastic body, or to the weight of the water on an overshot wheel, in its whole perpendicular descent. But this application of water to wheels, in this way, has hitherto proved impracticable, for whenever we attempt to confine the water, totally, from escaping sideways, we bring the principle of the hydrostatic paradox into action, which defeats the scheme.

To make this plain, let fig. 25, Pl. III. be a waterwheel; and, first, let us suppose the water to be brought to it by the penstock 41.6, to act by impulse on the float board, having liberty to escape every way as it strikes; then, by Art. 9, it will communicate but half its force. But if it be confined both at the sides and bottom, and can escape only upwards, to which the gravity will make some opposition, it will communicate more than half its force, and will not re-act back against the float C; but if we put soaling to the wheel, to prevent the water from escaping upwards, then the space between the floats will be filled as soon as the wheel begins to be retarded, and the paradoxical principle, Art. 48, is brought fully into action; namely: the pressure of water is every way equal; and it will press backwards against the bottom of the float C, with a force equal to its pressure on the top of the float b, and the wheel will immediately stop, and be held in equilibrio, and will not start again although all resistance be removed. There are many mills, where this principle is, in part, brought into action, which very much lessens their power.

ARTICLE 60.

OF THE MOTION OF BREAST AND PITCH-BACK WHEELS.

Many have been of opinion, that when water is put to act on a low breast wheel, as at a, (Pl. 3, fig. 25,) with 12 feet head, that then the four feet fall, below the point of impact a, is totally lost, because, say they, the impulse of the 12 feet head, will require the wheel to move with such velocity, to suit the motion of the water, as to move before the action of gravity, therefore, the water cannot act after the stroke; but, if they will consider well the principles of gravity acting on falling bodies, (Art. 10,) they will find, that if the velocity of a falling body be ever so great, the action of gravity to cause it to move faster, is still the same; so that, although an overshot wheel may move before the power of the gravity of the water thereon, yet no impulse downwards can give a wheel such velocity, as that the gravity of the water acting thereon can be thereby lessened.*

Hence, it appears, that when a greater head is used than that which is necessary to shoot the water fairly into the wheel, the impulse should be directed a little downward, as at D, (which is called pitch-back,) and it should have a circular sheeting, to prevent the water from leaving the wheel, because if it be shot horizontally on the top of a wheel, the impulse in that case will not give the water any greater velocity downwards, and, in this case, the fall would be lost, if the head were very great; and if the wheel moved to suit the velocity of the impulse, the water would be thrown out of the buckets, by the centrifugal force; and if we attempt to retard the wheel, so as to retain the water, the mill would be so ticklish and unsteady, that it would be almost impossible to attend it.

Hence may appear the reason why breast-wheels generally run quicker than overshots, although the fall, after the water strikes, be not so great.

1. There is generally more head allowed to breast-wheels than to overshots; and the wheel will incline to move with nearly 2-3ds the velocity of the water spouting from under the head. (Art. 41.)

2. If the water were permitted to fall freely after it issues from the gate, it would be accelerated by the fall, so that its velocity at the last point, would be equal to its velocity had it spouted from under a head equal to its whole perpendicular descent. This accelerated velocity of the water tends to accelerate the wheel; hence, to find the velocity of a breast-wheel, where the water strikes it in the direction of a tangent, as in fig. 31, 32, I deduce the following

* If gravity could be either decreased by velocity downwards, or increased by velocity upwards, then a vertical wheel without friction, either of gudgeons or air, would require a great force to continue its motion; because its velocity would decrease the gravity of its descending and increase that of its ascending side, which would immediately stop it; whereas, it is known, that it requires no power to continue its motion, but that which is necessary to overcome the friction of the gudgeons, &c.

THEOREM.

1. Find the difference of the velocity of the water under the head allowed to the wheel, above the point of impact, and the velocity of a body, having fallen the whole perpendicular descent of the water. Call this difference the acceleration by the fall: Then say, As the velocity a body would acquire in falling through the diameter of any overshot-wheel, is to the proper velocity of that wheel by the scale, (Art. 43,) so is the acceleration by the fall of the water before it strikes the wheel, to the acceleration of the wheel by its fall, after it strikes.

2. Find the velocity of the water issuing on the wheel; take ,577 of said velocity, to which add the accelerated velocity, and that sum will be the velocity of the breast-wheel.

This rule will hold nearly true, when the head is considerably greater than is assigned by the scale (Art. 43;) but as the head approaches that assigned by the scale, this rule will give the motion too quick.

EXAMPLE.

Given, a high breast-wheel, fig. 25, where the water is shot on at D, the point of impact—6 feet head, and 10 feet fall—required the motion of the circumference of the wheel, working to the best advantage, or maximum effect.

The velocity of a falling body, having 16 feet fall, the whole descent, } 32,4 feet.

Then, the velocity of the water, issuing on the wheel, 6 feet head, } 19,34, do.

Difference, - - - 13,06 do.

Then, as the velocity under a 16 feet fall (32,4 feet) is to the velocity of an overshot-wheel=8,76 feet, so is 13,06 feet, to the 16 feet diameter velocity accelerated, which is equal 3,5 feet, to which add ,577 of 19,34 feet (being 11,15 feet;) and this amounts to 14,65 feet per second, the velocity of the breast-wheel.

ARTICLE 61.

RULE FOR CALCULATING THE POWER OF ANY MILL-SEAT.

The only loss of power sustained by using too much head, in the application of water to turn a mill-wheel, is from the head producing only half its power. Therefore, in calculating the power of 16 cubic feet per second, on the different applications of fig. 25, Pl. III. we must add half the head to the whole fall, and count that sum the virtual perpendicular descent. Then, by the theorem in Art. 53, multiply the weight of the water per second by its perpendicular descent, and you have the true measure of its power.

But to simplify the rule, let us call each cubic foot 1, and the rule will then be—Multiply the cubic feet expended per second, by its virtual perpendicular descent in feet, and the product will be a true measure of the power per second. This measure must have a name, which I call Cuboch; that is, one cubic foot of water, multiplied by one foot descent, is one cuboch, or the unit of power.

EXAMPLES.

1. Given, 16 cubic feet of water per second to be applied by percussion alone, under 16 feet head, required the power per second.

Then, half 16=8×16=128 cubochs, for the measure of the power per second.

2. Given, 16 cubic feet per second, to be applied to a half breast of 4 feet fall and 12 feet head, required the power.

Then, half 12=6+4=10×16=160 cubochs, for the power.

3. Given, 16 cubic feet per second, to be applied to a pitch-back or high breast—fall 10, head 6 feet, required the power.

Then, half 6=3+10=13×16=208 cubochs, for the power per second.

4. Given, 16 cubic feet of water per second, to be applied as an overshot—head 4, fall 12 feet, required the power.

Then, half 4=2+12=14×16=224 cubochs, for the power.

The powers of equal quantities of water amounting to 16 cubic feet per second, the total perpendicular descents being equal, stand thus by the different modes of application:

The undershot, { 16 feet head,*
0 fall,
128 cubochs of power.

The half breast, { 12 feet head,
4 feet fall,
160 cubochs of power,

The high breast, { 6 feet head,
10 feet fall,
208 cubochs of power.

The overshot, { 4 feet head,
12 feet fall,
224 cubochs of power.

Ditto, { 2,5 feet head,
31,5 feet fall.
263 cubochs of power.

The last being the head necessary to shoot the water fairly into the buckets, may be said to be the best application. See Art. 43.

On these simple rules, and the rule laid down in Art. 43, for proportioning the head and fall, I have calcu-

* Water, by percussion, spends its force on the wheel in the following time, which is in proportion to the distance apart of the float-boards, and the difference of the velocity of the water and the wheel.

If the water runs with double the velocity of the wheel, it will spend all its force on the floats while the water runs to the distance of two float boards, and while the wheel runs to the distance of one; therefore, the water need not be kept to act on the wheel farther from the point of impact than the distance of about two float-boards.

But if the wheel run with two-thirds of the velocity of the water, then, while the wheel runs the distance of two floats, and while the water would have run the distance of three floats, it spends all its force; therefore, the water need be kept to act on the wheel the distance of three floats only past the point of impact.

If it be continued in action much longer, it will fall back, and react against the following bucket, and retard the wheel.

lated the following table, or scale, of the different quantities of water expended per second, with different perpendicular descents, to produce a certain power; in order to present at one view, the ratio of increase or decrease of quantity, as the perpendicular descent increases, or decreases.

A TABLE

Showing the quantity of water required with different falls, to produce by its gravity, 112 cubochs of power, which will drive a five feet stone about 97 revolutions in a minute, grinding about 5 bushels of wheat in an hour.

The virtual descent of the water, being half the head, added to all the fall, after it strikes the wheel.	Cubic feet of water required per second.	The virtual descent of the water, being half the head, added to all the fall, after it strikes the wheel.	Cubic feet of water required per second.
1	112	16	7
2	56	17	6.58
3	37.3	18	6.22
4	28	19	5.99
5	22.4	20	5.6
6	18.6	21	5.33
7	16	22	5.1
8	14	23	4.87
9	12.4	24	4.66
10	11.2	25	4.48
11	10.2	26	4.3
12	9.33	27	4.15
13	8.6	28	4
14	8	29	3.86
15	7.46	30	3.73

ARTICLE 62.

THEORY AND PRACTICE COMPARED.

I will here give a table of 18 mills in actual practice out of about 50 of which I have taken an account, in order to compare theory with practice, and in order to ascertain the power required on each superficial foot of the acting parts of the stone: But I must premise the following

THEOREMS.

1. To find the circumference of any circle, as of a mill-stone, by the diameter, or the diameter by the circumference; say,

As 7 is to 22, so is the diameter of the stone to the circumference; that is, multiply the diameter by 22, and divide product by 7, for the circumference; or, multiply the circumference by 7, and divide the product by 22, for the diameter.

2. To find the area of a circle, by the diameter: As 1, squared, is to ,7854, so is the square of the diameter to the area; that is, multiply the square of the diameter by ,7854, and, in a mill-stone, deduct 1 foot for the eye, and you have the area of the stone.

3. To find the quantity of surface passed by a millstone: The area, squared, multiplied by the revolutions of the stone, gives the number of superficial feet, passed in a given time.

OBSERVATIONS ON THE FOLLOWING TABLE OF EXPERIMENTS.

I have asserted, in Art. 44, that the head above the gate of a wheel, on which the water acts by its gravity, should be such, as to cause the water to issue on the wheel, with a velocity to that of the wheel, as 3 to 2. Compare this with the following table of experiments.

1. Exp. Overshot. Velocity of the water 12,9 feet per second, velocity of the wheel 8,2 feet per second, which is a little less than 2-3ds of the velocity of the water. This wheel received the water well. It is at Stanton, in Delaware state.

2. Overshot. Velocity of the water 11,17 feet per second, 2-3ds of which is 7,44 feet, velocity of the wheel 8,5 feet per second This received the water pretty well. It is at the above-mentioned place.

3. Overshot. Velocity of the water 12,16 feet per second, velocity of the wheel 10,2; throws out great part of the water by the back of the buckets, which strikes it and makes a thumping noise. It is allowed to run too fast; revolves faster than my theory directs. It is at Brandywine, in Delaware state.

4. Overshot. Velocity of the water 14,4 feet per second, velocity of the wheel 9,3 feet, a little less than 2-3ds of the velocity of the water. It receives the water very well; has a little more head than assigned by theory, and runs a little faster; it is a very good mill, situated at Brandywine, in the state of Delaware.

6. Undershot. Velocity of the wheel, loaded, 16, and when empty 24 revolutions per minute, which confirms the theory of motion for undershot wheels. See Art. 42.

7. Overshot. Velocity of the water 15,79 feet, velocity of the wheel 7,8 feet; less than 2-3ds of the velocity of the water; motion slower and head more than assigned by theory. The miller said the wheel ran too slowly, that he would have it altered; and that it worked best when the head was considerably sunk. This mill is at Bush, Hartford county, Maryland.

8. Overshot. Velocity of the water 14,96 feet per second, velocity of the wheel 8,8 feet, less than 2-3ds, very near the velocity assigned by the theory; but the head is greater, and the wheel runs best when the head is sunk a little; is counted the best mill, and is at the same place with the last mentioned.

9, 10, 11, 12. Undershot open wheels. Velocity of the wheels when loaded 20 and 40, and when empty 28 and 56 revolutions per minute, which is faster than my theory for the motion of undershot mills. Ellicott's mills, near Baltimore, in Maryland, serve to confirm the theory.

14. Overshot. Velocity of the water 16,2 feet, velocity of the wheel 9,1 feet, less than 2-3ds of the water, revolutions of the stone 144 per minute, the head nearly the same as by theory, the velocity of the wheel less, stone more. This shows the mill to be geared too high.

The wheel receives the water well, and the mill is counted a very good one, situated at Alexandria, in Virginia.

15. Undershot. Velocity of the water 24,3 per second, velocity of the wheel 16,67 feet, more than 2-3ds the velocity of the water. Three of these mills are in one house, at Richmond, Virginia—they confirm the theory of undershots, being very good mills.

16. Undershot. Velocity of the water 25,63 feet per second, velocity of the wheel 19,05 feet, being more than 2-3ds. Three of these mills are in one house, at Petersburg, in Virginia—they are very good mills, and confirm the theory. See Art. 43.

18. Overshot wheel. Velocity of the water 11,4 feet per second, velocity of the wheel 10,96 feet, nearly as fast as the water. The backs of the buckets strike the water, and drive a great part over; and as the motion of the stone is about right, and the motion of the wheel faster than assigned by the theory, it shows the mill to be too low geared, all which confirm the theory. See Art. 43.

In the following table I have counted the diameter of the mean circle to be 2-3ds of the diameter of the great circle of the stone, which is not strictly true. The mean circle, to contain half the area of any given circle, must be ,707 parts of the diameter of the said circle, differing but little from ,7, and somewhat exceeding 2-3ds.

Hence the following theorem for finding the mean circle of any stone.

THEOREM.

Multiply the diameter of the stone by ,707, and the product is the diameter of the mean circle.

EXAMPLE.

Given, the diameter of the stone 5 feet, required a mean circle that shall contain half its area.

Then, $5\times,707=3,535$ feet, the diameter of the mean circle.

ARTICLE 63.

FARTHER OBSERVATIONS ON THE FOLLOWING TABLE.

1. The mean power used to turn the 5 feet stones in the experiments (No. 1, 7, 14, 17,) is 87,5 cubochs of the measure established, Art. 61, and the mean velocity is 104 revolutions of the stones in a minute, the velocity of the mean circle being 18,37 feet per second, and their mean quantity ground is 3,8 lbs. per minute, which is 3,8 bushels per hour, and the mean power used to each foot of the area of the stone is 4,69 of the measure aforesaid, effected by 36582 superficial feet passing each other in a minute. Hence we may conclude,

1. That 87,5 cubochs of power per second will turn a 5 feet stone 104 revolutions in a minute, and grind 3,8 bushels in an hour.

2. That 4,69 cubochs of power are required to every superficial foot of a mill-stone, when its mean circle moves with a velocity of 18,37 feet per second. Or,

3. That for every 36582 feet of the face of stones that pass each other we may expect 3,8 lbs. will be ground, when the stones, grain, &c., are in the same state and condition as they were in the above experiments.

112   HYDRAULICS.   [Chap. 3.

A TABLE OF EXPERIMENTS ON EIGHTEEN MILLS IN PRACTICE.

Quantity ground per minute in pounds, or per hour in bushels.		3.5 2.5		3.75								4.5			3.5			
Superficial feet passed in a minute.		34594 21514		36435 35741 108091		36435		36435		36435		95264 49678 39558			74850 35741			
Velocity of the mean circle.	feet.	17.3 18.5		16.97 18.6		18.32 17.97 17.84		18.32		18.32		18.32		16.92 16.89 19.89 20.75 19.9			17.97		
Power required to each foot of face.		4.1 4.34		. .		4.9 5.9		.		.		.		4.67			5.15		
Area of the stones.	sup. ft.	18.63 13.13		. .		18.63 18.63 38.48		.		18.63		.		36.63 23.76 18.63			28.38 18.63		
Diameter of the stones in feet and inches.	ft. in.	3 2 5 4	6 4	6 4	3 4	4 4	4	5 5	7	5	10 4	5	10 6 6 5	4 5	6 5	4 4	4 4	8
Revolutions of the stones per minute.		99.7 124.8	122	122	104	108	126	105	103 73	105	105	105	71	88 114	113 95	103	124	116
Rounds in the trundles.		15 14	14	14	14	14	13	14 14	23	16	16	16	19	19 15	17 21	14	15	16
Cogs in the counter cog-wheels.		54 48	44	44	44	44	48	44 44	.	.	.	.	44 44	44 54	. .	44	48	48
Rounds in the wallowers.		27 22 24	24	24	24	24	22	22 22	.	.	.	.	25 24	24 25	. .	23	26	26
Number of cogs in the master-wheel.		88 88 72	66	72		78 72 84		42		84		42	96	96 60	44	66	72	
Velocity of the circumference per second.		8.2 8.5 10.2	9.3	9.8		loaded 7.8 unloaded 8.8		loaded 8.8 unloaded		loaded 7.8 unloaded		loaded unloaded	9.1	16.67 19.05	19.9	10.96		
Number of revolutions per minute.		8.5 9.1 13	12	105 16 24.5		9.5 10	20 28	40 56	20 28	40 56			8 9	32 45.5	16	14		
Diameter of the wheel.	feet.	18 18 15	.15	17.75 16.4 16.4 15				7.5		15		7.5	18.6	19.3 10	8	11 14		
Power per second, by simple theorem. Art. 61.	chs.	76 67	.	.		92 110		.		.		.	86	.	96			
Cubic feet expended per second, abating for friction by conjecture.	cub. ft.	3.8 3.3	.	.		.18 5.18 6.16		.		.		.	3.57	.	7.6			
Velocity of the water per second, by theory.	feet.	12.9 11.17 12.16 14.4		13.8		15.79 14.96 26.73		.		.		16.2	16.2 24.3 25.63	14	11.4			
Area of the gate, abating for contraction occasioned by friction.	feet.	.385 .325	.	.		.345 .425		.		.		.	.	.	.567			
Head above the centre of the gate.	feet.	2.67 1.9 2.2	3.1	3		3.83 3.5		.		.		4	4	. .	3	2		
Virtual or effective descent of the water.	feet.	20.2 19.2 16.2	16.6	19.25		17.8 17.8 11		.		.		20.6	21.5 9.5 10	5	12.5			
No. of experiments.		1 2 3	4	5		6 7 8 9		10		11		12	13	14 15 16	17	18		

In the 3d, 4th, 13th, and 18th experiments, in the above table, there are two pair of stones to one water-wheel, the gears, &c. of which are shown by the braces.

Observations continued from page 111.

As we cannot attain to a mathematical exactness in those cases, and as it is evident that all the stones in the foregoing experiments have been working with too little power, because it is known that a pair of good burr stones of 5 feet diameter will grind, sufficiently well, about 125 bushels in 24 hours—that is 5,2 bushels in an hour, which would require 6,4 cubochs of power per second—we may, for the sake of simplicity, say 6 cubochs, when 5 feet stones grind 5 bushels per hour. Hence we deduce the following simple theorem for determining the size of the stones to suit the power of any given seat, or the power required to any size of a stone.

THEOREM.

Find the power by the theorem, in Art. 61; then divide the power by 6, which is the power required, by 1 foot, and it will give you the area of the stone that the power will drive, to which add 1 foot for the eye, and divide by ,7854, and the quotient will be the square of the diameter: or, if the power be great, divide by the product of the area of any sized stones you choose, multiplied by 6, and the quotient will be the number of stones the power will drive: or, if the size of the stone be given, multiply the area by 6 cubochs, and the product is the power required to drive it.

EXAMPLE.

1. Given, 9 cubic feet per second, 12 feet perpendicular, virtual, or effective descent, required the diameter of the stone suitable thereto.

Then, by Art. 61, 9×12=108, the power, and 108÷6=18, the area, and 18×1÷,7854=24,2 the root of which is 4,9 feet, the diameter of the stone required.

5. The velocities of the mean circles of the stones in the table, are some below and some above 18 feet per second, the mean of them all being nearly 18 feet; therefore, I conclude that 18 feet per second is a good velocity in general, for the mean circle of any sized stone.

Of the different quantity of Surfaces that are passed by Mill-stones of different diameters with different velocities.

Supposing the quantity ground by mill-stones, and the power required to turn them, to be as the passing surfaces of their faces, each superficial foot that passes over another foot requires a certain power to grind a certain quantity: to explain this, let us premise,

1. The circumferences and diameters of circles are directly proportional. That is, a double diameter gives a double circumference.

2. The areas of circles are as the squares of their diameters. That is, a double diameter gives 4 times the area.

3. The square of the diameter of a circle, multiplied by ,7854, gives its area.

4. The square of the area of a mill-stone multiplied by its number of revolutions, gives the surface passed. Consequently,

5. In stones of unequal diameters revolving in equal times, their passing surfaces, quantity ground, and power required to drive them, will be as the squares of their areas, or as the biquadrate of their diameters. That is, a double diameter will pass 16 times the surface.*

6. If the velocity of their mean circles or circumferences be equal, their passing surfaces, quantity ground, and power required to move them, will be as the cubes of their diameters.†

7. If the diameters and velocities be unequal, their passing surfaces, and quantity ground, &c., will be as

---

* The diameter of a 4 feet stone squared, multiplied by ,7854, equal 12,56, its area; which squared is 157,75 feet, the surface passed at one revolution: and 8 multiplied by 8 equal 64, which multiplied by ,7854 equal 50,24 being the area of an 8 feet stone; which squared is 2524,04 the surface passed, which surfaces are as 1 to 16.

† Because the 8 feet stone will revolve only half as often as the 4 feet; therefore, their quantity of surface passed, &c., can only be half as much more as it was in the last case; that is, as 8 to 1.

the squares of their areas, multiplied by their revolutions.

8. If their diameters be equal, the quantity of surfaces passed, &c., are as their velocities or revolutions simply.

But we have been supposing theory and practice to agree strictly, which they will by no means do in this case. To the quantity ground and the proportion of power used by large stones more than by small ones, the ratio assigned by the theory will not apply; because the meal having to pass a greater distance through the stone, is operated upon oftener, which operation must be lighter, else it will be overdone; large stones may, therefore, be made to grind equal quantities with small ones, and with equal power, and to do it with less pressure; therefore, the flour will be better.* See Art. 111.

From these considerations, added to experiments, I conclude, that the power required and quantity ground, will be nearly as the area of the stones, multiplied into the velocity of the mean circles, or, which is nearly the same, as the squares of their diameters. But if the velocities of their mean circles or circumferences be equal, then it will be as their areas simply.

On these principles I have calculated the following table, showing the power required, and quantity ground, both by theory, and, what I suppose to be, the most correct practice.

---

* A French author (M. Fabre) says, that he has found by experiments, that, to produce the best flour, a stone 5 feet diameter should revolve between 48 and 61 times in a minute. This is much slower than the practice in America, but we may conclude that it is best to err on the side of a slower than of a faster motion than that of common practice; especially when the power is too small for the size of the stone.

A TABLE
OF THE
AREA OF MILL-STONES.
OF
DIFFERENT DIAMETERS,

And of the power required to move them with a mean velocity of 18 feet per second, &c.

Diameter of the stone in feet and parts.	Area of the stone in feet and parts, deducting 1 foot for the eye.	Power required to drive the stone, with mean velocity, 18 feet per second, allowing 6 cubchs to each foot of its area.	Circumference of the mean circle to contain half the area of the stone.	Revolutions of the stone per minute, with 18 feet velocity of mean circle per second.	Number of superficial feet passed per minute, being the square of the area of the stones multiplied by the number of revolutions.	Quantity ground in lbs. per minute or bushels per hour, supposing it to be as the number of superficial feet passed.	Power required, supposing it to be as the number of superficial feet passed.	Quantity ground, supposing it to be simply as the area of the stone with equal velocity.	Quantity ground, supposing it to be as the squares of the diameter of the stone, which appears to come nearest the true quantity.
feet.	s. t.	cuhs.	feet.		sup. ft.	lbs.	cuhs.	lbs.	lbs.
3.5	8.62	51.72	7.777	138.8	10312	1.49	33.1	2.3	2.45
3.75	9.99	59.94							2.8
4.	11.56	69.36	8.888	121.5	16236	2.3	52	3.1	3.2
4.25	13.18	79.							3.6
4.5	14.9	89.4	9.99	108.1	23999	3.46	77	4.	4.05
4.75	16.71	100.26							4.5
5.	18.63	111.78	11.09	97.4	34804	5.	111.78	5.	5.
5.25	20.64	123.84							5.53
5.5	22.76	136.5							6.05
5.75	24.96	153.7							6.6
6.	27.27	163.6	13.37	80.7	60012	8.6	192	7.3	7.2
6.25	29.67	178.							7.8
6.5	32.18	196.							8.4
6.75	34.77	208.6.							9.1
7.	37.48	225.	15.55	69.4	97499	14.06	313	10	9.8
1.	2	3	4	5	6	7	8	9	10

NOTE. One foot is deducted for the eye in each stone, and the reason why, in the 7th column, the quantity ground is not exactly as the cubes of the diameter of the stone, and, in the 9th column, not exactly as the squares of its diameter, is the deduction for the eye, which being equal in each stone, destroys the proportion.

The engine of a paper-mill, roll 2 feet diameter, 2 feet long, revolving 160 times in a minute, requires equal power with a 4 feet stone, grinding 5 bushels an hour.

I have now laid down, in Art. 61, 62, and 63, a theory for measuring the power of any mill-seat, and for ascertaining the quantity of that power that mill-stones of different diameters will require, by which we can find the diameter of the stones to suit the power of the seat; and have fixed on six cubochs of that power per second to every superficial foot of the mill-stone, as requisite to move the mean circle of the stone 18 feet per second, when in the act of grinding with moderate and sufficient feed; and have allowed the passing of 34804 feet per minute to grind 5 lbs. in the same time, which is the effect of the five feet stone in the table, by which, if right, we can calculate the quantity that a stone of any other size will grind with any given velocity.

I have chosen a velocity of 18 feet per second for the mean circle of all stones, which is slower than the common practice; but not too slow for making good flour. See Art. 111. Here will appear the advantage of large stones over small ones; for if we will make small stones grind as fast as large ones, we must give them such velocity as to heat the meal.

But I must here inform the reader, that the experiments, from which I have deduced the quantity of power to each superficial foot to be six cubochs, have not been sufficiently exact to be relied on; but it will be easy for every intelligent mill-wright to make accurate experiments to satisfy himself as to this point.*

---

* After having published the first edition of this work, I have been informed, that, by accurate experiments made at the expense of the British government, it was ascertained that the power produced by 40,000 cubic feet of water descending 1 foot, will grind and bolt 1 bushel of wheat. If this be true, then to find the quantity that any stream will grind per hour, multiply the cubic feet of water that it affords per hour, by the virtual descent, (that is, half of the head above the wheel, added to the fall after it enters an overshot-wheel,) and divide that product by 40,000, and the quotient will be the answer in bushels per hour that the stream will grind.

EXAMPLE.

Suppose a stream afford 32,000 cubic feet of water per hour, and the total fall 19,28 feet; then, by the table for overshot-mills, Art. 73, the wheel should be 16 feet diameter, head above the wheel 3,28 feet. Then half 3,28=1,64, which added to 16= 17,64 feet virtual descent, and 17,64×32000=563480, which divided by 40,000, gives 14,08 bushels per hour the stream will grind.

ARTICLE 64.

OF CANALS FOR CONVEYING WATER TO MILLS.

In digging canals, we must consider that water will come to a level on its surface, whatever may be the form of the bottom. If we have once determined on the area of the section of the canal necessary to convey a sufficient quantity of water to the mill, we need only to keep to that area in the whole distance, without paying much regard to the depth or width, if there be rocks in the way. Much expense may be oftentimes saved, by making the canal deep where it cannot easily be made wide enough, and wide where it cannot easily be made sufficiently deep. Thus, suppose we had determined it to be 4 feet deep, and 6 feet wide, then the area of its section will be 24.—Let fig. 36, Plate IV. represent a canal, the line A B the level or surface of the water, C D the side, E F the bottom, A C the width 6 feet, A E the depth, 4 feet. Then, if there be rocks at G, so that we cannot, without great expense, obtain more than 3 feet width, but can go 8 feet deep at a small expense: then $8 \times 3 = 24$, the section required. Again, suppose a flat rock to be at H, so that we cannot, without great expense, obtain more than 2 feet depth, but can, with small expense, obtain 12 feet width: then $2 \times 12 = 24$, the section required; and the water will come on equally well, even if it were not more than ,5 of a foot deep, provided it be proportionably wide. One disadvantage, however, arises in having canals very shallow in some places, because the water in dry seasons may be too low to rise over them; but if the water were always to be of one height, the disadvantage would be but trifling. The current will keep the deep places open, light sand or mud will not settle in them. This will seem paradoxical to some, but the experiment has been tried, and the fact established.

ARTICLE 65.

OF THE SIZE AND FALL OF CANALS.

As to the size and fall necessary to convey any quantity of water required to a mill, I do not find any rule laid down for either. But in order to establish one, let us consider, that the size depends entirely upon the quantity of water and the velocity with which it is to pass: therefore if we can determine on the velocity, which I will suppose to be from 1 to 2 feet per second —but the slower the better, as there will be the less fall lost—we can find the size of the canal by the following

THEOREM.

Divide the quantity required in cubic feet per second by the velocity in feet per second, and the quotient will be the area of the section of the canal. Divide that area by the proposed depth, and the quotient is the width: or, divide by the width, and the quotient is the depth.

PROBLEM I.

Given, a 5 feet mill-stone, its mean circle to be moved with a velocity of 18 feet per second, on a seat of 10 feet virtual, or effective, descent, required the size of the canal, with a velocity of 1 foot per second.

Then, by theorem in Art. 63: The area of the stone 18,63 feet, multiplied by 6 cubochs of power, is equal 111,78 cubochs for the power (in common practice say 112 cubochs) which divided by 10 the fall, quotes 11,178 cubic feet required per second, which divided by 1, the velocity proposed per second, gives 11,178 feet, the area of the section, which divided by the depth proposed, 2 feet, gives 5,58 feet for the width.

PROBLEM II.

Given, a mill-stone 6 feet diameter to be moved with a velocity of 18 feet per second of its mean circle, to be turned by an undershot-wheel on a seat of 8 feet perpendicular descent, required the power necessary per second to drive them, and the qantity of water per second to produce said power, likewise the size of the canal to convey the water with a velocity of 1,5 feet per second.

Then, by Art. 61, 8 feet perpendicular descent, on the undershot principle, is only=4 feet virtual or effective descent: and the area of the stone by the table (Art. 63) =27,27 feet ×6 cubochs=163,62 cubochs for the power per second, which divided by 4, the effective descent=40,9 cubic feet, the quantity required per second, which divided by the velocity proposed 1,5 feet per second=27,26, for the area of the section of the canal, which divided by 2,25 feet, the depth of the canal proposed=12 feet, the width.*

As to the fall necessary in the canal, I may observe, that the fall should be in the bottom of the canal, and none on the top, which should be all the way on a level with the water in the dam, in order that when the gate is shut down at the mill, the water may not overflow the banks, but stand at a level with the water in the dam; that is, as much fall as there is to be in the whole length of the canal, so much deeper must the canal be at the mill than at the dam. From many observations I conclude that about 3 inches to 100 yards will be sufficient, if the canal be long, but more will be requisite if it be short, and the head apt to run down when water is scarce, for the shallower the water, the greater must be the velocity, and the more fall is required. —A French author, M. Fabre, allows one inch to 500 feet.

---

* An acre of a mill-pond contains 43560 cubic feet of water, for every foot of its depth.

Suppose your pond contain 3 acres, and is 3 feet deep, then 43560, multiplied by 3, is equal 130680, which multiplied by 3, is equal 392040 cubic feet, its contents, which divided by the cubic feet your mill uses per second (say 10) is equal 39204 seconds, or 10 hours, the time the pond will keep the mill going.

ARTICLE 66.

OF AIR-PIPES TO PREVENT TIGHT TRUNKS FROM BURSTING WHEN FILLED WITH WATER.

When water is to be conveyed under ground, or in a tight trunk below the surface of the water in the reservoir, to any considerable distance, there must be air-pipes (as they have been called) to prevent the trunk from bursting. To understand their use, let us suppose a trunk 100 feet long, and 16 feet below the surface of the water; to fill which, a gate is to be drawn at one end, of equal size with the trunk. Then, if the water meet no resistance in passing to the other end, it acquires great velocity, which is suddenly to be stopped when the trunk is full. This great column of water in motion, in this case, would strike with a force equal to that of a solid body of equal weight and velocity, the shock of which would be sufficient to break any trunk that ever was made of wood. Many having thought the use of these pipes to be to let out the air, have made them too small; so that they would vent the air fast enough to let the water in with considerable velocity, but would not admit the water fast enough to check its motion gradually; in which case they are worse than useless; for if the air cannot escape freely, the water cannot enter freely, and the shock will be decreased by its resistance.

Whenever the air has been compressed in the trunk by the water coming in, it has made a great blowing noise in escaping through the crevices, and, therefore, has been viewed as the cause of the bursting of the trunk; whereas it acted, by its elastic principle, as a great preventive against it. For I apprehend, that if we were to pump all the air out of a trunk, 100 feet long, and 3 by 3 feet wide, and to let the water in with full force, it would burst, were it as strong as a cannon of cast metal; because, in that case, there would be 900 cubic feet of water, equal to 56250 lbs. pressed on by the weight of the atmosphere, with a velocity of 47 feet per second, to be suddenly stopped, the shock of which would be almost irresistible.

I consider it best, therefore, to make an air-pipe of the full size of the trunk, every 20 or 30 feet; but this will depend much on the depth of the trunk below the surface of the reservoir, and upon other circumstances.

Having now said what was necessary, in order the better to understand the theory of the power and principles of mechanical engines, and of water acting on water-wheels, upon different principles; and, for establishing true theories of the motion of the different kinds of water-wheels; I here quote many of the celebrated Mr. Smeaton's experiments, that the reader may compare them with the theories proposed, and judge for himself.

ARTICLE 67.

SMEATON'S EXPERIMENTS.

"*An experimental Inquiry, read in the Philosophical Society in London, May 3d, and 10th, 1759, concerning the Natural Powers of Water, to turn Mills and other Machines, depending on a circular Motion, by James Smeaton, F. R. S.*

"What I have to communicate on this subject was originally deduced from experiments made on working models, which I look upon as the best means of obtaining the outlines in mechanical inquiries. But in this case it is necessary to distinguish the circumstances in which a model differs from a machine in large: otherwise a model is more apt to lead us from the truth than towards it. Hence the common observation, that a thing may do very well in a model that will not do in large. And, indeed, though the utmost circumspection be used in this way, the best structure of machines cannot be fully ascertained, but by making trials with them of their proper size. It is for this purpose that, though the models referred to, and the greatest part of the following experiments, were made in the years 1752 and 1753, yet I deferred offering them to the society till I had an opportunity of putting the deductions made therefrom in real practice, in a variety of cases and for various purposes; so as to be able to assure the society, that I have found them to answer.

PART I.

CONCERNING UNDERSHOT WATER-WHEELS.

" Plate XII. is a view of the machine for experiments on water-wheels, wherein

ABCD is the lower cistern or magazine for receiving the water after it has left the wheel, and for supplying

DE the upper cistern or head, wherein the water being raised to any height by a pump, that height is shown by

FG a small rod divided into inches and parts, with a float at the bottom to move the rod up and down, as the surface of the water rises and falls.

HI is a rod by which the sluice is drawn, and stopped at any height required, by means of

K, a pin or peg, which fits several holes placed in the manner of a diagonal scale upon the face of the rod HI.

GL is the upper part of the rod of the pump for drawing the water out of the lower cistern, in order to raise and keep up the surface thereof to its desired height in the head DE, thereby to supply the water expended by the aperture of the sluice.

MM is the arch and handle of the pump, which is limited in its stroke by

N, a piece for stopping the handle from raising the piston too high, that also being prevented from going too low, by meeting the bottom of the barrel.

O is a cylinder upon which the cord winds, and which being conducted over the pulleys P and Q, raises

R, the scale into which the weights are put for trying the power of the water.

W the beam which supports the scale that is placed 15 or 16 feet higher than the wheel.

XX is the pump-barrel 5 inches diameter and 11 inches long. Y is the piston, and Z is the fixed valve.

GV is a cylinder of wood fixed upon the pump-rod, and reaches above the surface of the water; this piece of wood being of such thickness that its section is half the area of the pump-barrel, will cause the water to rise in the head as much while the piston is descending as while it is rising, and will thereby keep the gauge-rod FG more equally to its height.

a a shows one of the two wires that serve as a director to the float. b is the aperture of the sluice. c a is a cant-board for canting the water down the opening c d into the lower cistern. c e is a sloping board for bringing back the water that is thrown up by the wheel.

There is a contrivance for engaging and disengaging the scale and weight instantaneously from the wheel, by means of a hollow cylinder on which the cord winds by slipping it on the shaft; and when it is disengaged it is held to its place by a ratchet-wheel; for without this, experiments could not be made with any degree of exactness.

The apparatus being now explained, I think it necessary to assign the sense in which I use the term power.

The word power is used in practical mechanics, I apprehend, to signify the exertion of strength, gravity, impulse, or pressure, so as to produce motion.

The raising of a weight, relative to the height to which it can be raised in a given time, is the most proper measure of power. Or, in other words, if the weight raised be multiplied by the height to which it can be raised in a given time, the product is the measure of the power raising it; and, consequently, all those powers are equal. But note, all this is to be understood in case of slow or equable motion of the body raised; for in quick, accelerated, or retarded motions, the vis inertia of the matter moved will make a variation.

In comparing the effects produced by water-wheels with the powers producing them; or, in other words, to know what part of the original power is necessarily lost in the application, we must previously know how much of the power is spent in overcoming the friction of the machinery and the resistance of the air; also what is the real velocity of the water at the instant it strikes the wheel, and the real quantity of water expended in a given time.

From the velocity of the water at the instant that it strikes the wheel, given; the height of the head productive of such velocity can be deduced, from acknowledged and experienced principles of hydrostatics: so that by multiplying the quantity or weight of water really expended in a given time, by the height of head so obtained, which must be considered as the height from which that weight of water had descended, in that given time, we shall have a product equal to the origin 1 power of the water, and clear of all uncertainty that would arise from the friction of the water in passing small apertures, and from all doubts, arising from the different measure of spouting waters, assigned by different authors.

On the other hand, the sum of the weights raised by the action of this water, and of the weight required to overcome the friction and resistance of the machine, multiplied by the height to which the weight can be raised in the given time, the product will be the effect of that power; and the proportion of the two products will be the proportion of the power to the effect: so that by loading the wheel with different weights successively, we shall be able to determine at what particular load and velocity of the wheel the effect is a maximum.

To determine the Velocity of the Water striking the Wheel.

"First, let the wheel be put in motion by the water, but without any weight in the scale: and let the number of turns in a minute be 60: now, it is evident, that were the wheel free from friction and resistance, that 60 times the circumference of the wheel would be the space through which the water would have passed in a minute with that velocity wherewith it struck the wheel. But the wheel being encumbered with friction, and resistance, and yet moving 60 turns in a minute, it is plain, that the velocity of the water must have been greater than 60 circumferences, before it met with the wheel. Let the cord now be wound round the cylinder, but contrary to the usual way, and put as much weight in the scale as will, without any water, turn the wheel somewhat faster than 60 turns in a minute, suppose 63, and call this the counter-weight, then let it be tried again with the water assisted by this counter-weight, the wheel, therefore, will now make more than 60 turns in a minute, suppose 64, hence we conclude the water still exerts some power to turn the wheel. Let the weight be increased so as to make $64\tfrac{1}{2}$ turns in a minute without the water, then try it with the water and the weight as before, and suppose it now make the same number of turns with the water, as without; namely, $64\tfrac{1}{2}$, hence it is evident, that in this case the wheel makes the same number of turns as it would with the water, if the wheel had no friction or resistance at all, because the weight is equivalent thereto; for if the counter-weight were too little to overcome the friction, the water would accelerate the wheel, and if too great it would retard it: for the water in this case becomes a regulator of the wheel's motion, and the velocity of its circumference becomes a measure of the velocity of the water.

In like manner, in seeking the greatest product or maximum of effect; having found by trials what weight gives the greatest product, by simply multiplying the weight in the scale, by the number of turns of the wheel, find what weight in the scale, when the cord is on the contrary side of the cylinder, will cause the wheel to make the same number of turns, the same way, without water: it is evident that this weight will be nearly equal to all friction and resistance taken together; and, consequently, that the weight in the scale, with twice* the weight of the scale, added to the back or counter-weight, will

---

* The weight of the scale makes part of the weight both ways, namely; both of the weight and counter-weight.

be equal to the weight that could have been raised, supposing the machine had been without friction or resistance, and which multiplied by the height to which it was raised, the product will be the greatest effect of that power.

The Quantity of Water expended is found thus:—

"The pump was so carefully made, that no water escaped back through the leathers, it delivered the same quantity each stroke, whether quick or slow, and by ascertaining the quantity of 12 strokes, and counting the number of strokes in a minute that was sufficient to keep the surface of the water to the same height, the quantity expended was found.

These things will be farther illustrated by going over the calculations of one set of experiments.

Specimen of a Set of Experiments.

The sluice drawn to the 1st hole.	
The water above the floor of the sluice,	30 inches.
Strokes of the pump in a minute,	$39\frac{1}{2}$
The head raised by 12 strokes,	21
The wheel raised the empty scale and made turns in a minute,	80
With a counter-weight of one lb. 8 oz. it made,	85
Ditto, tried with water,	86

No.	lbs. oz.	turns in a min.	product.
1	4 : 0	45	180
2	5 : 0	42	210
3	6 : 0	$36\frac{1}{4}$	$217\frac{1}{2}$
4	7 : 0	$33\frac{3}{4}$	$236\frac{1}{4}$
5	8 : 0	30	240 max.
6	9 : 0	$26\frac{1}{2}$	$238\frac{1}{2}$
7	10 : 0	22	220
8	11 : 0	$16\frac{1}{2}$	$181\frac{1}{2}$
9	12 : 0	* ceased working.	

* When the wheel moves so slowly as not to rid the water so fast as supplied by the sluice, the accumulated water falls back upon the aperture, and the wheel immediately ceases moving.

NOTE. This note of the author argues in favour of drawing the gate near the float.

Counter-weight for 30 turns without water 2 oz. in the scale.

N. B. The area of the head was 105,8 square inches, weight of the empty scale and pulley 10 ounces, circumference of the cylinder 9 inches, and circumference of the water-wheel 75 inches.

Reduction of the above Set of Experiments.

" The circumference of the wheel 75 inches, multiplied by 86 turns, gives 6450 inches for the velocity of the water in a minute, 1-60th of which will be the velocity in a second, equal to 107,5 inches, or 8,96 feet, which is due to a head of 15 inches,* and this we call the virtual or effective head.

The area of the head being 105,8 inches, this multiplied by the weight of water of one cubic inch, equal to the decimal of ,579 of the ounce avoirdupois, gives 61,26 ounces for the weight of as much water as is contained in the head upon one inch in depth, 1-10th of which is 3,83 lbs.; this, multiplied by the depth 21 inches, gives 80,43 lbs. for the value of 12 strokes, and by proportion 39½ (the number made in a minute) will give 264,7 lbs., the weight of water expended in a minute.

Now, as 264,7 lbs. of water may be considered as having descended through a space of 15 inches in a minute, the product of these two numbers 3970 will express the power of the water to produce mechanical effects; which are as follows:—

The velocity of the wheel at a maximum as appears above, was 30 turns in a minute; which, multiplied by 9 inches, the circumference of the cylinder, makes 270 inches: but as the scale was hung by a pulley and double line, the weight was only raised half of this, namely; 135 inches.

---

* This is determined by the common maxim of hydrostatics; that the velocity of spouting water is equal to the velocity that a heavy body would acquire in falling from the height of the reservoir; and is proved by the rising of jets to the height of their reservoirs nearly.

HYDRAULICS.

	lbs.	oz.
The weight in the scale at the maximum,	8	0
Weight of the scale and pulley,	0	10
Counter-weight, scale, and pulley,	0	12
Sum of the resistance,	lbs. 9	6, or 9,375 lbs.

Now, as 9,375 lbs. are raised 135 inches, these two numbers being multiplied together produce 1266, which expresses the effect produced at a maximum: so that the proportion of the power to the effect is as 3970: 1266, or as 10: 3,18.

But though this be the greatest single effect producible from the power mentioned, by the impulse of the water upon an undershot wheel; yet, as the whole power of the water is not exhausted thereby, this will not be the true ratio between the power and the sum of all the effects producible therefrom: for, as the water must necessarily leave the wheel with a velocity equal to the circumference, it is plain that some part of the power of the water must remain after leaving the wheel.

The velocity of the wheel at a maximum is 30 turns a minute, and, consequently, its circumference moves at the rate of 3,123 feet per second, which answers to a head of 1,82 inches: this being multiplied by the expense of water in a minute; namely, 264,7 lbs. produces 481 for the power remaining: this being deducted from the original power, 3970, leaves 3489, which is that part of the power that is spent in producing the effect 1266; so that the power spent, 3489, is to its greatest effect 1266, as 10 : 3,62, or as 11 : 4.

The velocity of the water striking the wheel 86 turns in a minute, is to the velocity at a maximum 30 turns a minute, as 10 : 3,5, or as 20 to 7, so that the velocity of the wheel is a little more that 1-3d of the velocity of the water.

The load at a maximum has been shown to be equal to 9 lbs. 6 oz. and that the wheel ceased moving with 12 lbs. in the scale; to which, if the weight of the scale be added, namely; 10 oz.,* the proportion will be nearly as 3 to 4, between the load at a maximum and that by which the wheel is stopped.†

It is somewhat remarkable, that, though the velocity of the wheel in relation to the water turns out greater than 1-3d of the velocity of the water, yet the impulse of the water in case of the maximum is more than double of what is assigned by theory; that is, instead of 4-9ths of the column, it is nearly equal to the whole column.‡

It must be remembered, therefore, that in the present case, the wheel was not placed in an open river, where the natural current, after it has communicated its impulse to the float, has room on all sides to escape, as the theory supposes; but in a conduit or race, to which the float being adapted, the water cannot otherwise escape than by moving along with the wheel. It is observable, that a wheel working in this manner, as soon as the water meets the float, it, receiving a sudden check, rises up against the float, like a wave against a fixed object, insomuch, that when the sheet of water is not a quarter of an inch thick before it meets the float, yet this sheet will act upon the whole surface of a float, whose height is three inches; consequently, were the float no higher than the thickness of the sheet of water, as the theory also supposes, a great part of the force would be lost by the water dashing over it.

---

* The resistance of the air in this case ceases, and the friction is not added, as 12 lbs. in the scale was sufficient to stop the wheel after it had been in full motion, and, therefore, somewhat more than a counterbalance for the impulse of the water.

† I may here observe, that it is probable, that if the gate of the sluice had been drawn as near the float-boards as possible, [as is the practice in America, where water is applied to act by impulse alone,] that the wheel would have continued to move until loaded with $1\frac{1}{2}$ times the weight of the maximum load; namely, 9 lbs. 6 oz. multiplied by $1\frac{1}{2}$, equal to 14 lbs. 1 oz. It would then have agreed with the theory established Art. 41. This, perhaps, escaped the notice of our author.

‡ This observation of the author I think a strong confirmation of the truths of the theory established Art. 41; where the maximum velocity is made to be ,577 parts of the velocity of the water, and the load to be 2-3ds the greatest load: For if the gate had been drawn near the floats, the greatest load would probably have been 14 lbs. 1 oz., or as 3 to 2 of the maximum load.

In confirmation of what is already delivered, I have subjoined the following table, containing the result of 27 experiments made and reduced in the manner above specified. What remains of the theory of undershot wheels, will naturally follow from a comparison of the different experiments together.

A TABLE OF EXPERIMENTS.
NO. I.

Number.	Height of the water in the cistern.	Turns of the wheel, unloaded.	Virtual head deduced therefrom.	Turns at a maximum.	Load at the equilibrium.	Load at the maximum.	Water expended in a minute.	Power.	Effect.	Ratio of the power and effect.	Ratio of the velocities of the water and wheel.	Ratio of the load at the equilibrium to the load at the maximum.	Experiments.
	in.		inch.		lb. oz.	lb. oz.	lbs.						
1	33	88	15.85	30	13 10	10 9	275	4358	1411	10:3.24	10:3.4	10:7.75	At the 1st hole.
2	30	86	15.	30	12 10	9 6	264.7	3970	1266	10:3.2	10:3.5	10:7.4	
3	27	82	13.7	28	11 2	8 6	243	3329	1044	10:3.15	10:3.4	10:7.5	
4	24	78	12.3	27.7	9 10	7 5	235	2890	901.4	10:3.12	10:3.55	10:7.53	
5	21	75	11.4	25.9	8 10	6 5	214	2439	735.7	10:3.02	10:3.45	10:7.32	
6	18	70	9.95	23.5	6 10	5 5	199	1970	561.8	10:2.85	10:3.36	10:8.02	
7	15	65	8.54	23.4	5 2	4 4	178.5	1524	442.5	10:2.9	10:3.6	10:8.3	
8	12	60	7.29	22	3 10	3 5	161	1173	328	10:2.8	10:3.77	10:9.1	
9	9	52	5.47	19	2 12	2 8	134	733	213.7	10:2.9	10:3.65	10:9.1	
10	6	42	3.55	16	1 12	1 10	114	404.7	117	10:2.82	10:3.8	10:9.3	
11	24	84	14.2	30.75	13 10	10 14	342	4890	1505	10:3.07	10:3.66	10:7.9	At the 2d.
12	21	81	13.5	29	11 10	9 6	297	4009	1223	10:3.01	10:3.62	10:8.05	
13	18	72	10.5	26	9 10	8 7	285	2993	975	10:3.25	10:3.6	10:8.75	
14	15	69	9.6	25	7 10	6 14	277	2659	774	10:2.92	10:3.62	10:9.	
15	12	63	8.0	25	5 10	4 14	234	1872	549	10:3.94	10:3.97	10:8.7	
16	9	56	6.37	23	4 0	3 13	201	1280	390	10:3.05	10:4.1	10:9.5	
17	6	46	4.25	21	2 8	2 4	167.5	712	212	10:2.98	10:4.55	10:9.	
18	15	72	10.5	29	11 10	9 6	357	3748	1201	10:3.23	10:4.02	10:8.05	3d.
19	12	66	8.75	26.75	8 10	7 6	330	2887	878	10:3.05	10:4.05	10:8.1	
20	9	58	6.8	24.5	5 8	5 0	255	1734	541	10:3.01	10:4.22	10:9.1	
21	6	48	4.7	23.5	3 2	3 0	228	1064	317	10:2.99	10:4, 9	10:9.6	
22	12	68	9.3	27	9 2	8 6	359	3338	1006	10:3.02	10:3.97	10:9.17	4th.
23	9	58	6.8	26.25	6 2	5 13	332	2257	686	10:3.04	10:4.52	10:9.5	
24	6	48	4.7	24.5	3 12	3 8	262	1231	385	10:3.13	10:5.1	10:9.35	
25	9	60	7.29	27.3	6 12	6 6	355	2588	783	10:3.03	10:4.55	10:9.45	5th.
26	6	50	5.03	24.6	4 6	4 1	307	1544	456	10:2.92	10:4.9	10:9.3	
27	6	50	5.03	26	4 15	4 9	360	1811	534	10:2.95	10:5.2	10:9.25	6
1	2	3	4	5	6	7	8	9	10	11	12	13	

Maxims and Observations deduced from the foregoing Table of Experiments.

"Max. I. That the virtual or effective head being the same, the effect will be nearly as the quantity of water expended.

This will appear by comparing the contents of the columns 4, 8, and 10, in the foregoing sets of experiments, as, for

Example I. taken from No. 8 and 25; namely:—

No.	Virtual head.	Water expended.	Effect.
8	7,29	161	328
25	7,29	355	785

Now the heads being equal, if the effects be proportioned to the water expended, we shall have by maxim 1, as 161 : 355 :: 328 : 723; but 723 falls short of 785, as it turns out in experiment, according to No. 25, by 62. The effect, therefore, of No. 25, compared with No. 8, is greater than according to the present maxim, in the ratio of 14 to 13.*

The foregoing example, with four similar ones, may be seen at one view in the following table.

---

* If the true maximum velocity of the wheel be ,577 of the velocity of the water, and the true maximum load be 2-3ds of the whole column, as shown in Art. 42: then the effect will be to the power in the ratio of 100 to 38, or as 10 to 3,8, a little more then appears by the table of experiments in columns 9 and 10: the difference is owing to the disadvantageous application of the water on the wheel in the model.

TABLE OF EXPERIMENTS.

NO. II.

Proportional variation.	14 : 13	121 : 122	38 : 39	18 : 17	178 : 177
Variation.	+62	11−	18−	21+	3+
COMPARISON. Effect.	328 } 161 : 355 :: 328 : 723 785	975 } 285 : 357 :: 975 : 1221 1210	541 } 255 : 332 :: 541 : 704 686	317 } 228 : 262 :: 317 : 364 385	450 } 307 : 360 :: 450 : 531 584
Expense of water.	161 355	285 357	255 332	228 262	307 360
Virtual head.	7.29 7.29	10.5 10.5	6.8 6.8	4.7 4.7	5.08 5.08
No. Table I. Examples.	1st. { 8 25	2d. { 13 18	3d. { 22 23	4th. { 21 24	5th. { 26 27

Chap. 3.] HYDRAULICS. 135

By this table of experiments, it appears that some fall short of, and others exceed, the maximum, and all agree as nearly as can be expected in an affair where so many different circumstances are concerned; therefore, we may conclude the maxim to be true.

Max. II. That the expense of the water being the same, the effect will be nearly as the height of the virtual or effective head.

This also will appear by comparing the contents of columns 4, 8, and 10, in any of the sets of experiments.

Example I. of No. 2 and No. 24.

No.	Virtual head.	Expense.	Effect.
2	15	264,7	1266
24	4,7	262	385

Now, as the expenses are not quite equal, we must proportion one of the effects accordingly, thus:—

By maxim I.   262 : 264,7 :: 385 : 389
And by max. II.   15 : 4,7 :: 1266 : 397
                                    ———
                        Difference, 8

The effect, therefore, of No. 24, compared with No. 2, is less than according to the present maxim, in the ratio of 49 : 50.

Max. III. That the quantity of water expended being the same, the effect is nearly as the square root of its velocity.

This will appear by comparing the contents of columns 3, 8, and 10, in any set of experiments; as for

Example I. of No. 2, with No. 24; namely:—

No.	Turns in a minute.	Expense.	Effect.
2	86	264,7	1266
24	48	262	385

The velocity being as the number of turns, we shall have

By maxim I.   262 : 264,7 :: 385:389

And by max. III. $\left\{\begin{array}{c} 86^3 \quad\; 48^2 \\ 7396 : 2304 \end{array}\right\}$ ::1266:394

                                                 Difference, 5

The effect of No. 24, compared with No. 2, is less than by the present maxim in the ratio of 78 : 79.

Max. IV. The aperture being the same, the effect will be nearly as the cube of the velocity of the water.

This also will appear by comparing the contents of columns 3, 8, and 10, as, for

Example No. 1, and No. 10; namely:—

No.	Turns.	Expense.	Effect.
1	88	275	1411
10	42	114	117

Lemma. It must here be observed, that, if water pass out of an aperture in the same section, but with different velocities, the expense will be proportional to the velocity; and, therefore, conversely, if the expense be not proportional to the velocity, the section of water is not the same.

Now, comparing the water discharged with the turns of No. 1 and 10, we shall have 88:42::275:131,2; but the water discharged by No. 10 is only 114 lbs., therefore, though the sluice was drawn to the same height in No. 10 as in No. 1, yet the section of the water passing out, was less in No. 10 than No. 1, in the proportion of 114 to 131,2; consequently, had the effective aperture or section of the water been the same in No. 10 as in No. 1, so that 131,2 lbs. of water had been discharged instead of 114 lbs. the effect would have been increased in the same proportion; that is,

By lemma   88 : 42 :: 275:131,2

By maxim I,   114 : 131,2 :: 117:134,5

And by max. IV. $\left\{\begin{array}{c} 88^3 \;\;:\;\; 42^3 \\ 681472 : 74088 \end{array}\right\}$ :: 1411:153,5

                                            Difference, 19

The effect, therefore, of No. 10, compared with No. 1, is less than ought to be, by the present maxim, in the ratio of 7:8.

OBSERVATIONS.

"Observ. 1st. On comparing columns 2 and 4, table I., it is evident, that the virtual head bears no certain proportion to the head of water, but that when the aperture is greater, or the velocity of the water issuing therefrom less, they approach nearer to a coincidence: and, consequently, in the large opening of mills and sluices, where great quantities of water are discharged from moderate heads, the head of water and virtual head determined from the velocity will nearer agree, as experience confirms.

Observ. 2d. Upon comparing the several proportions between the powers and effects in column 11th, the most general is that of 10 to 3; the extremes are 10 to 3,2 and 10 to 2,8; but as it is observable, that where the quantity of water or the velocity thereof is great, that is, where the power is greatest, the 2d term of the ratio is greatest also, we may, therefore, well allow the proportion subsisting in large works as 3 to 1.

Observ. 3d. The proportion of velocities between the water and wheel in column 12 is contained in the limits of 3 to 1 and 2 to 1; but as the greater velocities approach the limits of 3 to 1, and the greater quantity of water approaches to that of 2 to 1, the best general proportion will be that of 5 to 2.*

Observ. 4th. On comparing the numbers in column 13, it appears, that there is no certain ratio between

* I will here observe, that Mr. Smeaton may be mistaken in his conclusion, that the best general ratio of the velocity of the water to that of the wheel will be as 5 to 2; because, we may observe, that, in the first experiment, where the virtual head was 15,85 inches, and the gate drawn to the 1st hole, the ratio is as 10 : 3,4. But in the last experiment, where the head was 5,03 inches, and the gate drawn to the 6th hole, the ratio is as 10 : 5,2; and that the 2d term of the ratio increases, gradually, as the head decreases, and quantity of water increases: therefore, we may conclude, that, in the large openings of mills, the ratio may approach to 3 to 2; which will agree with the practice and experiments of many able mill-wrights of America, and many experiments I have made on mills. And as it is better to give the wheel too great than too little velocity, I conclude, the wheel of an undershot mill must have nearly two-thirds of the velocity of the water to produce a maximum effect.

the load that the wheel will carry at its maximum, and what will totally stop it; but that they are contained within the limits of 20 to 19 and of 20 to 15; but as the effect approaches nearest to the ratio of 20 to 15 or of 4 to 3, when the power is greatest, whether by increase of velocity or quantity of water, this seems to be the most applicable to large works: but as the load that a wheel ought to have in order to work to the best advantage, can be assigned by knowing the effect it ought to produce, and the velocity it ought to have in producing it, the exact knowledge of the greatest load that it will bear is of less consequence in practice.*

It is to be noted, that in almost all of the examples under the three last maxims (of the four preceding) the effect of the less power falls short of its due proportion to the greater, when compared by its maxim. And hence, if the experiments be taken strictly, we must infer that the effects increase and diminish in a higher ratio than those maxims suppose; but as the deviations are not very considerable, the greatest being about 1-8th of the quantity in question, and as it is not easy to make experiments of so compound a nature with absolute precision, we may rather suppose that the less power is attended with some friction, or works under some disadvantage, not accounted for: and, therefore, we may conclude, that these maxims will hold very nearly, when applied to works in large.

After the experiments above-mentioned were tried, the wheel which had 24 floats was reduced to 12, which caused a diminution in the effect on account of a greater quantity of water escaping between the floats and the floor; but a circular sweep being adapted thereto, of such a length that one float entered the curve before the preceding one quitted it, the effect came so near to the former, as not to gives hopes of increasing the effect by increasing the number of floats past 24 in this particular wheel.

---

* Perhaps the author is here again deceived by the imperfection of the model; for had the water been drawn close to the float, the load that would totally stop the wheel would always be equal to the column of water acting on the wheel. See the note page 70. The friction of the shute and air, destroyed great part of the force of his small quantity of water.

ARTICLE 68.

PART II.

CONCERNING OVERSHOT WHEELS.

"In the former part of this essay, we have considered the impulse of a confined stream, acting on undershot wheels; we now proceed to examine the power and application of water, when acting by its gravity on overshot wheels.

It will appear in the course of the following deductions, that the effect of the gravity of descending bodies is very different from the effect of the stroke of such as are non-elastic, though generated by an equal mechanical power.

The alterations of the machinery already described, to accommodate the same for experiments on overshot wheels, were principally as follows:—

Plate XII. The sluice I b being shut down, the rod H I was taken off. The undershot water-wheel was taken off the axis, and instead thereof, an overshot wheel of the same size and diameter was put in its place. Note, this wheel was 2 inches deep in the shroud or depth of the bucket, the number of buckets was 36.

A trunk for bringing the water upon the wheel was fixed according to the dotted lines f g, the aperture was adjusted by a shuttle which also closed up the outer end of the trunk, when the water was to be stopped.

Specimen of a Set of Experiments.

Head 6 inches—$14\frac{1}{2}$ strokes of the pump in a minute, 12 ditto $=80$ lbs.*—weight of the scale (being wet) $10\frac{1}{2}$ ounces.

Counter-weight for 20 turns, besides the scale, 3 ounces.

No.	wt. in the scale.	turns.	product.	Observations.
1	0	60	——	⎫ threw most part of
2	1	56	——	⎬ the water out of
3	2	52	——	⎭ the wheel.
4	3	49	147	⎫ received the water
5	4	47	188	⎭ more quietly.
6	5	45	225	
7	6	$42\frac{1}{2}$	255	
8	7	41	287	
9	8	$38\frac{1}{2}$	308	
10	9	$36\frac{1}{2}$	$328\frac{1}{2}$	
11	10	$35\frac{1}{2}$	355	
12	11	$32\frac{3}{4}$	$360\frac{1}{2}$	
13	12	$31\frac{1}{4}$	375	
14	13	$28\frac{1}{2}$	$370\frac{1}{2}$	
15	14	$27\frac{1}{2}$	385	
16	15	26	390	
17	16	$24\frac{1}{2}$	392	
18	17	$22\frac{3}{4}$	$386\frac{3}{4}$	
19	18	$21\frac{3}{4}$	$391\frac{1}{2}$	
20	19	$20\frac{3}{4}$	$394\frac{1}{4}$	⎫ maximum.
21	20	$19\frac{3}{4}$	395	⎭
22	21	$18\frac{1}{4}$	$383\frac{1}{4}$	
23	22	18	396	worked irregularly.
24	23	overset by its load.		

* The small difference in the value of 12 strokes of the pump from the former experiments, was owing to a small difference in the length of the stroke, occasioned by the warping of the wood.

Reduction of the preceding Specimen.

"In these experiments the head being 6 inches, and the height of the wheel 24 inches, the whole descent will be 30 inches: the expense of water was $14\frac{1}{2}$ strokes of the pump in a minute, whereof 12 contained 80 lbs., therefore, the water expended in a minute, was $96\frac{2}{3}$ lbs. which multiplied by 30 inches, give the power =2900.

If we take the 20th experiment for the maximum, we shall have $20\frac{3}{4}$ turns in a minute, each of which raised the weight $4\frac{1}{2}$ inches, that is, 93,37 inches in a minute. The weight in the scale was 19 lbs., the weight of the scale $10\frac{1}{2}$ oz., the counter-weight 3 oz. in the scale, which, with the weight of the scale $10\frac{1}{2}$ oz., make in the whole $20\frac{1}{2}$ lbs. which is the whole resistance or load; this, multiplied by 93,37 makes 1914, for the effect.

The ratio, therefore, of the power and effect will be as 2900 : 1914, or as 10 : 6,6, or as 3 to 2, nearly.

But, if we compute the power from the height of the wheel only, we have $96\frac{2}{3}$ lbs. ×24 inches=2320 for the power, and this will be to the effect as 2320 : 1914, or as 10 : 8,2, or as 5 to 4, nearly.

The reduction of this specimen is set down in No. 9 of the following table, and the rest were deduced from a similar set of experiments, reduced in the same manner.

TABLE III.

CONTAINING THE RESULT OF 16 SETS OF EXPERIMENTS ON OVERSHOT WHEELS.

Number.	Whole descent.	Water expended per minute.	Turns at the maximum per minute.	Weight raised at the maximum.	Power of the whole descent.	Power of the wheel.	Effect.	Ratio of the whole power and effect.	Ratio of the power of the wheel and effect.	Mean ratio.
	inchs.	lbs.		lbs.						
1	27	30	19	6 1-2	810	720	556	10 : 6.9	10 : 7.7	Medium. 10 : 8.1
2	27	56 2-3	16 1-4	14 1-2	1530	1360	1060	10 : 6.9	10 : 7.8	
3	27	56 2-3	20 3-4	12 1-2	1530	1360	1167	10 : 7.6	10 : 8.4	
4	27	63 1-3	20 1-2	13 1-2	1710	1524	1245	10 : 7.3	10 : 8.2	
5	27	76 2-3	21 1-2	15 1-2	2070	1840	1500	10 : 7.3	10 : 8.2	
6	28 1-2	73 1-3	18 3-4	17 1-2	2090	1764	1476	10 : 7	10 : 8.4	10 : 8.2
7	28 1-2	96 2-3	20 1-4	20 1-2	2755	2320	1868	10 : 6.8	10 : 8.1	
8	30	90	20	19 1-2	2700	2160	1755	10 : 6.5	10 : 8.1	10 : 8.2
9	30	96 2-3	20 3-4	20 1-2	2900	2320	1914	10 : 6.6	10 : 8.2	
10	30	113 1-3	21	23 1-2	3400	2720	2221	10 : 6.5	10 : 8.2	
11	33	56 2-3	20 1-4	13 1-2	1870	1360	1230	10 : 6.6	10 : 9.	10 : 8.5
12	33	106 2-3	22 1-4	21 1-2	3520	2560	2153	10 : 6.1	10 : 8.4	
13	33	146 2-3	23	27 1-2	4840	3520	2846	10 : 5.9	10 : 8.1	
14	35	65	19 3-4	16 1-2	2275	1560	1466	10 : 6.5	10 : 9.4	10 : 8.5
15	35	120	21 1-2	25 1-2	4200	2880	2467	10 : 5.9	10 : 8.6	
16	35	163 1-2	25	26 1-2	5728	3924	2981	10 : 5.2	10 : 7.6	
1	2	3	4	5	6	7	8	9	10	11

OBSERVATIONS AND DEDUCTIONS FROM THE FOREGOING EXPERIMENTS.

I. *Concerning the Ratio between the Power and Effect of Overshot Wheels.*

" The effective power of the water must be reckoned upon the whole descent, because it must be raised to that height in order to be in a condition of producing the same effect a second time.

The ratios between the powers so estimated, and the effects at a maximum deduced from the several sets of experiments, are exhibited at one view in column 9 of table III.; and, hence, it appears, that those ratios differ from that of 10 to 7,6 to that of 10 to 5,2; that is, nearly from 4 : 3 to 4 : 2. In those experiments where the heads of water and quantities expended are least, the proportion is nearly as 4 to 3; but where the heads and quantities are greatest, it approaches nearer to that of 4 to 2, and by a medium of the whole the ratio is that of 3 : 2, nearly. We have seen before, in our observations upon the effects of undershot wheels, that the general ratio of the power to the effect, when greatest, was as 3 : 1. *The effect, therefore, of overshot wheels, under the same circumstances of quantity and fall, is, at a medium, double to that of the undershot: and a consequence thereof, that non-elastic bodies, when acting by their impulse or collision communicate only a part of their original power:* the other part being spent in changing their figure in consequence of the stroke.*

The powers of water computed from the height of the wheel only, compared with the effects as in column 10, appear to observe a more constant ratio: for, if we take the medium of each class, which is set down in column 11, we shall find the extremes to differ no more than from the ratio of 10 : 8.1 to that of 10 : 8,5, and as the second term of the ratio gradually increases from 8,1 to 8,5 by an increase of head from 3 inches to 11, the ex-

---

* These observations of the author agree with the theory, Art. 41—42. I may add, that non-elastic bodies, when acting by impulse or collision, communicate only half of their original power, by the laws of motion.

cess of 8,5 above 8,1 is to be imputed to the superior impulse of the water at the head of 11 inches, above that of 3 inches; so that if we reduce 8,1 to 8, on account of the impulse of the 3 inch head, we shall have the ratio of the power computed upon the height of the wheel only, to the effect at a maximum, as 10 : 8, or as 5 : 4, nearly. And from the equality of the ratio, between power and effect, subsisting where the constructions are similar, we must infer that the effects as well as the powers, are, as the quantities of water and perpendicular heights, multiplied together respectively.

II. *Concerning the most proper Height of the Wheel in Proportion to the whole Descent.*

"We have already seen, in the preceding observation, that the effect of the same quantity of water, descending through the same perpendicular space, is double, when acting by its gravity upon an overshot wheel, to what the same produces when acting by its impulse, upon an undershot. It also appears, that, by increasing the head from 11 to 3 inches, that is, the whole descent, from 27 to 35, or in the ratio of 7 to 9, nearly, the effect is advanced no more than in the ratio of 8,1 to 8,4; that is, as 7 : 7,26, and, consequently, the increase of the effect is not 1-7th of the increase of the perpendicular height. Hence, it follows, that the higher the wheel is in proportion to the whole descent, the greater will be the effect; because it depends less upon the impulse of the head, and more upon the gravity of the water in the buckets: and, if we consider how obliquely the water issuing from the head must strike the buckets, we shall not be at a loss to account for the little advantage that arises from the impulse thereof; and shall immediately see of how little consequence this impulse is to the effect of an overshot wheel. However, as every thing has its limits, so has this: for thus much is desirable, that the water should have somewhat greater velocity than the circumference of the wheel, in coming thereon; otherwise the wheel will not only be retarded by the buckets striking the water, but thereby dashing a part of it over, so much of the power is lost.

The velocity that the circumference of the wheel ought to have, being known, the head requisite to give the water its proper velocity is easily computed by the common rules of hydrostatics, and it will be found much less than what is commonly practised.

III. *Concerning the Velocity of the Circumference of the Wheel in order to produce the greatest effect.*

" If a body be let fall freely from the surface of the head to the bottom of the descent, it will take a certain time in falling, and in this case the whole action of gravity is spent in giving the body a certain velocity: but, if this body in falling be made to act upon some other body, so as to produce a mechanical effect, the falling body will be retarded; because, a part of the action of gravity is then spent in producing the effect, and the remainder only giving motion to the falling body: and, therefore, the slower a body descends, the greater will be the portion of the action of gravity applicable to the producing of a mechanical effect. Hence, we are led to this general rule, that the less the velocity of the wheel, the greater will be the effect thereof. A confirmation of this doctrine, together with the limits it is subject to in practice, may be deduced from the foregoing specimen of a set of experiments.

From these experiments it appears, that when the wheel made about 20 turns in a minute, the effect was nearly upon the greatest; when it made 30 turns, the effect was diminished about 1-20th part; but, that, when it made 40, it was diminished about 1-4th: when it made less than $18\frac{1}{4}$ its motion was irregular; and when it was loaded so as not to admit its making 18 turns, the wheel was overpowered by its load.

It is an advantage in practice, that the velocity of the wheel should not be diminished farther than what will procure some solid advantage in point of power; because, as the motion is slower, the buckets must be made larger;

and the wheel being more loaded with water, the stress upon every part of the work will be increased in proportion: the best velocity for practice, therefore, will be such as when the wheel here used made about 30 turns in a minute; that is, when the velocity of the circumference is a little more than 3 feet in a second.

Experience confirms, that this velocity of 3 feet in a second, is applicable to the highest overshot wheels as well as the lowest; and all other parts of the work being properly adapted thereto, will produce very nearly the greatest effect possible. However, this also is certain, from experience, that high wheels may deviate farther from this rule, before they will lose their power, by a given aliquot part of the whole, than low ones can be admitted to do; for a wheel of 24 feet high may move at the rate of 6 feet per second without losing any considerable part of its power: and, on the other hand, I have seen a wheel of 33 feet high that has moved very steadily and well, with a velocity but little exceeding 2 feet."*

[Mr. Smeaton has also made a model of a wind-mill, and a complete set of experiments on the power and effect of the wind, acting on wind-mill sails of different constructions. But as the accounts thereof are quite too long for the compass of my work, I, therefore, extract little more than a few of the principal maxims deduced from his experiments, which, I think, may not only be of use to those who are concerned in building wind-mills, but may, also, serve to confirm some principles deduced from his experiments on water-mills.]

* Probably this wheel was working a forge or furnace bellows, which have deceived many by their slow regular motion.

ARTICLE 69.

PART III.

OF THE CONSTRUCTION AND EFFECTS OF WIND-MILL SAILS.*

" In trying experiments on wind-mill sails, the wind itself is too uncertain to answer the purpose; we must, therefore, have recourse to artificial wind.

This may be done two ways; either by causing the air to move against the machine, or the machine to move against the air. To cause the air to move against the machine in a sufficient column, with steadiness and the requisite velocity, is not easily put in practice: To carry the machine forward in a right line against the air, would require a larger room than I could conveniently meet with. What I found most practicable, therefore, was to carry the axis whereon the sails were to be fixed progressively round in the circumference of a large circle. Upon this idea the machine was constructed.†

Specimen of a Set of Experiments.

Radius of the sails,	21 inches.
Length of do. in cloth,	18
Breadth of do.,	5,6
‡ { Angle at the extremity,	10 degs.
Do. at the greatest inclination,	25
20 turns of the sails raised the weight,	11,3 inch.
Velocity of the centre of the sails in the circumference of the great circle in a second, in which the machine was carried round,	6 feet.
Continuance of the experiment,	52 seconds.

---

* Read May 31st and June 14th, 1759, in the Philosophical Society of London

† I decline giving any description or draught of this machine, as I have not room; but I may say, that it was constructed so as to wind up a weight, (as did the other model) in order to find the effect of the power. I also insert a specimen of a set of experiments, which I fear will not be well understood for want of a full explanation of the machine.

‡ In the following experiments, the angle of the sails is accounted from the plane of their motion; that is, when they stand at right angles to the axis, their angle is denoted ° deg.; this notation being agreeable to the language of practitioners, who call the angle so denoted the weather of the sail; which they denominate greater or less, according to the quantity of the angle.

No.	Weight in the scale.	turns.	Product.
1	0 lbs.	108	0
2	6	85	510
3	$6\frac{1}{2}$	81	$526\frac{1}{2}$
4	7	78	546
5	$7\frac{1}{2}$	73	$547\frac{1}{2}$ maxim.
6	8	65	520
7	9	0	0

The product is found by simply multiplying the weight in the scale by the number of turns.

By this set of experiments it appears, that the maximum velocity is 2-3ds of the greatest velocity, and that the ratio of the greatest load to that of the maximum is, as 9 to 7,5, but, by adding the weight of the scale and friction to the load, the ratio turns out to be as 10 : 8,4 or 5 to 4, nearly. The following table is the result of 19 similar sets of experiments.

By the following table it appears that the most general ratio between the velocity of the sails unloaded and when loaded to a maximum, is 3 to 2, nearly.

And the ratio between the greatest load and the load at a maximum (taking such experiments where the sails answered best,) is, at a medium, about as 6 to 5, nearly.

And that the kind of sails used in the 15th and 16th experiments are best of all, because they produce the greatest effect or product, in proportion to their quantity of surface, as appears in column 12.

TABLE IV.

Containing Nineteen Sets of Experiments on Wind-mill Sails of various Structures, Positions, and Quantities of Surface.

The kind of sails made use of	Number.	Angle at the extremities.	Greatest angle.	Turns of the sails, unloaded.	Turns at a maximum.	Load at a maximum.	Greatest load.	Product.	Quantity of surface.	Ratio of the greatest velocity to the velocity at a maximum.	Ratio of the greatest load to the load at a maximum.	Ratio of the surface to the product.
		deg.	deg.			lbs.	lbs.		sq.in			
I.	1	35°	35°	66	42	7. 56	12. 59	318	404	10:7	10:6	10: 7. 9
II.	2	12	12		70	6. 3	7. 56	441	404		10: 8.3	10: 10. 1
	3	15	15	105	69	6. 72	8. 12	464	404	10: 6.6	10: 8.3	10: 10.15
	4	18	18	96	66	7. 0	9. 81	462	404	10: 7	10: 7.1	10: 10.15
III.	5	9	26. 5		66	7. 0		462	404			10: 11. 4
	6	12	29. 5		70. 5	7. 35		518	404			10: 12. 8
	7	15	32. 5		63. 5	8. 3		527	404			10: 13. 0
IV.	8	0	15	120	93	4. 75	5. 31	442	404	10: 7.7	10: 8.9	10: 11. 0
	9	3	18	120	79	7. 0	8. 12	553	404	10: 6.6	10: 8.6	10: 13. 7
	10	5	20		78	7. 5	8. 12	585	404		10: 9.2	10: 14. 5
	11	7.5	22. 5	113	77	8. 3	9. 81	639	404	10: 6.8	10: 8.5	10: 15. 8
	12	10	25	108	73	8. 69	10. 37	634	404	10: 6.8	10: 8.4	10: 15. 7
	13	12	27	100	66	8. 41	10. 94	580	404	10: 6.6	10: 7.7	10: 14. 4
V.	14	7. 5	22. 5	123	75	10. 65	12. 59	799	505	10: 6.1	10: 8.5	10: 15. 8
	15	10	25	117	74	11. 08	13. 69	820	505	10: 6.3	10: 8.1	10: 16. 2
	16	12	27	114	66	12. 09	14. 23	799	505	10: 5.8	10: 8.4	10: 15. 8
	17	15	30	96	63	12. 09	14. 78	762	505	10: 6.6	10: 8.2	10: 15. 1
VI.	18	12	22	105	64. 5	16. 42	27. 87	1059	854	10: 6.1	10: 5.9	10: 12. 4
	19	12	22	99	64. 5	18. 06		1165	1146	10: 5.9		10: 10. 1
	1	2	3	4	5	6	7	8	9	10	11	12

I. Plain sails at an angle of 55 degrees.
II. Plain sails weathered according to common practice.
III. Weathered according to Maclaurin's theorem.
IV. Weathered in the Dutch manner, tried in various positions.
V. Weathered in the Dutch manner, but enlarged towards the extremities.
IV. 8 sails, being sectors of ellipses in their best positions.

TABLE V.

Containing the result of 6 sets of experiments, made for determining the difference of effect according to the different velocity of the wind.

Ratio of the greatest load to the load at a maximum.		10:8.3 10:9.1		10:8.5 10:8.7	14
Ratio of the greatest velocity to the velocity at a maximum.		10:6.9 10:5.9		10:6.7 10:6.2	13
Ratio of the two products.		10:27.3	10:27.8	10:26	12
Product of the lesser load and greater velocity.		805	832	795	11
Turns of the sails therewith.		180	180	158	10
Maximum load for half the velocity.		4.47	4.62	5.08	9
Product.		295 2003	300 2278	307 2047	8
Greatest load.	lbs.	5.37 18.06		5.87 21.34	7
Load at the maximum.	lbs.	4.47 16.42	4.62 17.52	5.03 18.61	6
Turns of the sails at a maximum.		66 122	65 130	61 110	5
Turns of the sails unloaded.		96 207		91 178	4
Velocity of the wind in a second.	in. ft.	4½ 9 4 8	4½ 9 4 8	4½ 9 4 8	3
Angle at the extremity.	degr.	5 5	7.5 7.5	10 10	2
Number.		1 2	3 4	5 6	1

N. B. The sails were of the same size and kind as those of Nos. 10, 11, and 12, table IV. Continuance of the experiment one minute.

Concerning the Effects of Sails according to the different Velocity of the Wind.

"From the foregoing table the following maxims are deduced.

Maxim I. The velocity of wind-mill sails, whether unloaded or loaded, so as to produce a maximum, is nearly as the velocity of the wind, their shape and position being the same.

This appears by comparing the respective numbers of columns 4 and 5, table V., wherein those numbers 2, 4, and 6, ought to be double of No. 1, 3, and 5, and are as nearly so as can be expected by the experiments.

Maxim II. The load at the maximum is nearly, but somewhat less than, as the square of the velocity of the wind, the shape and positions of the sails being the same.

This appears by comparing No. 2, 4, and 6, in column 6, with 1, 3, and 5, wherein the former ought to be quadruple of the latter, (as the velocity is double,) and are as nearly so as can be expected.

Maxim III. The effects of the same sails at a maximum are nearly, but somewhat less than, as the cubes of the velocity of the wind.*

It has been shown, maxim I, that the velocity of sails at a maximum, is nearly as the velocity of the wind; and by maxim II., that the load at the maximum, is nearly as the square of the same velocity. If those two maxims would hold precisely, it would be a consequence that the effect would be in a triplicate ratio thereof. How this agrees with experiment will appear by comparing the products in column 8, wherein those of No. 2, 4, and 6, (the velocity of the wind being double,) ought to be octuple of those of No. 1, 3, and 5, and are nearly so.

Maxim IV. The load of the same sails at the maximum is nearly as the squares of, and their effects as the cubes of, their number of turns in a given time.

---

* This confirms the 7th law of spouting fluids.

This maxim may be esteemed a consequence of the three preceding ones."

These 4 maxims agree with and confirm the 4 maxims concerning the effects of spouting fluids acting on undershot mills; and, I think, sufficiently confirm as a law of motion, that the effect produced, if not the instant momentum of a body in motion, is as the square of its velocity, as asserted by the Dutch and Italian philosophers.

Smeaton says, that by several trials in large, he has found the following angles to answer as well as any:—

"The radius is supposed to be divided into 6 parts, and 1-6th, reckoning from the centre is called 1, the extremity being denoted 6.

No.	Angle with the axis.	Angle with the plane of motion.
1	72°	18°
2	71	19
3	72	18 middle.
4	74	16
5	77½	12½
6	83	7 extremity."

He seems to prefer the sails being largest at the extremities.

END OF PART FIRST.

THE

YOUNG MILL-WRIGHT'S

GUIDE.

PART THE SECOND.

INTRODUCTION.

WHAT has been said in the first part, was meant to establish theories, and to furnish easy rules. In this part I mean to show their practical application, in as concise a manner as possible, referring only to the articles in the first part, where the reasons and demonstrations are given.

This part is particularly intended for the help of young and practical mill-wrights, whose time will not admit of a full investigation of those principles and theories, which have been laid down; I shall, therefore, endeavour to reduce the substance of all that has been said, to a few tables, rules, and short directions, which, if found to agree with experience, will be sufficient for the practitioner.

CHAPTER IV.

OF THE DIFFERENT KINDS OF MILLS.

ARTICLE 70.

OF UNDERSHOT MILLS.

UNDERSHOT wheels move by the percussion or stroke of the water, and are only half as powerful as other wheels that are moved by the gravity of the water. See Art. 9. Therefore, this construction ought not to be adopted, except where there is but little fall, or great plenty of water. The undershot wheel, and all others that move by percussion, should move with a velocity nearly equal to two-thirds of the velocity of the water. See Art. 42, Fig. 28, Plate IV., represents this construction.

For a rule for finding the velocity of the water, under any given head, see Art. 51. Upon the principles, and by the rule, given in that article, is formed the following table of the velocity of spouting water, under different heads, from one to twenty-five feet high above the centre of the issue; to which is added the velocity of the wheel suitable thereto, and the number of revolutions a wheel of fifteen feet diameter (which I esteem a good size) will revolve in a minute; also, the number of cogs and rounds in the wheels, both for double and single gears, so as to produce about ninety-seven or one hundred revolutions per minute, for a five feet stone, which I think a good motion and size for a mill-stone, grinding for merchantable flour.

That the reader may fully understand how the following table is calculated, let him observe,

1. That, by Art. 42, the velocity of the wheel must be just 577 thousandth parts of the velocity of the water; therefore, if the velocity of the water, per second, be multiplied by ,577, the product will be the maximum velocity of the wheel, or velocity that will produce the greatest effect, which is the third column in the table.

2. The velocity of the wheel per second, multiplied by 60, produces the distance the circumference moves per minute, which, divided by 47,1 feet, the circumference of a 15 feet wheel, gives the number of revolutions of the wheel per minute, which is the fourth column.

3. That, by Art. 20 and 74, the number of revolutions of the wheel per minute, multiplied by the number of cogs in all the driving wheels, successively, and that product, divided by the product of the number of cogs in all the leading wheels, multiplied successively, the quotient is the number of revolutions of the stones per

minute, which is found in the ninth and twelfth columns.

4. The cubochs of power required to drive the stone being, by Art. 61, equal to 111,78, cubochs per second, which, divided by half the head of water, added to all the fall, (if any,) being the virtual or effective head by Art. 61, gives the quantity of water, in cubic feet, required per second, which is found in the thirteenth column.

5. The quantity required, divided by the velocity with which it is to issue, gives the area of the aperture of the gate, and is shown in the fourteenth column.

6. The quantity required, divided by the velocity proper for the water to move along the canal, gives the area of the section of the canal; as in the fifteenth column.

7. Having obtained their areas, it is easy, by Art. 65, to determine the width and depth, which may be varied to suit other circumstances.

THE MILL-WRIGHT'S TABLE

FOR

UNDERSHOT MILLS,

CALCULATED FOR A WATER-WHEEL OF FIFTEEN FEET, AND STONES OF FIVE FEET DIAMETER.

Head of water above the point of impact.	Velocity of the water per second at the point of impact.	Velocity of the wheel per second, loaded at the maximum.	Number of revolutions of the wheel of 15 feet diameter, per minute.	No. of cogs in the master cog-wheel.	Rounds in the wallower.	Cogs in the counter cog-wheel.	Rounds in the trundle.	Revolutions of the stone per minute.	Rounds in the trundle.	Cogs in the cog-wheel for single gear.	Rounds in the trundle.	Revolutions of the stone per minute.	Cubic feet of water required per second to drive a 5 feet stone 97 revolutions per minute.	Area of the gate to vent the water, or rather of a section of the column of water at place of impact.	Area of a section of the canal sufficient to bring on the water with 1.5 feet velocity.
feet.	feet.	feet.											cub. ft.	sup. ft.	sup. ft.
1	8.1	4.67	5.94	112	22	54	16	101.6					223.5	27.5	149.
2	11.4	6.57	8.36	96	23	54	19	99					111.78	9.8	74.5
3	14.	8.07	10.28	88	25	54	19	100.5					74.52	4.6	43.
4	16.2	9.34	11.19	78	23	48	20	97					55.89	3.45	37.26
5	18.	10.38	13.22	66	24	48	18	97	112	15	98.66		44.7	2.48	29.8
6	19.84	11.44	14.6	66	24	48	20	96.2	112	17	96.2		37.26	1.9	24.84
7	21.43	12.36	15.74	66	25	44	19	96.2	104	17	96.2		31.9	1.48	21.26
8	22.8	13.15	16.75	66	25	44	20	97.2	96	16	100.		27.94	1.22	18.6
9	24.3	14.02	17.86	66	26	42	19	100.2	96	17	100.8		24.84	1.02	16.56
10	25.54	14.73	18.78	60	25	44	20	99	96	18	100.		22.89	.9	15.26
11	26.73	15.42	19.7	60	26	44	20	100	96	19	99.5		20.32	.76	13.54
12	28.	16.16	20.5	60	27	44	20	100	96	20	98.4		18.63	.66	12.42
13	29.16	16.32	21.42	60	27	42	20	99.8	96	21	102.6		16.27	.56	10.8
14	30.2	17.42	22.19	60	28	42	20	99	88	20	97.63		15.94	.53	10.6
15	31.34	18.08	23.03	60	29	42	20	99	88	21	96.5		14.9	.47	9.93
16	32.4	18.69	23.8						88	21	99.7		13.97	.43	9.31
17	33.32	19.22	24.48						84	21	97.9		13.14	.39	8.76
18	34.34	19.81	25.23						80	21	96.1		12.42	.36	8.28
19	35.18	20.29	25.82						80	21	98.3		11.76	.33	7.84
20	36.2	20.88	26.6						78	21	98.3		11.17	.3	7.4
21	37.11	21.41	27.26						78	22	97.		10.64	.29	7.1
22	37.98	21.86	27.84						78	22	98.6		10.16	.26	6.77
23	38.79	22.38	28.5						72	21	97.7		9.72	.25	6.48
24	39.69	22.90	29.17						66	20	96.2		9.32	.23	6.21
25	40.5	23.36	29.75						60	18	99.		8.94	.22	5.96
1	2	3	4	5	6	7	8	9	10	11	12		13	14	15

It must be observed, that five feet fall is the least that a single gear can be built on, to keep the cog-wheel clear of the water, and give the stone sufficient motion.

Although double gear is calculated to fifteen feet fall, yet I do not recommend them above ten feet, unless for some particular convenience, such as for two pair of stones to one wheel, &c. &c. The number of cogs in the wheels is even, and is thus suited to eight, six, or four arms, so as not to pass through any of them, this being the common practice; but when the motion cannot be obtained without a trundle that will cause the same cogs and rounds to meet too often, such as 16 into 96, which will meet every revolution of the cog-wheel, or 18 to 96, which will meet every third revolution; I advise the putting in either of one more, or one less, as may best suit the motion, which will cause them to change oftener. See Art. 82.

It should be recollected, that the friction at the aperture of the gate will greatly diminish both the velocity and power of the water, where the head is great, if the gate be made of the usual form, that is, wide and shallow. Where the head is great, the friction will be great. See Art. 55: therefore, the wheel must be narrow, and the aperture of the gate of a square form, in order to avoid the friction and loss in a wide wheel, especially if it do not run very closely to the sheeting.

Use of the Table.

Having levelled your mill-seat carefully, and finding such fall and quantity of water as determines you to make choice of an undershot wheel; for instance, suppose 6 feet fall, and about 45 cubic feet of water per second, which you may find in the way directed in Art. 53; cast off about one foot for fall in the tail-race below the bottom of the wheel, if subject to back-water, which leaves you 5 feet head; then look for 5 feet head in the first column of the table, and against it are all the calculations for a 15 feet water-wheel and 5 feet stones; in the thirteenth column you have 44,7 cubic feet of water, which shows you have enough for a pair of five feet stones; and the velocity of the water will be 18 feet per second, the velocity of the wheel 10,38 feet per second, and it will revolve 13,22 times per minute. If you choose double gear, then 66 cogs in the master cog-wheel, 24 rounds in the wallower, 48 cogs in the counter cog-wheel, and 18 rounds in the trundle, will give the stone 97 revolutions in a minute; if single gear, 112 cogs and 15 rounds give 98,66 revolutions in a minute; it will require 44,7 cubic feet of water per second; the size of the gate must be 2,48 feet, which will be about 4 feet wide, and ,62 feet, or about $7\frac{1}{4}$ inches deep; the size of the canal must be 29,8 feet; that is, about 3 feet deep and 9,93 or nearly 10 feet wide. If you choose single gear, you must make your water-wheel much smaller, say $7\frac{1}{2}$, the half of 15 feet, then the cog-wheel must have half the number of cogs, the trundle-head the same, the spindle will be longer, and the husk lower; the mill will then be full as good as with double gear: in the case supposed, however, a cog-wheel of 66 cogs would not answer, because it would reach the water; but where the head is 10 or 12 feet, it will do very well.

If you choose stones, or water-wheels, of other sizes, it will be easy, by similar rules, to proportion the whole to suit, seeing you have the velocity of the periphery of a wheel of any size.*

* One advantage large wheels have over small ones is, that they cast off the back water much better. The buckets of the low wheel will lift the water much more than those of the high wheel; because the nearer the water rises to the centre of the wheel, the nearer the buckets approach the horizontal or lifting position.

To make a wheel cast off back-water, some mill-wrights fix the sheeting below the wheel, with joints and hinges, so that the end down stream can be raised so as to shoot the water, as it leaves the wheel, on to the surface of the back-water, and thus roll it from the wheel, it is thought that it will drive off the back-water much better.

Plate IV. fig. 28, shows an undershot wheel. Some mill-wrights prefer to slant the forebay under the wheel, as in the figure, that the gate may be drawn near the floats; because they think that the water acts with more power near the gate, than at a distance; which appears to be the case, when we consider, that the nearer we approach the gate, the nearer the column of water approaches to what is called a perfectly definite quantity. See Art. 59.

Others, again, say, that it acquires equal power of descending the shute. (It will certainly acquire equal velocity abating only for the friction of the shute and the air.) When the shute has a considerable descent, the greater the distance from the gate, the greater the velocity and power of the water; but where the descent of the shute is not sufficient to overcome the friction of the air, &c., then

Observations on the Table.

1. The table is calculated for an undershot wheel constructed, and the water shot on, as in Plate IV. fig. 28. The head is counted from the point of impact I., and the motion of the wheel at a maximum, about ,58 of the velocity of the water; but when there is plenty of water, and great head, the wheel will run best at about ,66 or two-thirds of the velocity of the water; therefore, the stones will incline to run faster than in the table, in the ratio of 58 to 66, nearly; for which reason, I have set the motion of 5 feet stones under 100 revolutions in a minute, which is slower than common practice; they will incline to run between 96 and 110 revolutions.

2. I have taken half of the whole head above the point of impact, for the virtual or effective head, by Art. 53, which I apprehend will be too little in very low heads, and, perhaps, too much in high ones. As the principle of non-elasticity does not seem to operate against the power so much in low as in high heads; therefore, if the head be only 1 foot, it may not require 223,5 cubic feet of water per second, and if 20 feet, may require more than 11,17 cubic feet of water per second, the quantities given in the table.

ARTICLE 71.

OF TUB MILLS.

A tub mill has a horizontal water-wheel, that is acted on by the percussion of the water altogether; the shaft is

the nearer the gate, the greater the velocity and power of the water; which argues in favour of drawing the gate near the floats. Yet, where the fall is great, or water plenty, and the expense of a deep penstock considerable, the small difference of power is not worth the expense of thus obtaining it. In these cases, it is best to have a shallow penstock, and a long shute to convey the water down to the wheel, drawing the gate at the top of the shute: this is frequently done to save expense in building saw-mills, with flutter-wheels, which are small undershot wheels, fixed on a crank shaft, and made so small as to obtain a sufficient number of strokes of the saw in a minute, say about 120. This wheel is to be of such a size as is calculated to suit the velocity of the water at the point of impact, so as to make that number of revolutions (120) in a minute.

Thomas Ellicott's method of shooting the water on an undershot wheel, where the fall is great, is shown in plate 13, fig. 6.

vertical, carrying the stone on the top of it, and serves in place of a spindle; the lower end of this shaft is set in a step fixed in a bridge-tree, by which the stone is raised and lowered, as by the bridge-tree of other mills; the water is shot on the upper side of the wheel, in the direction of a tangent with its circumference. See fig. 29, Plate IV., which is a top view of the tub-wheel, and fig. 39, which is a side view of it, with the stone on the top of the shaft, bridge-tree, &c. The wheel runs in a hoop, like a mill-stone hoop, projecting so far above the wheel as to prevent the water from shooting over the wheel, and whirls it about until it strikes the buckets, because the water is shot on in a deep narrow column, 9 inches wide and 18 inches deep, to drive a 5 feet stone, with 8 feet head—the whole of this column cannot enter the buckets until a part has passed half way round the wheel, so that there are always nearly half the buckets struck at once: the buckets are set obliquely, that the water may strike them at right angles. See Plate IV., fig. 30. As soon as it strikes, it escapes under the wheel in every direction, as in fig. 29.*

* NOTE. That in Plate IV. fig. 30, I have allowed the gate to be drawn inside of the penstock, and not in the shute near the wheel, as is the common practice; because the water will leak out much alongside of the gate, if drawn in the shute. But here we must consider that the gate must always be full drawn, and the quantity of water regulated by a regulator in the shute near the wheel; so that the shute will be perfectly full, and pressed with the whole weight of the head, else a great part of the power may be lost.

To show this more plainly, suppose the long shute A, from the high head (shown by dotted lines) of the undershot mill, fig. 28, be made tight by being covered at top, then, if we draw the gate A, but not fully, and if the shute at bottom be large enough to vent all the water that issues through the gate when the shute is full to A, then it cannot fill higher than A; therefore, all that part of the head above A is lost, it being of no other service than to supply the shute, and keep it full to A, and the head from A to the wheel, is all that acts on the wheel.

Again, when we shut the gate, the shute cannot run empty, because it would leave a vacuum in the head of the shute at A; therefore, the pressure of the atmosphere resists the running out of the water from the shute, and whatever head of water is in the shute, when the gate is shut, will balance its weight of the pressure of the atmosphere, and prevent it from acting on the lower side of the gate, which will cause it to be very hard to draw. For, suppose 11 feet head of water to be in the shute when the gate was shut, its pressure is equal to about 5 lbs. per square inch; then, if the gate be 48 by 6 inches, which is equal to 288 inches, this multiplied by 5, is equal to 1440 lbs. the additional pressure on the gate.

Again, if the gate be full drawn, and the shute be not much larger at the up-

The disadvantages of these wheels are,

1. Under the best construction the water does not act to advantage on them; and it is in general necessary to make them so small, in order to give velocity to the stone, that the buckets take up a third part of their diameter.

2. The water acts with less power than on undershot wheels, as it is less confined at the time of striking the wheel, and its non-elastic principle operates more fully.

3. If the head be low, it is with difficulty we can put a sufficient quantity of water to act on them so as to drive them with sufficient power; I, therefore, advise to let the water strike on them in two places; as in Plate IV. fig. 29; the apertures need then only be about 6 by 13 inches each, instead of 9 by 18; they will then operate to more advantage, as nearly all the buckets will be acted on at once.

Their advantages are,

Their exceeding simplicity and cheapness, having no cogs nor rounds to be kept in repair, their wearing parts are few, and have but little friction; the step-gudgeon runs under water, therefore, if well fixed, it will not get out of order in a long time; they will move with sufficient velocity and power with 9 or 10 feet total fall, if there be plenty of water; and, if they be well fixed, they will not require much more water than undershot wheels; they are, therefore, preferable in all seats which have a surplus of water, and above 8 feet fall.

In order that the reader may fully understand how the following table for tub-mills is calculated, let him consider,

1. That the tub-wheel moves altogether by percussion, the water flying clear of the wheel the instant it

per than the lower end, these defects will cause much loss of power. To remedy all this, put the gate H at the bottom of the shute to regulate the quantity of water by, and make a valve at A to shut on the inside of the shute, like the valve of a pair of bellows, which will close when the gate A is drawn, and open when the gate shuts, and let air into the shute; this plan will do better for saw-mills with flutter-wheels, or tub-mills, than long open shutes, as by it we avoid the friction of the shute and the resistance of the air.

To understand what is here said, the reader must be acquainted with the theory of the pressure of the atmosphere, vacuums, &c. See these subjects touched on in Art. 56.

strikes, and that it is better, (by Art. 70,) for such wheels to move faster than the calculated maximum velocity; therefore, instead of ,577, we will allow them to move ,66 velocity of the water; then multiplying the velocity of the water by ,66, gives the velocity of the wheel, at the centre of the buckets; which constitutes the 3d column in the table.

2. The velocity of the wheel per second, multiplied by 60, and divided by the number of revolutions the stone is to make in a minute, gives the circumference of the wheel at the centre of the buckets; which circumference, multiplied by 7, and divided by 22, give the diameter from the centre of the buckets, to produce the number of revolutions required; which are contained in the 4th, 5th, 6th, and 7th columns.

3. The cubochs of power required, by Art. 63, to drive the stone, divided by half the head, give the cubic feet of water required to produce said power; which are found in the 8th and 10 columns.

4. The cubic feet of water, divided by the velocity, will give the sum of the apertures of the gates; which are shown in the 9th and 11th columns.

5. The cubic feet of water, divided by 1,5 feet, the velocity of the water in the canal, gives the area of a section of the canal; which is shown in the 12th and 13th columns.

6. For the quantity of water, aperture of gate, and size of canal, for 5 feet stones, see table for undershot mills, in Art. 70.

OF TUB MILLS.

THE MILL-WRIGHT'S TABLE

FOR

TUB MILLS.

Head of water above the point of impact or set of the wheel.	Velocity of the water per second.	Velocity of the wheel, counted at the centre of the buckets, and being 66 velocity of the water.	Diameter of the wheel, to the centre of the buckets, for a stone 4 feet diameter, 122 revolutions in a minute.	Ditto for a 5 feet stone, to revolve 98 times in a minute.	Ditto for a 6 feet stone, to revolve 81 times in a minute.	Ditto for a 7 feet stone, to revolve 70 times in a minute.	Cubic feet of water, per second, required to drive the 4 feet stones.	Sum of the areas of the apertures of the gate for a 4 feet stone.	Cubic feet of water required, per second, for a 6 feet stone.	Sum of the areas of the apertures for a 6 feet stone.	Area of a section of the canal sufficient to bring the water to 4 feet stones, with a velocity of 1.5 feet per second.	Ditto for 6 feet stones.
ft.	feet.	feet.	feet.	feet.	feet.	feet.	cub. ft.	su.ft.	cub. ft.	su.ft.	sup.ft.	sup.ft.
8	22.8	15.04	2.17	2.73	3.3	3.9	17.34	.76	40.9	1.79	11.56	27.3
9	24.3	16.03	2.5	3.12	3.68	4.37	15.41	.64	36.35	1.5	10.3	24.23
10	25.54	16.85	2.63	3.28	3.97	4.59	13.87	.54	32.72	1.28	9.25	21.7
11	26.73	17.64	2.75	3.44	4.15	4.8	12.61	.47	29.74	1.11	8.4	19.83
12	28.	18.48	2.9	3.6	4.34	4.9	11.56	.41	27.26	.97	7.7	18.17
13	29.16	19.24	3.01	3.74	4.53	5.24	10.67	.36	25.17	.86	7.1	16.8
14	30.2	19.93	3.12	3.9	4.7	5.43	9.9	.33	23.36	.77	6.6	15.56
15	31.34	20.68	3.24	4.03	4.87	5.67	9.24	.29	21.93	.7	6.16	14.62
16	32.4	21.38	3.34	4.12	5.01	5.83	8.67	.27	20.45	.6	5.71	13.6
17	33.32	21.99	3.43	4.25	5.18	5.95	8.16	.24	19.24	.57	5.44	12.15
18	34.34	22.66	3.54	4.41	5.32	6.13	7.7	.22	18.18	.52	5.13	12.12
19	35.18	23.21	3.63	4.52	5.47	6.33	7.3	.2	17.	.48	4.9	11.33
20	36.2	23.89	3.71	4.62	5.49	6.47	6.93	.19	16.36	.45	4.62	10.9
1	2	3	4	5	6	7	8	9	10	11	12	13

Use of the Table for Tub-Mills.

Having levelled your mill-seat, and found that you have above 8 feet fall, and plenty of water, and wishing to build a mill on the simplest, cheapest, and best construction to suit your seat, you will, of course, make choice of a tub-mill.

Cast off 1 foot for fall in the tail-race, below the bottom of the wheel, if it be subject to back-water, and 9 inches for the wheel; then suppose you have 9 feet left for head above the wheel; look in the table, against 9 feet head, and you have all the calculations necessary for 4, 5, 6, and 7 feet stones, the quantity of water required to drive them, the sum of the areas of the apertures, and the areas of the canals.

If you choose stones of any other size, you can easily proportion the parts to suit, by the rules by which the table is calculated.

Let it be recollected, however, that it is a very common error, to build tub mills in situations where they must fail during a dry season. They are suited to those places only where water runs to waste during the whole year. There are hundreds of such mills in the United States which are useless at the season when they are most needed, whilst a well-constructed overshot, breast, or pitchback wheel, might be kept constantly running.

ARTICLE 72.

OF BREAST MILLS.

Breast wheels, which have the water shot on them in a tangential direction, are acted on by the principles both of percussion and of gravity; all that part above the point of impact, called head, acts by percussion, and all that part below said point, called fall, acts by gravity.

We are obliged, in this structure of breast mills, to use more head than will act to advantage; because we cannot strike the water on the wheel, in a true tangential direction, higher than I, the point of impact, as shown in Plate IV., fig. 31, which is a breast wheel, with 12 feet perpendicular descent, 6,5 feet of which are above the point I., as head, and 5,5 feet below, as fall. The upper end of the shute that carries the water down to the wheel, must project some inches above the point of the gate when full drawn, otherwise the water will strike towards the centre of the wheel; and it must not project too high, or else the water in the penstock will not come fast enough into the shute when the head sinks a little. The bottom of the penstock is a little below the top end of the shute, to leave room for stones and gravel to settle, and prevent them from getting into the gate.

We might lay the water on higher, by setting the top of the penstock close to the wheel, and using a sliding gate at bottom, as shown by the dotted lines; but this is not approved of in practice. See Ellicott's mode, Plate XV., fig. 1.

But if the water in the penstock be nearly as high as the wheel, it may be carried over; as shown by the upper dotted lines, and shot on backwards, making that part next the wheel, the shute to guide the water into the wheel, and the gate very narrow or shallow, allowing the water to run over the top of it when drawn; by this method (called pitch-back) the head may be reduced to the same as it is for an overshot wheel; and then the motion of the circumference of the wheel will be equal to the motion of an overshot wheel, whose diameter is equal to the fall below the point of impact, and their powers will be equal.

This structure of a wheel, Plate IV. fig. 31, I view as a good one for the following reasons; namely:—

1. The buckets, or floats, receive the percussion of the water at right angles, which is the best direction possible.

2. It prevents the water from flying towards the centre of the wheel without re-acting against the bottom of the buckets, and retains it in the wheel, to act by its gravity in its descent, after the stroke.

3. It admits air, and discharges the water freely, with-

out lifting it at bottom; and this is an important advantage, because, if the buckets of a wheel be tight, and the wheel wade a little in back-water, they will lift the water to a considerable distance as they empty; the pressure of the atmosphere then prevents the water from leaving the buckets freely, and it requires a great force to lift them out of the water with the velocity of the wheel; this may be proved by dipping a common water-bucket into water, and lifting it out, bottom up, with a quick motion; you have to lift not only the water in the bucket, but it appears to suck much more up after it; which is the effect of the pressure of the atmosphere. See Art. 56. This shows the necessity of air-holes to let air into the buckets, that the water may have liberty to escape freely.

Its disadvantages are,

1. It loses much water, if it be not kept closely to the sheeting. And,

2. It requires too great a part of the total fall to be used as head, which is a loss of power; one foot fall being equal in power to two feet head.

Plate IV. fig. 32, is a draught, showing the position of the shute for striking the water on a wheel in a tangent, for all the total perpendicular descents from 6 to 15 feet; the points of impact are numbered inside the fig. with the number of the total fall, for each respectively. The top of the shute is only about 15 inches from the wheel, in order to set the point of impact as high as possible, allowing 3 feet above the upper end of the shute to the top of the water in the penstock, which is little enough, when the head is to be often run down any considerable distance; but where the stream is steady, being always nearly the same height in the penstock, 2 feet would be sufficient, especially in the greatest total falls. Where the quantity is less, raising the shute 1 foot would also raise the point of impact nearly the same, and increase the power, because 1 foot fall is equal in power to 2 feet head, by Art. 61.

On these principles, to suit the applications of water as represented by fig. 32, I have calculated the following table for breast mills. And, in order that the reader may fully understand the principles on which it is calculated, let him consider as follows:—

1. That all the water above the point of impact, called head, acts wholly by percussion, and all below said point, called fall, acts wholly by gravity, (see Art. 60:) these form the 2d and 3d columns.

2. That half the head, added to the whole fall, constitutes the virtual or effective descent, by Art. 61; which is given in the 4th column.

3. That if the water were permitted to descend freely down the circular sheeting, after it passed the point of impact, its velocity would be accelerated, by Art. 60, so as to be, at the lowest point, equal to the velocity of water spouting from under a head equal to the whole descent; the maximum velocity of this wheel will, consequently, be compounded of the velocity to suit the head, and the acceleration after it passes the point of impact. Therefore, to find the velocity of this wheel, I first multiply the velocity of the head, in column 5, by ,577, (as for undershot mills,) which gives the velocity suitable to the head; I then, (by the rule for determining the velocity of overshots,) say, as the velocity of water descending 21 feet, equal to 37,11 feet per second, is to the velocity of the wheel ,10 feet per second, so is the acceleration of velocity, after it passes the point of impact, to the accelerated velocity of the wheel; and these two velocities added, give the velocity of the wheel; which is shown in the 6th column.

4. The velocity of the wheel per second, multiplied by 60, and divided by the circumference of the wheel, gives the revolutions per minute: see 7th column.

5. The number of cogs in the cog-wheel, multiplied by the number of revolutions, per minute, of the water-wheel, and divided by the rounds in the trundle-head, will give the number of revolutions of the stone per minute; and if we divide by the number of revolutions the stone is to have, it gives the rounds in the trundle, and, when fractions arise, take the nearest whole number; see columns 8, 9, and 10.

6. The cubochs of power required to turn the stone, by Art. 63, divided by the virtual descent, give the cubic feet of water required per second; column 11.

7. The cubic feet of water, divided by the velocity allowed to it in the canal, suppose 1,5 feet per second, give the area of a section of the canal; column 12.

8. If the mill is to be double geared, take the revolutions of the wheel from column 7 of this table, and look in column 4 of the undershot table, Art. 70, for the number of revolutions nearest to it, and against that number you have the gearing that will give a 5 feet stone the right motion.

THE MILL-WRIGHT'S TABLE

FOR

BREAST MILLS.

Calculated for a water-wheel fifteen feet, and stones five feet diameter: the water being shot on in the direction of a tangent, to the circumference of the wheel.

Total perpendicular descent or fall of the water from the top of the water in the penstock, to ditto in tail-race.	Head above the point of impact.	Fall below the point of impact.	Virtual or effective descent, being half the head added to the fall.	Velocity of the water per second at the point of impact.	Velocity of the circumference of the wheel per second.	Number of revolutions of a wheel fifteen feet diameter, per minute.	Cogs in the cog-wheel, for single gear.	Rounds in the trundle-head.	Revolutions of the stone per minute.	Cubic feet of water required per second.	Area of a section of the canal, allowing the velocity of the water in it to be 1.5 feet per second.
feet.	feet.	feet.	feet.	feet.	feet.	No.	No.	N.	No.	cu. ft.	sup. ft.
6	4.5	1.5	3.75	17.13	10.61	13.5	112	15	100.8	29.8	19.25
7	5.	2.	4.5	18.	11.3	14.4	112	16	100.8	24.83	16.55
8	5.5	2.5	5.25	18.99	12.07	15.3	104	16	99.4	21.29	14.19
9	5.9	3.1	6.05	19.48	12.53	16.	104	16	102.7	18.45	12.3
10	6.2	3.8	6.9	20.16	13.07	16.6	96	16	99.6	16.2	10.8
11	6.5	4.5	7.75	20.64	13.53	17.	96	16	102.	14.42	9.61
12	6.8	5.3	8.7	21.11	14.03	17.81	96	17	100.5	12.73	8.49
13	6.8	6.2	9.6	21.11	14.35	18.28	96	18	97.5	11.63	7.75
14	6.9	7.1	10.55	21.3	14.41	18.35	96	18	97.8	10.59	7.06
15	7.	8.	11.5	21.13	14.76	18.56	96	18	98.4	9.72	6.48
1	2	3	4	5	6	7	8	9	10	11	12

Use of the Table for Breast Mills.

Having a seat with above 6 feet fall, but not enough for an overshot mill, and the water being scarce, so that you wish to make the best use of it, should lead you to the choice of a breast mill.

Cast off about 1 foot for fall in the tail race below the bottom of the wheel, if much subject to back-water, and suppose you have then 9 feet total descent; look for 9 feet in the first column of the table, and against it you have it divided into 5,9 feet head above, and 3,1 feet fall below, the point of impact, which is the highest point that the water can be fairly struck on the wheel; leaving the head 3 feet deep above the shute; which is equal to 6,5 feet virtual or effective descent; the velocity of the water striking the wheel will be 18,99 feet, and the velocity of the wheel 12,07 feet per second; it will revolve 16 times in a minute; and, if single geared, 104 cogs and 16 rounds, gives the stone 99,4 revolutions in a minute, requiring 21,29 cubic feet of water per second; the area of a section of the canal must be 14,19 feet, or about 3 feet deep, and 5 feet wide. If the stones be of any other size, it is easy to proportion the gearing to give them any required number of revolutions.*

If you wish to proportion the size of the stones to the power of your seat, multiply the cubic feet of water your stream affords per second, by the virtual descent in column 4, and that product is the power in cubochs; then look in the table, in Art. 63, for the size of the stone that most nearly suits that power.

For instance, suppose your stream afford 14 cubic feet of water per second, then 14 multiplied by 6,05 feet virtual descent, produce 84,7 cubochs of power; which, in the table in Art. 63, comes nearest to 4,5 feet for the diameter of the stones: but by the rules laid down in Art. 63, the size may be found more exactly.

---

* The mill-wright will do well to examine with attention, the article in the appendix, written by the late *W. Parkin*, a practical and scientific workman, whose suggestions are of the utmost importance, as they may lead to the correction of errors, which the editor is assured, from his own observations, are almost universal, the too great velocity, and the too little capacity of water-wheels.

Six cubochs of power are required to every superficial foot of the stones.

ARTICLE 73.

OF OVERSHOT MILLS.

Fig. 33, Plate IV, represents an overshot wheel; the water is laid on at the top, so that the upper part of the column will be in the direction of a tangent, with the circumference of the wheel, but so that all the water may strike within the circle of the wheel.

The gate is drawn about 30 inches behind the perpendicular line from the centre of the wheel, and the point at the shute ends at said perpendicular with a direction a little downwards, which gives the water a little velocity downwards to follow the wheel; for if it be directed horizontally, the head will give it no velocity downwards, and if the head be great, the parabolic curve, which the spouting water forms, will extend beyond the outside of the circle of the wheel, and it will incline to fly over. See Art. 44 and 60.

The head above the wheel acts by percussion, as on an undershot wheel, and we have shown, Art. 43, that the head should be such as to give to the water a velocity of 3, for 2 of the wheel. After the water strikes the wheel it acts by gravity; therefore, to calculate the power, we must take half the head and add it to the fall, for the virtual descent, as in breast-mills.

The velocity of overshot wheels is as the square roots of their diameters. See Art. 43.

On these principles, I have calculated the following table for overshot wheels; and, in order that the reader may understand it fully, let him consider well the following premises:—

1. That the velocity of the water spouting on the wheel must be one and a half times the velocity of the wheel, by Art. 43: then, to find the head that will give said velocity, say, as the square of 16,2 feet per second, is to 4 feet, the head that gives that velocity, so is the

square of the velocity required, to the head that will give the velocity sought for: but to this head, so found, we must add a little, by conjecture, to overcome the friction of the aperture. See Art. 55.

In this table, I have added to the heads of wheels of from 9 to 12 feet diameter ,1 of a foot, and from 12 to 20 feet I have added one-tenth more, for every foot increase of diameter, and from 20 to 30 feet I have added ,05 more to every foot diameter's increase; which gives a 30 feet wheel 1,5 feet additional head, while a 9 feet wheel has only one-tenth of a foot, to overcome the friction. The reason of this great difference will appear when we consider that the friction increases as the aperture decreases, and as the velocity increases: still this depends much on the form of the gate, for if that be nearly square, there will be but little friction; but if very oblong, say 24 inches by half an inch, then it will be very great.

The heads thus found, compose the 3d column.

2. The head, added to the diameter of the wheel, makes the total descent, as in column 1.

3. The velocity of the wheel per second, taken from the table in Art. 43, multiplied by 60, and divided by the circumference of the wheel, gives the number of revolutions of the wheel per minute; as in column 4.

4. The number of revolutions of the wheel per minute, multiplied by the number of cogs in all the driving wheels successively, and that product divided by the product of all the leading wheels, gives the number of revolutions of the stone per minute, and is found in column 9, double gear, for 5 feet stones; and in column 12, single gear, for 6 feet stones.

5. The cubochs of power required to drive the stone, by table in Art. 63, divided by the virtual or effective descent, which is half the head added to the fall, or the diameter of the wheel, gives the cubic feet of water required per second to drive the stone, and is column 13.

6. The cubic feet required, divided by the velocity you intend the water to have in the canal, gives the area of a section of the canal. The width multiplied by the depth, must always produce this area. See Art. 64.

Chap. 4.]   OF OVERSHOT MILLS.   173

7. The number of cogs in the wheel, multiplied by the quarters of inches in the pitch, produces the circumference of the pitch circle; which multiplied by 7, and divided by 22, gives the diameter in quarters of inches, which reduced to feet and parts, forms column 15. The reader may here at once observe how near the cog-wheel, in the single gear, will be to the water; that is, how near it is, in size, to the water-wheel.

Use of the Table.

Having with care levelled the seat on which you mean to build, and found, that after deducting 1 foot for fall below the wheel, and a sufficiency for the sinking of the head race, according to its length and size, and having a total descent remaining sufficient for an overshot wheel, suppose 17 feet; then, on looking in the first column of the table, for the descent nearest to it, we find 16,74 feet, and against it a wheel 14 feet diameter; head above the wheel 2,7 feet; revolutions of the wheel per minute 11,17; double gears, to give a 5 feet stone 98,7 revolutions per minute; single gears, to give a 6 feet stone 76,6 revolutions per minute; the cubic feet of water required for a 5 feet stone 7,2 feet per second, and the area of a section of the canal 5 feet; about 2 feet deep, and 2,5 feet wide.

If it be determined to proportion the size of the stones exactly to suit the power of the seat, it may be done as directed in Art. 63. All the rest can be proportioned by the rules by which the table is calculated.

THE MILL-WRIGHT'S TABLE

FOR

OVERSHOT MILLS.

CALCULATED FOR FIVE FEET STONES, DOUBLE GEAR, AND SIX FEET STONES, SINGLE GEAR.

Total descent of the water, which is in this table made to suit the diameter of the wheel and head above it.	Diameter of the wheel.	Head above the wheel, allowing for the friction of the aperture, so as to give the water velocity 3 for 2 of the wheel.	Number of revolutions of the wheel per minute.	Double gear, 5 feet stones.					Single gear 6 ft. stones.			Cubic feet of water required per second, for five feet stones	Area of a section of the canal, allowing the velocity of the water in it to be 1 foot per second.	Diameter of the pitch circle of the great cog-wheels for single gear, pitch 4½ inches.	
				No. of cogs in master cog-wheel.	Rounds in the wallower.	Cogs in the counter cog-wheel.	Rounds in the trundle.	Revolutions of the stone per minute.	Cogs in the cog-wheel.	Rounds in the trundle.	Revolutions of the stone per minute.				
feet.	ft.	feet.										cu. ft.	sup. ft.	feet.	inches.
10.51	9	1.51	14.3	54	21	44	16	102.9	60	11	78.	11.46	11.46	6:9 0-4	12-22
11.74	10	1.74	13.	54	21	48	18	98.	60	10	78.	10.3	10.3		
12.94	11	1.94	12.6	60	21	48	19	96.	66	11	75.6	9.34	9.34	7:5 1-4	
14.2	12	2.2	12.	66	23	48	17	97.	66	10	79.2	8.53	8.53		
15.47	13	2.47	11.54	66	21	48	17	99.3	84	12	80.7	7.92	7.92	9:5 1-2	
16.74	14	2.74	11.17	72	23	48	17	98.7	96	14	76.6	7.2	7.2	10:9 3-4	6-22
17.99	15	2.99	10.78	78	23	48	18	98.3	96	13	81.9	6.77	6.77		
19.28	16	3.28	10.4	78	23	48	17	99.5	120	16	76.	6.4	6.4	13:6 1-4	2-22
20.5	17	3.5	10.1	78	21	48	18	96.6	120	15	80.8	6.	6.		
21.8	18	3.8	9.8	84	24	48	17	97.	128	16	78.4	5.56	5.56	14:5 0-4	8-22
23.03	19	4.03	9.54	84	23	48	17	98.3	128	15	81.4	5.32	5 32		
24.34	20	4.34	9.3	88	23	48	17	100.	128	15	79.3	5.04	5.04		
25.54	21	4.54	9.1	88	23	48	17	98.3	128	15	77.6	4.81	4.81		
26.86	22	4.86	8.9	96	24	48	17	100.5	128	14	81.4	4.57	4.57		
27.99	23	4.99	8.7	96	25	54	18	100.2				4.34	4.34		
29.27	24	5.27	8.5	96	25	54	17	103.				4.19	4.19		
30.45	25	5.45	8.3	96	25	54	17	101.				4.	4.		
31.57	26	5.57	8.19	96	25	54	17	99.6				3.82	3.82		
32.77	27	5.77	8.03	104	25	54	18	100.2				3.7	3.7		
33.96	28	5.96	7.93	104	25	54	18	99.				3.6	3.6		
35.15	29	6.15	7.75	112	26	54	18	100.1				3.4	3.4		
36.4	30	6.4	7.63	112	26	54	18	98.6				3.36	3.36		
1	2	3	4	5	6	7	8	9	10	11	12	13	14	15	

Observations on the Table.

1. It appears that single gearing does not well suit this construction; because, where the water wheels are low, their motion is so slow that the cog-wheels, (if made large enough to give sufficient motion to the stone, without having the trundle too small, see Art. 23,) will touch the water. And again, when the water-wheels are above 20 feet high, the cog-wheels require to be so high, in order to give motion to the stone without having the trundle too small, that they become unwieldy; the husk also is too high, and the spindle so short as to be inconvenient. Single gearing, therefore, seems to suit overshot wheels only, where their diameters is between 12 and 18 feet; and even then the water-wheel will have to run rather too fast, or the trundle be too small; and the stones should be, at least, 6 feet in diameter.

2. I have, in the preceding tables, supposed the water to pass along the canal with 1,5 feet velocity per second; but being of opinion, that 1 feet per second is nearer the proper motion, that is, about 20 yards per minute; the cubic feet required per second, will, in this case, be the area of a section of the canal, as given in column 14 of this table.

3. Although I have calculated this table for the velocities of the wheels to vary as the square roots of their diameters, which makes a 30 feet wheel move 11,99 feet per second, and a 12 feet wheel to move 7,57 feet per second, yet they will do to have equal velocity, and head, which is the common practice among mill-wrights. But, for the reasons I have mentioned in Art. 43, I prefer giving them the velocity and head assigned in the table, in order to obtain steady motion.

4. Many have been deceived, by observing the exceedingly slow and steady motion of some very high overshot wheels, working forge or furnace bellows, concluding therefrom, that they will work as steadily with a very slow, as with any quicker motion, not considering, perhaps, that it is the principle of the bellows that re-

gulates the motion of the wheel, which is different from any other resistance, for it soon becomes perfectly equable; therefore, the motion will be uniform, which is not the case with mills of any kind.

5. An opinion is sometimes entertained, that water is not well applied by an overshot wheel, because, it is said, those buckets which nearly approach a line drawn perpendicularly through the centre, either above or below, act on too short a lever. To correct this erroneous idea, I have divided the fall of the overshot wheel, fig. 33, Plate IV., into feet, shown by dotted lines. Now, by Art. 53 and 54, every cubic foot of water on the wheel produces an equal quantity of power in descending each perpendicular foot, called a cuboch of power; and that because where the lever is shortest, the greatest quantity of water is contained within the foot perpendicular; or, in other words, each cubic foot of water is a much longer time, and passes a greater distance, in descending a perpendicular foot, than where the lever is longest: this exactly compensates for the deficiency in the length of lever. See this demonstrated, Art. 54. It is true that the effect of the lower foot is, in practice, entirely lost, by the running of the water out of the buckets.

Of Mills moved by Reaction.

We have now treated of the four different kinds of mills that are in general use. There is another, the invention of, or rather an improvement made by, the late ingenious James Rumsey, which moves by the reaction of water.*

---

* This is sometimes known by the name of Barker's mill; several of which have been built in different places; but it is believed that they have all been abandoned, as they have not in practice answered the expectations which had been entertained respecting them. A modification of this mode of applying the power of water, has, of late years, been extensively used in the United States, and been made the subject of several patents. These wheels will be noticed in the appendix.—EDITOR.

CHAPTER V.

ARTICLE 74.

RULES AND CALCULATIONS.

The fundamental principle, on which are founded all rules for calculating the motion produced by a combination of wheels, and for calculating the number of cogs to be put in them, to produce any motion that is required, has been given in Art. 20; and is as follows:—

If the revolutions that the first moving wheel makes in a minute be multiplied by the number of cogs in all the driving wheels successively, and the product noted; and the revolutions of the last leading wheel be multiplied by the number of cogs in all the leading wheels successively, and the product noted; these products will be equal in all possible cases. Hence, we deduce the following simple rules:—

1st. For finding the motion of the mill-stone; the revolutions of the water-wheel, and the cogs in the wheels, being given:—

RULE.

Multiply the revolutions of the water-wheel per minute, by the number of cogs in all the driving wheels successively, and note the product; and multiply the number of cogs or rounds in all the leading wheels successively, and note the product; then divide the first product by the last, and the quotient is the number of revolutions of the stone per minute.

EXAMPLE.

Given, the revolutions of the water-wheel
per minute, - - - - - 10,4
No. of cogs in the master cog-wheel - 78 ⎫ Drivers.
No. of do. in the counter cog-wheel - 48 ⎭
No. of rounds in the wallower - - 23 ⎫ Leaders.
No. of do. in the trundle - - 17 ⎭

Then 10,4 the revolutions of the water-wheel, multiplied by 78, the cogs in the master-wheel, and 48, the cogs in the counter-wheel, are equal to 38937,6; and 23 rounds in the wallower, multiplied by 17 rounds in the trundle, are equal to 391, by which we divide 38937,6, and it gives 99,5, the revolutions of the stone per minute; which are the calculations for a 16 feet wheel, in the overshot table.

2d. For finding the number of cogs to be put in the wheels, to produce any number of revolutions required to the mill-stone, or to any wheel.

RULE.

Take any suitable number of cogs for all the wheels, except one; then multiply the revolutions of the first mover per minute, by all the drivers, except the one wanting (if it be a driver,) and the revolutions of the wheel required, by all the leaders, and divide the greatest product by the least, and it will give the number of cogs required in the omitted wheel, to produce the desired revolutions.

Note. If any of the wheels be for straps, take their diameters in inches and parts, and multiply and divide with them, as with the cogs.

EXAMPLE.

Given, the revolutions of the water-wheel   10,4
And the cogs in the master wheel   -   78 ⎱
Ditto in the counter wheel   -   -   48 ⎰ Drivers.
Rounds in the wallower   -   -   -   23

The number of the trundle is required, to give the stone 99 revolutions.

Then 10,4, multiplied by 78 and 48, is equal to 38937,6; and 99, multiplied by 23, is equal to 2277, by which divide 38937,6 and it gives 16,66; instead of which, I take the nearest whole number, 17, for the rounds in the trundle, and find, by rule 1st, that it produces 99,5 revolutions, as required.

For the exercise of the inexperienced, I have constructed fig. 7, Plate XI.; which I call the circle of motion, and which serves to prove the fundamental principle on which the rules are founded; the first shaft being, also, the last of the circle.

A	is a cog-wheel of	20	cogs, and is a	driver.
B	do.	24	-	leader.
C	do.	24	-	driver.
D	do.	30	-	leader.
E	do.	25	-	driver.
F	do.	30	-	leader.
G	do.	36	-	driver.
H	do.	20	-	leader.

But if we trace the circle the backward way, the leaders become drivers.

I	is a strap-wheel	$14\frac{1}{2}$	inches diameter,		driver.
K	do.	30	do.	-	leader.
L	cog-wheel	12	cogs	-	driver.
M	do.	29	do.	-	leader.

MOTION OF THE SHAFTS.

The upright shaft, and first driver, AH 36 revs. in a min.
BC 30 do.
DE 24 do.
FG 20 do.
HA 36 do.
M 4 do. which is the shaft of a hopper-boy.

If this circle be not so formed, as to give the first and last shafts (which are here the same) exactly the same motion, one of the shafts must break as soon as they are put in motion.

The learner may exercise the rules on this circle, until he can form a similar circle of his own; and then he need never be afraid to undertake to calculate any other combination of motion.

I omit showing the work for finding the motion of the several shafts in this circle, and the wheels to produce said motion; but leave it for the practice of the learner, in the application of the foregoing rules.

EXAMPLES.

1st. Given, the first mover AH 36 revolutions per minute, and first driver A 20 cogs, leader B 24; required, the revolutions of shaft BC. Answer, 30 revolutions per minute.

2dly. Given, first mover 36 revolutions per minute, drivers 20—24—25, and leaders 24—30—30; required, the revolutions of the last leader. Answer, 20 revolutions per minute.

3dly. Given, first mover 20 revolutions per minute, and first driver, strap-wheel, $14\frac{1}{2}$ inches, cog-wheel 12, and leader, strap-wheel, 30 inches, cog-wheel 29; required, the revolutions of the last leader, or last shaft. Answer, 4 revolutions.

4thly. Given, first mover 36 revolutions, driver A 20, C 24, leader B 24, D 30; required, the number of leader F, to produce 20 revolutions per minute. Answer, 30 cogs.

5thly. Given, first mover 36 revolutions per minute, driver A 20, C 24, E 25, driver pulley $14\frac{1}{2}$ inches diameter, L 12, and leader B 24, D 30, F 30, M 29; required the diameter of the strap-wheel K, to give the shaft 4, four revolutions per minute. Answer, 30 inches diameter.

The learner may, for exercise, work the above questions, and every other than he can propose on the circle.

ARTICLE 75.

The following are the proportions for finding the circumference of a circle, its diameter being given, or the diameter by the given circumference; namely:

As 1 is to 3,1416, so is the diameter to the circumference; and as 3,1416 is to 1, so is the circumference to the diameter: Or, as 7 is to 22, so is the diameter to the circumference; and as 22 is to 7, so is the circumference to the diameter. The last proportion makes the diameter a little too large; it, therefore, suits mill-wrights best for finding the pitch circle; because the sum of the distances, from centre to centre, of all the cogs in a wheel, makes the circle too short, especially where the number of cogs is few, because the distance is taken in straight lines, instead of on the circle. In a wheel of 6 cogs only, the circle will be so much too short, as to give the diameter $\frac{2}{22}$ parts of the pitch or distance of the cogs too short. Hence, we deduce the following

RULES FOR FINDING THE PITCH CIRCLE.

Multiply the number of cogs in the wheel, by the quarters of inches in the pitch, and that product by 7, and divide by 22, and the quotient is the diameter in quarters of inches, which is to be reduced to feet.

EXAMPLE.

Given, 84 cogs 4½ inches pitch; required the diameter of the pitch circle.

Then, by the rule, 84 multiplied by 18, and by 7, is equal to 10584; which, divided by 22, is equal to $481\frac{2}{22}$ quarter inches, equal to 10 feet $\frac{12}{22}$ inches, for the diameter of the pitch circle required.

ARTICLE 76.

A true and expeditious method of finding the diameter of the pitch circle, is to find it in measures of the pitch itself that you use.

RULE.

Multiply the number of cogs by 7, and divide by 22, and you have the diameter of the pitch circle, in measures of the pitch, and 22d parts of said pitch.

EXAMPLE.

Given, 78 cogs; required, the diameter of the pitch circle. Then, by the rule,

$$\begin{array}{r}78\\7\\\hline\end{array}$$

22)546(24$\frac{18}{22}$ ⎱ Measures of the pitch for the diameter
   44   ⎰   of the circle required.

$$\begin{array}{r}\hline 106\\88\\\hline 18\end{array}$$

Half of which diameter, 12$\frac{9}{22}$ of the pitch, is the radius, or half diameter, by which the circle is to be swept.

To use this rule, set a pair of compasses to the pitch, and screw them fast, so as not to be altered until the wheel is pitched; divide the pitch into 22 equal parts; then step 12 steps, on a straight line with the pitch compasses, and 9 of these equal parts of the pitch, make the radius that is to describe the circle.

To save the trouble of dividing the pitch for every wheel, the workman may mark the different pitch which he commonly uses, on the edge of his two-foot rule, (or make a little rule for the purpose,) and carefully divide them there, where they will always be ready for use. See plate IV. fig. 35.

By these rules, I have calculated the following table of the radii of pitch circles of the different wheels commonly used, from 6 to 136 cogs.

A TABLE

OF THE

PITCH CIRCLES OF THE COG-WHEELS

COMMONLY USED,

From 6 to 136 cogs, both in measures of the pitch, and in feet, inches, and parts.

Cogs in the wheel. No.	Radius of the pitch circle in measures of the pitch and 22 parts and tenths of parts of said pitch. Pitch. 22 parts.	Radius of the pitch circle of the wheels in column 1, taken in inches, quarters, and 22 parts of a quarter, when the pitch is 2¼ inches, for bolting gears, &c. inches. quarters. 22 parts.	Cogs in the wheel. No.	Radius of the pitch circle in measure of the pitch and 22 parts of said pitch. Pitch. 22 parts.	Radius of the pitch circle of the wheels in the 4th column taken in feet, inches, quarters, and 22 parts of a quarter, when the pitch is 4¼ inches, for large gears, &c. feet. inches. quarters. 22 parts.	Ditto, when the pitch is 4¾ inches. feet. inches. quarters. 22 parts.
6	1 —	2:2: 0	33	5 5 1-2	1:10:1: 51-2	1:11:2:11
7	1 3.5	2:3:12	34	5 9	1:10:3:21	2: 0:1: 8
8	1 6.7	3:1: 3	35	5 12 1-2	1:11:2:14 1-2	2: 1:0: 5
9	1 10.2	3:2:13	36	5 16	2: 0:1: 8	2: 1:3: 2
10	1 13.6	4:0: 3	37	5 19 1-2	2: 1:0: 11-2	2: 2:1:21
11	1 17.1	4:1:17	38	6 1	2: 1:2:17	2: 3:0:10
12	1 20.5	4:3: 5	39	6 4 1-2	2: 2:1:10 1-2	2: 3:3:15
13	2 1.9	5:0:17	40	6 8	2: 3:0: 4	2: 4:2:12
14	2 5.3	5:2: 8	42	6 15	2: 4:1:13	2: 6:0: 6
15	2 8.8	5:3:20	44	7 —	2: 5:3: 0	2: 7:2: 0
16	2 12.2	6:1:11	48	7 14	2: 8:1:18	2:10:1:10
17	2 15.7	6:3: 2	52	8 4	2:11:0:14	3: 1:0:20
18	2 19.1	7:0:15	54	8 11	3: 0:2: 1	3: 2:2:14
19	3 0.6	7:2: 6	56	8 20	3: 1:3:10	3: 4:0: 8
20	3 4.1	7:3:18	60	9 13	3: 4:2: 6	3: 6:3:18
21	3 7.5	8:1: 9	66	10 11	3: 8:2:11	3:11:1: 0
22	3 11.	8:3: 0	72	11 10	4: 0:2:16	4: 3:2: 4
23	3 14.5	9:0:13	78	12 9	4: 4:2:21	4: 7:3: 8
24	3 18.	9:2: 4	84	13 8	4: 8:3: 4	5: 0:0:12
25	3 21.5	9:3:17	88	14 0	4:11:2: 0	5: 3:0: 0
26	4 3.	10:1: 8	90	14 7	5: 0:3: 9	5: 4:1:16
27	4 6.5	10:2:21	96	15 6	5: 4:3:14	5: 8:2:20
28	4 10.	11:0:12	104	16 13	5:10:1: 6	6: 2:1:19
29	4 13.5	11:2: 3	112	17 18	6: 3:2:20	6: 8:0:16
30	4 17.	11:3:16	120	19 2	6: 9:0:12	7: 1:3:14
31	4 20.5	12:1: 7	128	20 8	7: 2:2: 4	7: 7:2:12
32	6 2.	12:2:20	136	21 14	7: 7:3:18	8: 1:1:10
1	2	3	4	5	6	7

Use of the foregoing Table.

Suppose you are making a cog-wheel with 66 cogs; look for the number in the 1st or 4th column, and against it, in the 2d or 5th column, you find 10, 11; that is, 10 steps of the pitch (you use) in a straight line, and 11 of 22 equal parts of said pitch added, make the radius that is to describe the pitch circle.

The 3d, 6th, and 7th columns, contain the radius in feet, inches, quarters, and 22 parts of a quarter; which may be made use of in roughing out timber, and fixing the centres that the wheels are to run in, so that they may gear to the right depth; but on account of the difference in the parts of the same scales or rules, and the difficulty of setting the compasses exactly, they can never be true enough for the pitch circles.

RULE COMMONLY PRACTISED.

Divide the pitch into 11 equal parts, and take in your compasses 7 of those parts, and step, on a straight line, counting 4 cogs for every step, until you come up to the number in your wheel; if there be an odd one at last, take ¼ of a step—if 2 be left, take ½ of a step—if 3 be left, take ¾ of a step, for them; and these steps, added, make the radius or sweep-staff of the pitch circle: but on account of the difficulty of making these divisions sufficiently exact, there is little truth in this rule—and where the number of cogs is few, it will make the diameter too short, for the reason formerly mentioned.

The following geometrical rule is more true, and, in some instances, more convenient.

RULE.

Draw the line AB, Plate IV. fig. 34, and draw the line 0,22 at random; then take the pitch in your compasses, and beginning at the point 22, step 11 steps towards A, and 3½ steps to point X, towards O, draw the line AC through the point X; draw the line DC parallel to AB, and, without having altered your compasses, begin at point O, and step both ways, as you did on AB; then, from the respective points, draw the cross lines parallel to O,22; and the distance from the point, where they cross the line AC, to the line AB, will be the radius of the pitch circles for the number of cogs respectively, as in the figure. If the number of cogs be odd, say 21, the radius will be between 20 and 22.

This will also give the diameter too short, if the wheels have but few cogs; but, where the number of cogs is above twenty, the error is imperceptible.

All these rules are founded on the proportion, that, as 22 is to 7, so is the circumference to the diameter.

ARTICLE 77.

CONTENTS OF GARNERS, HOPPERS, &C. IN BUSHELS.

A Table of English dry Measure.

Solid Inches.				
33.6	Pint.			
268.8	8	Gall.		
537.6	16	2	Peck.	
2150.4	64	8	4	Bushel.

The bushel contains 2150,4 solid inches. Therefore, to measure the contents of any garner, take the following

RULE.

Multiply its length in inches, by its breadth in inches, and that product by its height in inches, and divide the last product by 2150,4, and it will give the bushels it contains.

But to shorten the work decimally, because 2150,4 solid inches, make 1,244 solid feet, multiply the length, breadth, and height, in feet, and decimal parts of a foot, by each other, and divide by 1,244, and it will give the contents in bushels.

EXAMPLE.

Given, a garner 6,25 feet long, 3,5 feet wide, 10,5 feet high; required its contents in bushels. Then, 6,25 multiplied by 3,5 and 10,5, is equal to 229,687; which, divided by 1,244, gives 184 bushels and 6 tenths.

To find the contents of a hopper, take the following

RULE.

Multiply the length by the width at the top, and that product by one-third of the depth, measuring to the very point, and divide by the contents of a bushel, either in inches or decimals, and the quotient will be the contents in bushels.

EXAMPLE.

Given, a hopper, 42 inches square at the top, and 24 inches deep; required, the contents in bushels.

Then 42 multiplied by 42, and that product by 8, is equal to 14112 solid inches: which, divided by 2150,4, the solid inches in a bushel, gives 6,56 bushels, or a little more than 6½ bushels.

To make a garner to hold any given quantity, having two of its sides given, pursue the following

RULE.

Multiply the contents of 1 bushel by the number of bushels the garner is to hold; then multiply the given sides into each other, and divide the first by the last product, and the quotient will be the side wanted, in the same measure by which you have wrought in.

EXAMPLE.

Given, two sides of a garner 6,25 by 10,5 feet; required, the other side, to hold 184,6 bushels.

Then, 1,244 multiplied by 184,6 is equal to 229,642; which, divided by the product of the two sides 65,625, the quotient is 3,5 feet for the side wanted.

To make a hopper to hold any given quantity, having the depth given.

RULE.

Divide the inches contained in the bushels it is to hold, by 1-3d the depth in inches; and the quotient will be the square of one of the sides, at the top, in inches.

Given, the depth 24 inches; required, the sides to hold 6,56 bushels.

Then, 6,56 multiplied by 2150,4 equal to 14107,624; which, divided by 8, gives 1764, the square root of which is 42 inches; which is the length of the sides of the hopper wanted.

CHAPTER VI.

ARTICLE 78.

OF THE DIFFERENT KINDS OF GEARS, AND FORMS OF COGS.

In order to conceive a just idea of the most suitable form or shape of cogs in cog-wheels, we must consider that they describe, with respect to the pitch circles, a figure called an Epicycloid.

And when one wheel works in cogs set in a straight line, such as the carriage of a saw-mill, the cogs or rounds, moving out and in, form a curve called a Cycloid.

To describe this figure, let us suppose the large circle in Plate V, fig. 37, to move on the straight line from O to A; then the point O, in its periphery, will describe the arch ODA, which is called a Cycloid; and by the way in which the curve joins the line, we may conceive what should be the form of the point of the cog.

Again, suppose the small circle to run round the large one; then the point o, in the small circle, will describe the arch O b C, called an Epicycloid; by which we may conceive what should be the form of the point of the cogs. But, in common practice, we generally let the cogs extend but a short distance past the pitch circle; so that their precise form is not so important.

ARTICLE 79.

OF SPUR GEARS.

The principle of spur gears, is that of two cylinders rolling on each other, with their shafts or axes truly parallel. Here the touching parts move with equal velocity, and have, therefore, but little friction; but to prevent these cylinders from slipping, we are obliged to indent, or to set cogs in, them.

It appears to me that, in this kind of gear, the pitch of the driving wheel should be a little larger than that of the leading wheel, for the following reasons:—

1. If there is to be any slipping, it will be much easier for the driver to slip a little past the leader, than for the cogs to have to force the leader a little before the driver; which would be very hard on them.

2. If the cogs should bend any, by the stress of the work, as they assuredly do, this will cause those that are coming into gear to touch too soon, and rub hard at entering.

3. It is much better for cogs to rub hard as they are going out of gear, than as they are coming in; because then they work with the grain of the wood; whereas, at entering they work against it, and would wear much faster.

The advantage of this kind of gear is, that we can make the cogs as wide as we please, so that their bearing may be so large that they will not cut, but only polish each other, and wear smooth; therefore, they will last a long time.

Their disadvantages are,

1st. That if the wheels be of different sizes, and the pitch circles are not made to meet exactly, they will not run smoothly. And,

2dly. We cannot, conveniently, change the direction of the shafts.

Fig. 38, Plate V. shows two spur wheels working into each other; the dotted lines show the pitch circles, which must always meet exactly. The ends of the cogs are made circular, as is commonly done; but, if they were made true epicycloids, adapted to the size of the wheels, they would work with less friction, and, consequently, be much better.

Fig. 39, is a spur and face wheel, or wallower, whose pitch circles should always meet exactly.

The rule for describing the sides of the cogs, so as nearly to approach the figure of an epicycloid, is as follows; namely: Describe a circle a little inside of the pitch circle, for the point of your compasses to be set in, so as to describe the sides of the cogs, (as the four cogs at A, Plate V. fig. 38—39,) as near as you can to the curve of the epicycloid that is formed by the little wheel moving round the great one; the greater the difference between the great and small wheels, the greater distance must this circle be within the pitch circle: in doing this properly, much will depend upon the judgment of the workman.*

ARTICLE 80.

OF FACE GEARS.

The principle of face gears, is that of two cylinders rolling with the side of one on the end of the other, their axes being at right angles. Here, the greater the

---

* The following is Mr. Charles Taylor's rule for ascertaining the true cycloidical or epicycloidical form for the point of cogs:—

Make a segment of the pitch circle of each wheel, which gear into each other; fasten one to a plain surface, and roll the other round it as shown, Plate V. fig. 37, and, with a point in the moveable segment, describe the epicycloid o b c; set off at the end o one-fourth part of the pitch for the length of the cog outside of the pitch circle. Then fix the compasses at such an opening, that with one leg thereof, in a certain point, (to be found by repeated trials,) the other leg will trace the epicycloid from the pitch circle to the end of the cog: preserve the set of the compasses, and through the point where the fixed leg stood, sweep a circle from the centre of the wheel, in which set one point of the compasses to describe the point of all the cogs of that wheel whose segment was made fast to the plane.

If the wheels be bevel gear, this rule may be used to find the true form of both the outer and inner ends of the cogs, especially if the cogs be long, as the epicycloid is different in different circles. In making cast-iron wheels, it is absolutely necessary to attend to forming the cogs to the true epicycloidical figure, without which they will grind and wear rapidly.

The same rule serves for ascertaining the cycloidal form of a right line of cogs, such as those of a saw-mill carriage, &c., or of cogs set inside of a circle or hollow cone. Where a wheel works within a wheel, the cogs require a very different shape.

bearing, and the less the diameter of the wheels, the greater will be the friction; because the touching parts move with different velocities—therefore, the friction will be great.

The advantages of this kind of gear are,

1st. Their cogs stand parallel to each other; therefore, moving them a little out of or in gear, does not alter the pitch of the bearing parts of the cogs, and they will run smoother than spur gears, when their centres are out of place.

2dly. They serve for changing the direction of the shafts.

Their disadvantages are,

1st. The smallness of the bearing, so that they wear out very fast.*

2dly. Their great friction and rubbing of parts.

The cogs for small wheels are generally round, and put in with round shanks. Great care should be taken in boring the holes for the cogs, with a machine, to direct the auger straight, that the distance of the cogs may be equal, without dressing. And all the holes of all the small wheels in a mill should be bored with one auger, and made of one pitch; then the miller may keep by him a quantity of cogs ready turned to a gauge, to suit the auger; and, when any fail, he can put in new ones, without much loss of time.

Fig. 40, Plate V. represents a face cog-wheel working into a trundle; showing the necessity of having the corners of the sides of the cogs sniped, or worked, off in a cycloidal form, to give liberty for the rounds to enter between the cogs, and pass out again freely. To describe the sides of the cogs of the right shape to meet the rounds when they get fairly into gear, as at c, there must be a circle described on the ends of the cogs, a little outside of the pitch circle, for the point of the compasses to be set in, to scribe the ends of the cogs; for, if the point be set in the pitch circle, it will leave the inner

---

* If the bearing of the cogs be small, and the stress so great that they cut one another, they will wear exceedingly fast; but if it be so large, and the stress so light, that they only polish one another, they will last very long.

corners too full, and make the outer ones too scant. The middle of the cog is to be left straight, or nearly so, from bottom to top, and the side nearly flat, at the distance of half the diameter of the round, from the end, the corners only being worked off to make the ends of the shape in the figure; because, when the cog comes fully into gear, as at c, the chief stress is there, and there the bearing should be as large as possible. The smaller the cog-wheel, the larger the trundle, and the wider the cogs, the more will the corners require to be worked off. Suppose the cog-wheel to turn from 40 to b, the cog 40, as it enters, will bear on the lower corner, unless it be sufficiently worked off; when it comes to c, it will be fully in gear, and if the pitch of the cog-wheel be a little larger than that of the trundle, the cog a will bear as it goes out, and let c fairly enter before it begins to bear.

Suppose the plumb line A B to hang directly to the centre of the cog-wheel, the spindle is, by many millwrights, set a little before the line or centre, that the working round, or stave, of the trundle may be fair with said line, and meet the cog fairly as it comes to bear; by this means, also, the cogs enter with less, and go out with more friction. Whether there be any real advantage in thus setting the spindle foot before the centre plumb line, does not seem to be determined.

ARTICLE 81.

OF BEVEL GEARS.

The principle of bevel gears, is that of two cones rolling on the surface of each other, their vertexes meeting in a point, as at A, fig. 41, Plate V. Here the touching surfaces move with equal velocities in every part of the cones; therefore, there is but little friction. These cones, when indented, or fluted, with teeth diverging from the vertex to the base, to prevent them from slipping, become bevel gear; and as these teeth are very small at the point or vertex of the cone, they may be cut off 2 or 3 inches from the base, as 19 and 25, at B; they then have the appearance of wheels.

To make these wheels of a suitable size for any number of cogs you choose to have to work into one another, take the following

RULE.

Draw lines to represent your shafts, in their proper direction, with respect to each other, to intersect A; then take from any scale of equal parts, as feet, inches, or quarters, as many parts as your wheels are to have cogs, and at that distance from the respective shafts, draw the dotted lines a, b, c, d, for 21 and 20 cogs; and from where they cross at e, draw e A. On this line, which makes the right bevel, the pitch circles of the wheels will meet, to contain that proportion of cogs of any pitch.

Then, to determine the size of the wheels to suit any particular pitch, take from the table of pitch circles, the radius in measures of the pitch, and apply it to the centre of the shaft, and the bevel line A e, taking the distance at right angles with the shaft; and it will show the point in which the pitch circles will meet, to suit that particular pitch.

By the same rule, the sizes of the wheels at B and C are found.

Wheels of this kind, when made of cast iron, answer exceedingly well.

The advantages of this kind of gear are,

1. They have very little friction, or sliding of parts.
2. We can make the cogs of any width of bearing we choose; therefore, they will wear a great while.
3. By them we can set the shafts in any direction desired, to produce the necessary movements.

Their disadvantage is,

They require to be kept exactly of the right depth in gear, so that the pitch circles meet constantly, else they will not run smooth, as is the case with spur gears.

The universal joint, as represented fig. 43, may be

applied to communicate motion, instead of bevel gear, where the motion is to be the same, and the angle not more than 30 or 40 degrees. This joint may be constructed by a cross, as in the figure, or by four pins, fastened at right angles on the circumference of a hoop or solid ball. It may sometimes serve to communicate the motion, instead of two or three face wheels. The pivots, at the end of the cross, play in the ends of the semicircles. It is best to screw the semicircles to the blades, that they may be taken apart.

ARTICLE 82.

OF MATCHING WHEELS TO MAKE THE COGS WEAR EVEN.

Great care should be taken in matching or coupling the wheels of a mill, that their number of cogs be not such that the same cogs will often meet; because, if two soft ones meet often, they will both wear away faster than the rest, and destroy the regularity of the pitch; whereas, if they are continually changing, they will wear regular, even if they be, at first, a little irregular.

For finding how often wheels will revolve before the same cogs meet again, take the following

RULE.

1. Divide the cogs in the greater wheel by the cogs in the lesser; and if there be no remainder, the same cogs will meet once every revolution of the great wheel.

2. If there be a remainder, divide the cogs in the lesser wheel by the said remainder; and if it divide them equally, the quotient shows how often the great wheel will revolve before the same cogs meet.

3. But if it will not divide equally, then the great wheel will revolve as often as there are cogs in the small wheel, and the small wheel as often as there are cogs in the large wheel, before the same cogs meet:

they never can be made to change more frequently than this.

EXAMPLE.

Given, wheels of 13 and 17 cogs; required, how often each will revolve before the same cogs meet again.

Then 13)17(1
     13
     ―
   4)13(3
    12   Answer,
    ―   Great wheel 13, and
    1   Small wheel 17 revolutions.

ARTICLE 83.

THEORY OF ROLLING SCREENS AND FANS, FOR SCREENING AND FANNING THE WHEAT IN MILLS.

Let fig. 42, Plate V. represent a rolling screen and fan, fixed for cleaning wheat in a merchant-mill. DA the screen, AF the fan, AB the wind tube, 3 feet deep from a to b, and 4 inches wide, in order that the grain may have a good distance to fall through the wind, to give time and opportunity for the light parts to be carried forward, away from the heavy parts. Suppose the tube to be of equal depth and width for the whole of its length, except where it communicates with the tight boxes or garners under it; namely: C for the clean wheat, S for the screenings and light wheat, and c for the cheat, chaff, &c. Now, it is evident, that if wind be driven into the tube at A, and if it can no where escape, it will pass on to B, with the same force as at A, let the tube be of any length or direction; and any thing which it will move at A, it will carry out at B, if the tube be of an equal size all the way.

It is also evident, that if we shut the holes of the fan at A and F, and let no wind into it, none can be forced into the tube; hence, the best way to regulate the blast is, to fix shutters sliding at the air holes, to give more or less feed, or air, to the fan, so as to produce a blast sufficient to clean the grain.

The grain enters, in a small stream, into the screen at D, where it passes into the inner cylinder. The screen consists of two cylinders of sieve wire; the inmost one has the meshes so open as to pass all the wheat through it to the outer one, retaining only the white caps, large garlic, and every thing larger than the grain of the wheat, which falls out at the tail A.

The outer cylinder is so close in the meshes, as to retain all good wheat, but to sift out the cheat, cockle, small wheat, garlic, and every thing less than good grains of wheat; the wheat is delivered out at the tail of the outer cylinder, which is not quite as long as the inner one, whence it drops into the wind tube at a; and as it falls from a to b, the wind carries off every thing lighter than good wheat; namely: cheat, chaff, light garlic, dust, and light, rotten grains of wheat; but, in order to effect this completely, it should fall, at least, 3 feet through the current of wind.

The clean wheat falls into the funnel b, and thence into the garner C, over the stones. The light wheat, screenings, &c. fall into garner S, and the chaff settles into the chaff room c. The current slackens in passing over this room, and drops the chaff, but resumes its full force as soon as it is over, and carries out the dust through the wall at B. To prevent the current from slackening too much, as it passes over S and c, and under the screen, make the passages, where the grain comes in and goes out, as small as possible, not more than half an inch wide, and as long as necessary. If the wind escapes any where but at B, it defeats the object, and carries the dust into the mill. Valves may be fixed to shut the passages by a weight or spring, so that the weight of the wheat, falling on them, will open them just enough to let it pass, without suffering any wind to escape.

The fan is to be so set as to blow both the wheat and screenings, and carry out the dust. It is to be recollected, also, that the wind cannot escape into the garners or screen-room, if they are tight; for as soon as they are full no more can enter.

By careful attention to the foregoing principle, we may fix fans to answer our purposes.

The principal things to be observed in fixing screens and fans, are,

1. Give the screen 1 inch to the foot fall, and between 15 and 18 revolutions in a minute.

2. Make the fan blow strong enough, let the wings be 3 feet wide, 20 inches long, and revolve 140 times in a minute.

3. Regulate the blast by giving more or less feed of wind.

4. Leave no place for the wind to escape but at the end through the wall.

5. Wherever you want it to blow hardest, there make the tube narrowest.

6. Where you want the chaff and cheat to fall, there widen the tube sufficiently.

7. Make the fans blow both the wheat and screenings, and carry the dust clear out of the mill.

8. The wind tube may be of any length, and either crooked or strait, as may best suit; but no where smaller than where the wheat falls.

CHAPTER VII.

ARTICLE 84.

OF GUDGEONS, THE CAUSE OF THEIR HEATING AND GETTING LOOSE, AND REMEDIES THEREFOR.

The cause of gudgeons heating, is the excessive friction of their rubbing parts, which generates the heat in proportion to the weight that presses the rubbing surfaces together, and the velocity with which they move.

The cause of their getting loose is, their heating, and burning the wood, or drying it, so that it shrinks in the bands, and gives the gudgeons room to work.

To avoid these effects,

1. Increase the surface of contact, or rubbing parts, and, if possible, decrease their velocity; so much heat will not then be generated.

2. Conduct the heat away from the gudgeon as fast as it is generated.

To increase the surface of contact, without increasing the velocity, lengthen the neck or bearing part of the gudgeon. If the length be doubled, the weight will be sustained by a double surface, and the velocity remain the same; there will not then be so much heat generated: and, even supposing the same quantity of heat generated, there will be a double surface exposed to air, to convey it away, and a double quantity of matter, in which it will be diffused.

To convey the heat away as fast as generated, cause a small quantity of water to drop slowly on the gudgeon. A small is better than a large quantity; it should be just sufficient to keep up the evaporation, and not destroy the polish made by the grease, which it will do if the quantity be too great; and this will let the box and gudgeon come into contact, which will cause both to wear rapidly away.

The large gudgeons, for heavy wheels, are usually made of cast iron. Fig. 6, Plate XI., is a perspective view of one of the best form; a a a a, are four wings, at right angles with each other, extending from side to side of the shaft. These wings are larger, every way, at the end that is farthest in the shaft, than at the outer end, for convenience in casting them, and, also, that the bands may drive on tight, one over each end of the wings. Fig. 4 is an end view of the shaft, with the gudgeon in it and a band on the end; these bands, being put on hot, become very tight as they cool, and, if the shaft be dry, will not get loose, but will do so, if green: but, by driving a few wedges along side of each wing, it can be easily fastened, by any ordinary hand, without danger of moving it from the centre.

One great use of these wings is, to convey away the heat from the gudgeon to the bands, which are in contact with the air; by thus distributing it through so much metal, with so large a surface exposed to the air, the heat is carried off as fast as generated, and, therefore can never accumulate to a degree sufficient to burn loose, as it is apt to do in common gudgeons of wrought iron.

These gudgeons should be made of the best hard metal, well refined, in order that they may wear well, and not be subject to break; but of this there is but little danger, if the metal be good. I propose, sometimes, to have wings cast separate from the neck, as represented in fig. 4, Plate XI.; where the inside light square shows a mortise for the steeled gudgeon, fig. 8, to be fitted into, with an iron key behind the wings, to draw the gudgeon tight, if ever it should work loose; when thus made, it may be taken out, at any time, to repair.

This plan would do well for step gudgeons for heavy upright shafts, such as those of tub-mills.

When the neck is cast with the wings, the square part in the shaft need not be larger than the light square, representing the mortise.*

---

* Grease of any kind, used with a drill, in boring cast iron, prevents it from cutting, but will, on the contrary, make it cut wrought iron, or steel, much faster. This quality in cast iron renders it most suitable for gudgeons, and may be the principal cause why cast iron gudgeons have proved much better than any one expected. Several of the most experienced and skilful mill-wrights and millers assert that they have found cast gudgeons to run on cast boxes better than on stone or brass. In one instance they have carried heavy overshot wheels, which turned seven feet mill-stones, they have run for ten years, doing much work; and have hardly worn off the sand marks.

CHAPTER VIII.

ARTICLE 85.

ON BUILDING MILL-DAMS, LAYING FOUNDATIONS, AND BUILDING MILL-WALLS.

THERE are several points to be attained, and dangers to be guarded against, in building mill-dams.

1. Construct them so, that the water, in tumbling over them, cannot undermine their foundations at the lower side.*

2. And so that heavy logs, or large pieces of ice, floating down, cannot catch against any part of them, but will slide easily over.†

---

* If you have not a foundation of solid rocks, or of stones, so heavy that the water will never move them, there should be such a foundation made with great stones, not lighter than mill stones, (if the stream be heavy, and the fall great,) well laid, as low and as close as possible, with their up-stream end lowest, to prevent any thing from catching under them. But if the bottom be sand or clay, make a foundation of the trunks of long trees, laid close together on the bottom of the creek, with their but ends down stream, as low and as close as possible, across the whole tumbling space. On these the dam may be built, either of stone or wood, leaving 12 or 15 feet below the breast or fall, for the water to fall upon. See fig. 3, Plate X., which is a front view of a log dam, showing the position of the logs; also, of the stones in the abutments.

† If the dam be built of timber and small stones, &c., make the breast perpendicular, with straight logs, laid close one upon another, putting the largest, longest, and best logs on the top; make another wall of logs 12 or 15 feet up-stream, laying them close together, to prevent lamprey eels from working through them; they are not to be so high as the other, by 3 feet; tie these walls together, at every 6 feet, with cross logs, with the buts down stream, dove-tailed and bolted strongly to the logs of the lower wall, especially the upper log, which should be strongly bolted down to them. The spaces between these log walls, are to be filled up with stones, gravel, &c. Choose a dry season for this work; then the water will run through the lower part while you build the upper part tight.

To prevent any thing from catching against the top log, flag the top of the dam with broad or long stones, laying the down-stream end on the up-stream side of the log, to extend a little above it, the other end lowest, so that the next tier of stones will lap a little over the first; still getting lower, as you advance up-stream. This will glance logs, &c. over the dam, without their catching against any thing. If suitable stones cannot be had, I would recommend strong plank or small logs, laid close together, with both ends pinned to the top logs of the wall, the up-stream end being 3 feet lower than the other: But if plank is to be used, there need only be a strong frame raised on the foundation logs, to support the plank or the timber it is pinned to. See a side view of this frame, fig. 45, Plate IV. Some plank the breast to the front posts, and fill the hollow space

3. Build them so that the pressure or force of the current of the water will press their parts more firmly together.*

4. Give them a sufficient tumbling space to vent all the water in time of freshets.†

5. Make the abutments so high, that the water will not overflow them in time of freshets.

6. Let the dam and mill be a sufficient distance apart; so that the dam will not raise the water on the mill, in time of high floods.‡

with stone and gravel; but this may be omitted, if the foundation logs are sufficiently long up-stream, under the dam, to prevent the whole from floating away. First, stone, and then gravel, sand, and clay, are to be filled in above this frame, so as to stop the water. If the abutments be well secured, the dam will stand well.

A plank laid in a current of water, with the up-stream end lowest, set at an angle of $22\frac{1}{2}$ degrees, with the horizon or current of the water, will be held firmly to its place by the force of the current, and, in this position, it requires the greatest force to remove it; and the stronger the current, the firmer it is held to its place;—this points out the best position for the breast of dams.

* If the dam be built of stone, make it in the form of an arch or semicircle, standing up-stream, and endeavour to fix strong abutments on each side, to support the arch; then, in laying the stones, put the widest end up-stream, and the more they are forced down-stream, the tighter they will press together. All the stones of a dam should be laid with their up-stream ends lowest, and the other end lapped over the preceding, like the shingles or tiles of a house, to glance every thing smoothly over, as at the side 3, of fig. 3, Plate X. The breast may be built up with stone, either on a good rock or log foundation, putting the best in front, leaning a little up-stream, and on the top lay one good log, and another 15 feet up-stream on the bottom, to tie the top log to, by several logs, with good buts, down-stream, dove-tailed and bolted strongly, both at bottom and top of the top and up-stream logs; fill in between them with stone and gravel, laying large stones slanting next the top log, to glance any thing over it. This will be much better than to build all of stone; because if one at top gives way, the breach will increase rapidly, and the whole go down to the bottom.

† If the tumbling space be not long enough, the water will be apt to overflow the abutments; and if they be of earth or loose stones, they will be broken down, and, perhaps, a very great breach made. If the dam be of logs, the abutments will be best made of stone, laid as at the side 3, in fig. 3; but if stone is not to be had, they must be made of wood, although it will be subject to rot soon, being above water.

‡ I have, in many instances, seen a mill set so close to the dam, that the pierhead, or forebay, was in the breast, so that, in case of a leak or breach about the forebay, or mill, there is no chance of shutting off the water, or conveying it another way; but all must be left to its fate. Such mills are frequently broken down, and carried away; even the mill stones are sometimes carried a considerable distance down the stream, buried under the sand, and never found.

The great danger from this error will appear more plainly, if we suppose six mills on one stream, one above the other, each at the breast of the dam, and a great flood to break the first or uppermost dam, say through the pierhead, carrying with it the mill, stones, and all; this so increases the flood, that it overflows the next dam, which throws the water against the mill, and it is taken away; the

ARTICLE 86.

ON BUILDING MILL-WALLS.

The principal things to be considered in building mill-walls, are,

1. To lay the foundations with large, good stones, so deep as to be out of danger of being undermined, in case of such an accident as the water breaking through at the mill.*

2. Set the centre of gravity, or weight of the wall, on the centre of its foundation.†

water of these two dams has now so augmented the flood, that it carries every mill before it until it comes to the dam of the sixth, which it sweeps away also; but suppose this dam to be a quarter of a mile above the mill, which is set well into the bank, the extra water that is thrown into the canal, runs over at the waste left in its banks for the purpose; and the water having a free passage by the mill, does not injure it, whereas, had it been at the breast of the dam, it must have gone away with the rest. A case, similar to this, actually happened in Virginia, in 1794; all the mills and d ms on Falling Creek, in Chesterfield county, were carried away at once, except the lowest, (Mr. Wardrope's,) whose dam, having broke the year before, was rebuilt a quarter of a mile higher up; by which means his mill was saved.

* If the foundation be not good, but abounding with quick-sands, the wall cannot be expected to stand, unless it be made good by driving piles until they meet the solid ground; on the top of these may be laid large, flat pieces of timber, for the walls to be built on; they will not rot under water, when constantly excluded from the air.

† It is a common practice to build walls plumb outside, and to batter them from the inside; which throws their centre of gravity to one side of their base. If, therefore, it settles any, it will incline to fall outwards. Mill-walls should be battered so much outside, as to be equal to the offset inside, to cause the whole weight to stand on the centre of the foundation, unless it stands against a bank, as the wall next the wagon, in Plate VIII. The bank is very apt to press the wall inwards, unless it stands battering. In this case, build the side against the bank plumb, even with the ground, and then begin to batter it inwards. The plumb rules should be made a little widest at the upper end, so as to give the wall the right inclination, according to its height; to do which, take a line, the length equal to the height of the wall, set one end, by a compass point, in the lower end of the plumb rule, and strike the plumb line; then move the other end just as much as the wall is to be battered in the whole height, and it will show the inclination of the side of the rule, that will batter the wall exactly right. The error of building walls plumb outside, is frequently committed in building the abutments of bridges; the consequence is, they fall down in a short time; because the earth between the walls is expanded a little by every hard frost, which forces the walls over.

3. Use good mortar, and it will, in time, become as hard as stone.*

4. Arch over all the windows, doors, &c.

5. Tie them well together by the timbers of the floors.

* Good mortar, made of pure, well-burnt limestone, properly made up with sharp, clean sand, free from any sort of earth, loam, or mud, will, in time, actually petrify, and turn to the consistence of a stone. It is better to put too much sand into your mortar than too little. Workmen choose their mortar rich, because it works pleasantly; but rich mortar will not stand the weather so well, nor grow so hard as poor mortar. If it were all lime, it would have little more strength than clay.

PART THE THIRD.

CHAPTER IX.

DESCRIPTION OF THE AUTHOR'S IMPROVEMENTS IN THE MACHINERY FOR MANUFACTURING GRAIN INTO MEAL AND FLOUR.

ARTICLE 87.

INTRODUCTORY REMARKS.

These improvements consist of the invention and application of the following machines; namely:—

1. The Elevator.
2. The Conveyer.
3. The Hopper-boy.
4. The Drill.
5. The Descender.

These five machines are variously applied, in different mills, according to their construction, so as to perform every necessary movement of the grain, and meal, from one part of the mill to another, or from one machine to another, through all the various operations from the time the grain is emptied from the wagoner's bag, or from the measure on board the ship, until it be completely manufactured into flour, either superfine or of other qualities, and separated, ready for packing into barrels, for sale or exportation. All which is performed by the force of the water, without the aid of manual labour, excepting to set the different machines in motion, &c. This lessens the labour and expense of attendance of flour mills, fully one-half. The whole, as applied, is represented in Plate VIII.

ARTICLE 88.

1. OF THE ELEVATOR.

THE elevator is an endless strap, revolving over two pulleys, one of which is situated at the place whence the grain or meal is to be hoisted, and the other where it is to be delivered; to this strap is fastened a number of small buckets, which fill themselves as they pass under the lower pulley, and empty themselves as they pass over the upper one. To prevent any waste of what may spill out of these buckets, the strap, buckets, and pulleys, are all enclosed, and work in tight cases, so that what spills will descend to the place from whence it was hoisted. AB, in fig. 1, Plate VI., is an elevator for raising grain, which is let in at A, and discharged at B, into the spouts leading to the different garners. Fig. 2 is a perspective view of the strap, with different kinds of buckets, and the various modes of fastening them to the strap.

2. OF THE CONVEYER.

The conveyer K I, Plate VI., fig. 1, is an endless screw of two continued spirals, put in motion in a trough; the grain is let in at one end, and the screw drives it to the other, or collects it to the centre, as at y, to run into the elevator, (see Plate VIII., 37—36—4, and 44—45) or it is let in at the middle, and conveyed each way, as 15—16, Plate VIII.

Fig. 3, Plate VI., is a top view of the lower pulley of a meal elevator in its case, and a meal conveyer in its trough, for conveying meal from the stones, into the elevator, as fast as ground. This conveyer is an eight-sided shaft, set on all sides with small inclining boards, called flights, for conveying the meal from one end of the trough to the other; these flights are set in spirally, as shown by the dotted line; but the flights being set across the spiral line, the principle of the machine is changed from a screw to that of a number of ploughs; which is found to answer better for conveying warm meal.

Besides these conveying flights, there are others sometimes necessary, which are called lifters; and set with their broadsides foremost, to raise the meal from one side of the shaft, and let it fall on the other side to cool; these are only used where the meal is hot, and the conveyer short; there may be half as many of these as of the conveying flights. See 21—22, in Plate VIII.; which is a conveyer, carrying the meal from three pair of stones to the elevator, 23—24.

3. OF THE HOPPER-BOY.

Fig. 12, Plate VII., is a hopper-boy; which consists of a perpendicular shaft, A B having a slow motion, (not above 4 revolutions in a minute,) carrying round with it the horizontal piece C D, which is called the arm; this, on the under side, is set full of small inclining boards, called flights, so as to gather the meal towards the centre, or to spread it from the centre to that part of the arm which passes over the bolting hopper; at this part, one board is set broadside foremost, as E, (called the sweeper,) which drives the meal before it, and drops it into the hoppers H H, as the arms pass over them. The meal is generally let fall from the elevator, at the extremity of the arm, at D, where there is a sweeper, which drives the meal before it, trailing it in a circle the whole way round, so as to discharge nearly the whole of its load, by the time it returns to be loaded again: the flights then gather it towards the centre, from every part of the circle; which would not be the case, if the sweepers did not lay it round; but the meal would, in this case, be gathered from one side only of the circle. These sweepers are screwed on the back of the arm, so that they may be raised or lowered, in order to make them discharge sooner or later, as may be found necessary.

The extreme flight at each end of the arms is put on with a screw, passing through its centre, so that they may be turned to drive the meal outwards; the use of which is, to spread the warm meal as it falls from the elevator, in a ring, round the hopper-boy, while it, at the same time, gathers the cooled meal into the bolting hopper; so that the cold meal may be bolted, and the warm meal spread to cool, by the same machine, at the same time, if the miller chooses so to do. The foremost edge of the arm is sloped up in order to make it rise over the meal, and its weight is nearly balanced by the weight w, hung to one end of a cord, passing over the pulley P, and to the stay iron F. About $4\frac{1}{2}$ feet of the lower end of the upright shaft is made round, passing loosely through a round hole in the flight arm, giving it liberty to rise and fall freely, to suit any quantity of meal under it. The flight arm is led round by the leading arm L M, there being a cord passed through the holes L M, at each end, and made fast to the flight arm D C. This cord is lengthened or shortened by a hitch stick N, with two holes for the cord to pass through, its end being passed through a hole at D, and fastened to the end of a stick; this cord must reeve freely through the holes at the end of the arms, in order that the ends may both be led equally. The flight arm falls behind the leader about 1-6th part of the circle. The stay-iron C F E, is formed into a ring at F, which fits the shaft loosely, keeps the arm steady, and serves for hanging the hands of an equal height, by means of the screws C E.

Fig. 13, Plate VII., is a perspective view of the under side of the flight arms. The arm a c, with flights and sweepers complete; s s s show the screws which fasten the sweepers to the arms. The arm c b, is to show the rule for laying out for the flights. When the sweeper at b is turned in the position of the dotted line, it drives the meal outwards. Fig. 14, Plate VII., represents a plate of metal on the bottom of the shaft, to keep the arm from the floor, and 15 is the step gudgeon.

4. OF THE DRILL.

The drill is an endless strap revolving over two pulleys, like an elevator, but set nearly horizontal, and instead of buckets, there are small rakes fixed to the strap, which draw the grain or meal along the bottom of the case. See G H, Plate VI., fig. 1. The grain is let in at H, and discharged at G. This can sometimes be applied at less expense than a conveyer; it should be set a little descending; it will move grain or meal with ease, and will answer well, even when a little ascending.

5. OF THE DESCENDER.

The descender is a broad, endless strap of very thin, pliant leather, canvass, flannel, &c. revolving over two pulleys, which turn on small pivots, in a case or trough, to prevent waste, one end of which is to be lower than the other. See E F, Plate VI., fig. 1. The grain or meal falls from the elevator on the upper strap, at E; and by its own gravity and fall, sets the machine in motion, which discharges the load over the lower pulley F. There are two small buckets to bring up what may spill or fall off the strap, and lodge in the bottom of the case.

This machine moves on the principles of an over-shot water-wheel, and will convey meal to a considerable distance, with a small descent. Where a motion can be readily obtained from the water, it is to be preferred, as, when working by itself, it is easily stopped, and is apt to be troublesome.

The crane spout is hung on a shaft to turn on pivots or a pin, so that it may turn every way, like a crane; into this spout the grain falls from the elevator, and it can be directed by it into any garner. The spout is made to fit close, and play under a broad board, and the grain is let into it through the middle of this board, near the pin; it will then always enter the spout. It is seen under B, Plate VI., fig. 1. L is a view of the under side of it, and M is a top view of it. The pin or shaft may reach down so low, that a man may stand on the floor and turn it by the handle x.

CHAPTER X.

ARTICLE 89.

APPLICATION OF THE FOREGOING MACHINES IN THE PROCESS OF MANUFACTURING WHEAT INTO SUPERFINE FLOUR.

PLATE VIII. is not meant to show the plan of a mill, but merely the application and use of the foregoing machines.

The grain is emptied from the wagon into the spout I, which is set in the wall, and conveys it into the scale 2, that is made to hold 10, 20, 30, or 60 bushels, at pleasure.

There should, for the convenience of counting, be weights of 60 lbs. each divided into 30, 15, and $7\frac{1}{2}$ lbs.; then each large weight would show a bushel of wheat, and the smaller ones, halves, pecks, &c., which any one could count with ease.

When the wheat is weighed, draw the gate at the bottom of the scale, and let it run into the garner 3; at the bottom of which there is a gate to let it into the elevator 4—5, which raises it to 5; the crane spout is to be turned over the great store garner 6, which communicates from floor to floor, to garner 7, over the stones 8, which may be intended for shelling or rubbing the wheat, before it is ground, to take off all dust that sticks to the grain, or to break smut, fly-eaten grain, lumps of dust, &c. As it is rubbed, it runs into 3 again; in its passage it goes through a current of wind, blowing into the tight room 9, having only the spout a, through the lower floor, for the wind to escape; all the chaff will settle in the room, but most of the dust will pass out with the wind at a. The wheat again runs into the elevator at 4, and the crane spout, at 5, is turned over the screen hoppers 10 or 11, and the grain lodged there, out of which it runs into the rolling screen 12, and descends through the current of wind made by the fan 13; the clean heavy grain descends, by 14, into the conveyer 15—16, which conveys it into all the garners over the stones 7—17—18, and these regularly supply the stones 8—19—20, keeping always an equal quantity in the hoppers, which will cause them to feed regularly; as it is ground, the meal falls to the conveyer 21—22, which collects it to the meal elevator at 23, and it is raised to 24, whence it gently runs down the spout to the hopper-boy at 25, which spreads and cools it sufficiently, and gathers it into the bolting hoppers, both of which it attends regularly; as it passes through the superfine cloths 26, the superfine flour falls into the packing chest 28, which is on the second floor. If the flour is to be loaded on wagons, it should be packed on this floor, that it may conveniently be rolled into them; but if the flour is to be put on board a vessel, it will be more convenient to pack on the lower floor, out of chest 29, and thence roll it into the vessel at 30. The shorts and bran should be kept on the second floor, that they may be conveyed by spouts into the vessel's hold, to save labour.

The rubbings which fall from the tail of the 1st reel 26, are guided into the head of the 2d reel 27; which is in the same chest, near the floor, to save both room and machinery. On the head of this reel is 6 or 7 feet of fine cloth, for tail flour; and next to it the middling stuff, &c.

The tail flour which falls from the tail of the 1st reel 26, and head of the 2d reel 27, and requires to be bolted over again, is guided by a spout, as shown by dotted line 21—22, into the conveyer 22—23, to be hoisted again with the ground meal; a little bran may be let in with it, to keep the cloth open in warm weather;—but if there be not a fall sufficient for the tail flour to run into the lower conveyer, there may be one set to convey it into the elevator, as 31—32. There is a little regulating board, turning on the joint x, under the tail of the first reels, to guide more or less with the tail flour.

The middlings, as they fall, are conveyed into the eye of either pair of mill-stones by the conveyer 31—32, and ground over with the wheat; this is the best way of grinding them, because the grain keeps them from being killed; there is no time lost in doing it, and they are regularly mixed with the flour. There is a sliding board set slanting, to guide the middlings over the conveyer, that the miller may take only such part, for grinding over, as he shall judge fit; a little regulating board stands between the tail flour and middlings, to guide more or less into the stones, or elevator.

The light grains of wheat, screenings, &c., after being blown by the fan 13, fall into the screenings garner 32; the chaff is driven farther on, and settles in the chaff-room 33; the greater part of the dust will be carried out with the wind through the wall. For the theory of fanning wheat, see Art. 83.*

To clean the Screenings.

Draw the little gate 34, and let them into the elevator at 4, to be elevated into garner 10; then draw gate 10, and shut 11 and 34, and let them pass through the rolling screen 12 and fan 13, and as they fall at 14, guide them down a spout (shown by dotted lines) into the elevator at 4, and elevate them into the screen-hopper 11; then draw gate 11, shut 10, and let them take the same course over again, and return into the garner 10, &c. as often as necessary; when cleaned, guide them into the stones to be ground.

The screenings of the screenings are now in garner 32, which may be cleaned as before, and an inferior quality of meal made out of them.

By these means the wheat may be effectually separated from the seed of weeds, &c., and these saved for food for cattle.

This completes the whole process from the wagon to the wagon again, without manual labour, except in packing the flour and rolling it in.

---

* The bolting reels may all be set in a line connected by jointed gudgeons, supported by bearers. The meal, as it leaves the tail of one reel, may be introduced into the head of the other, by an elevator bucket, fixed on the head of the reel, open at the side next the centre, so that it will dip up the meal, and, as it passes over the centre, drop it in. This improvement was made by Mr. Jonathan Ellicott; and by it, in many cases, many wheels and shafts, and much room may be saved.

ARTICLE 90.

OF ELEVATING GRAIN FROM SHIPS.

If the grain come to the mill by ships, No. 35, and require to be measured at the mill, then a conveyer, 35—4, may be set in motion by the great cog-wheel, and may be under or above the lower floor, as may best suit the height of the floor above high water. This conveyer must have a joint, as 36, in the middle, to give the end that lies on the side of the ship, liberty to rise and lower with the tide. The wheat, as measured, is poured into the hopper at 35, and is conveyed into the elevator at 4; which conveyer will so rub the grain as to answer the end of rubbing stones. And, in order to blow away the dust, when rubbed off, before it enters the elevator, part of the wind made by the fan 13, may be brought down by a spout, 13—36, and, when it enters the case of the conveyer, it will pass each way, and blow out the dust at 37 and 4.

In some instances, a short elevator may be used, with the centre of the upper pulley, 38, fixed immoveably, the other end resting on the deck, but so much aslant as to give the vessel liberty to raise and lower, the elevator will then slide a little on the deck. The case of the lower strap of this elevator must be considerably crooked, to prevent the points of the bucket from wearing by rubbing in their descent. The wheat, as measured, is poured into a hopper, which lets it in at the bottom of the pulley.

But if the grain is not to be measured at the mill, then fix the elevator 35—39, to take it out of the hold, and elevate it through any conveniently situated door. The upper pulley is fixed in a gate that plays up and down in circular rabbets, to raise and lower to suit the tide and depth of the hold, and to reach the wheat. 40 is a draft of the gate, and manner of hanging the elevator in it. (See particular description thereof, in the latter part of Article 95.)

This gate is hung by a stout rope, passing over a strong

pulley or roller 41, and thence round the axis of the wheel 42, round the rim of which wheel there is a rope, which passes round the axis of the wheel 43, round the rim of which is a small rope, leading down over the pulley P, to the deck, and fastened to the cleet q; a man, by pulling this rope, can hoist the whole elevator; because, if the diameter of the axis be 1 foot, and the wheel 4 feet, the power is increased 16 fold. The elevator is hoisted up, and rested against the wall, until the ship comes to, and is fastened steadily in the right place; then it is set in the hold on the top of the wheat, and the bottom being open, the buckets fill as they pass under the pulley; a man holds by the cord, and lets the elevator settle as the wheat sinks in the hold, until the lower part of the case rests on the bottom of the hold, it being so long as to keep the buckets from touching the vessel; by this time it will have hoisted 1, 2, or 300 bushels, according to the size of the ship and depth of the hold, at the rate of 300 bushels per hour. When the grain ceases running in of itself, the man may shovel it up, till the load is discharged.

The elevator discharges the wheat into the conveyer at 44, which conveys it into the screen-hoppers 10—11, or into any other, from which it may descend into the elevator 4—5, or into the rubbing-stones 8.

This conveyer may serve instead of rubbing-stones, and the dust rubbed off thereby may be blown out through the wall at p, by a wind-spout from the fan 13, into the conveyer at 45. The holes at 44 and 10—11 are to be small, to let but little wind escape any where, excepting through the wall, where it will carry off the dust.

A small quantity of wind might be let into the conveyer 15—16, to blow away the dust rubbed off by it.

The fan, to be sufficient for all these purposes, must be made to blow very strongly, and the strength of the blast may be regulated as directed by Art. 83.

ARTICLE 91.

A MILL FOR GRINDING PARCELS.

Here each person's parcel is to be stored in a separate garner, and kept separate through the whole process of manufacture, which occasions much labour; almost all of which is performed by the machines. See Plate VI., fig. 1; which is a view of one side of a mill, containing a number of garners holding parcels, and a side view of the wheat elevator.

The grain is emptied into the garner g, from the wagon, as shown in Plate VIII.; and by drawing the gate A, it is let into the elevator A B, and elevated into the crane-spout B, which, being turned into the mouth of the garner-spout B C, which leads over the top of a number of garners, and has, in its bottom, a little gate over each garner; these gates and garners are all numbered with the same numbers, respectively.

Suppose we wish to deposit the grain in the garner No. 2, draw the gate 3 out of the bottom, and shut it in the spout, to stop the wheat from passing along it, past the hole, so that it must all fall into the garner; and thus proceed for the other garners 3 4 5 6, &c. These garners are all made like hoppers, about 4 inches wide at the floor, and nearly the length of the garner; but as it passes through the next story, it is brought to the form of a spout, 4 inches square, leading down to the general spout K A, which leads to the elevator: in each of these spouts is a gate numbered with the number of its garner; so that when we want to grind the parcel in garner 2, we draw the gate 2 in the lower spout, to let the wheat run into the elevator at A, to be elevated into the crane-spout B, which is to be turned over the rolling-screen, as shown in Plate VIII.

Under the upper tier of garners, there is another tier in the next story, set so that the spouts from the bottom of the upper tier pass down the partitions of the lower tier, and the upper spouts of the lower tier pass between the partitions of the upper tier, to the garner-spout.

These garners, and the gates leading both into and out of them, are numbered as the others.

If it be not convenient to fix the descending spouts B C, to convey the wheat from the elevator to the garners, and K A to convey it from the garners to the elevator again, then the conveyers r s and I K may be used for said purposes.

To keep the parcels separate, there should be a crane-spout to the meal elevator; or any other method may be adopted, by which the meal of the second parcel may be guided to fall on another part of the floor, until the first parcel is all bolted, and the chests cleared out, when the meal of the second parcel may be guided into the hopper-boy.

I must here observe, that in mills for grinding parcels, the tail flour must be hoisted by a separate elevator to the hopper-boy, to be bolted over, and not run into the conveyer, as shown in Plate VIII; because then the parcels could not be kept separate.

The advantages of the machinery, applied to a mill for grinding parcels, are very great.

1. Because without them there is much labour in moving the different parcels from place to place; all which is here done by the machinery.

2. The meal, as it is ground, is cooled by the machinery, and bolted in so short a time, that, when the grinding is done, the bolting is, also, nearly finished. Therefore,

3. It saves room, because the meal need not be spread over the floor to cool, during 12 hours, as is usual; and but one parcel need be on the floor at one time.

4. It gives greater despatch, as the miller need never stop either stones or bolts, in order to keep parcels separate. The screenings of each parcel may be cleaned, as directed in Art. 89, with very little trouble; and the flour may be nearly packed before the grinding is finished; so that if a parcel of 60 bushels arrive at the mill in the evening, the owner may wait till morning, when he may have it all finished; he may use the offal for feed for his team, and proceed with his load to market.

ARTICLE 92.

A GRIST MILL FOR GRINDING VERY SMALL PARCELS.

Fig. 16, Plate VII., is a representation of a grist-mill, so constructed that the grist being put into the hopper, it will be ground and bolted and returned into the bags again.

The grain is emptied into the hopper at A, and as it is ground it runs into the elevator at B, and is elevated and let run into the bolting hopper down a broad spout at C, and, as bolted, it falls into the bags at d. The chest is made to come to a point like a funnel, and a division made to separate the fine and coarse, if wanted, and a bag put under each part; on the top of this division is set a regulating board on a joint, as x, by which the fine and coarse can be regulated at pleasure.

If the bran require to be ground over, (as it often does,) it is made to fall into a box over the hopper, and by drawing the little gate b, it may be let into the hopper as soon as the grain is all ground, and as it is bolted the second time, it is let run into the bag by shutting the gate b, and drawing the gate c.

If the grain be put into the hopper F, then as it is ground it falls into the drill, which draws it into the elevator at B, and it ascends as before.

To keep the different grists separate;—when the miller sees the first grist fall into the elevator, he shuts the gate B or d, and gives time for it all to get into the bolting reel; he then stops the knocking of the shoe by pulling the shoe line, which hangs over the pulleys p p, from the shoe to near his hand, making it fast to a peg; he then draws the gate B or d, and lets the second grist into the elevator, to fall into the shoe or bolting hopper, giving time for the first grist to be all in the bags, and the bags of the second grist to be put in their places; he then unhitches the line from the peg, and lets the shoe knock again, and begins to bolt the second grist.

If he does not choose to let the meal run immediately into the bags, he may have a box made with feet, to stand in the place of the bags, for the meal to fall in, out of which it may be taken and put into the bags, as fast as it is bolted, and mixed as desired; and as soon as the first parcel is bolted, the little gates at the mouth of the bags may be shut, while the meal is filled out of the box, and the second grist may be bolting.

The advantages of this improvement on a grist mill are,

1. It saves the labour of hoisting, spreading, and cooling the meal, and of carrying up the bran to be ground over, sweeping the chest, and filling up the bags.

2. It does all with great despatch, and little waste, without having to stop the stones or bolting-reel, to keep the grists separate, and the bolting is finished almost as soon as the grinding; therefore, the owner will be the less time detained.

The chests and spouts should be made steep, to prevent the meal from lodging in them; so that the miller, by striking the bottom of the chest, will shake out all the meal.

The elevator and drill should be so made as to clean out at one revolution. The drill might have a brush or two, instead of rakes, which would sweep the case clean at a revolution; and the shoe of the bolting hopper should be short and steep, so that it will clean out rapidly.

The same machinery may be used for merchant-work, by having a crane-spout at C, or a small gate, to turn the meal into the hopper-boy that tends the merchant bolt.

A mill, thus constructed, might grind grists in the day-time, and merchant-work at night.

A drill is preferable to a conveyer for grist mills, because it may be cleaned out much sooner and better. The lower pulley of the elevator is twice as large in diameter as the pulleys of the drill; the lower pulley of the elevator, and one pulley of the drill, are on the same shaft, close together; the elevator moves the drill, and the pulley of the drill being smallest, gives room for the meal to fall into the buckets of the elevator.

ARTICLE 93.

OF ELEVATING GRAIN, SALT, OR ANY GRANULAR SUBSTANCE FROM SHIPS INTO STORE-HOUSES, BY THE STRENGTH OF A HORSE.

Plate VII., fig. 17, represents the elevator, and the manner of giving it motion; the horse is hitched to the end of the sweep-beam A, by which he turns the upright shaft, on the top of which is the driving cog-wheel of 96 cogs, $2\frac{1}{2}$ inches pitch, to gear into the leading wheel of 20 cogs, on the same shaft with which is another driving wheel of 40 cogs, to gear into another leading wheel of 19 cogs, which is on the same shaft with the elevator pulley; then, if the horse make about 3 revolutions in a minute (which he will do if he walk in a circle of 20 feet diameter) the elevator pulley will make about 30 revolutions in a minute; and if the pulley be 2 feet in diameter, and a bucket be put on every foot of the strap, to hold a quart each, the elevator will hoist about 187 quarts per minute, or 320 bushels in an hour, 3840 bushels in 12 hours; and for every foot the elevator is high, the horse will have to sustain the weight of a quart of wheat, say 48 feet, which is the height of the highest store-houses, then the horse would have to move $1\frac{1}{2}$ bushels of wheat upwards, with a velocity equal to his own walk; which, I presume, he can do with ease, and overcome the friction of the machinery: From this will appear the great advantages of this application.

The lower end of the elevator should stand near the side of the ship, and the grain, salt, &c., be emptied into a hopper; the upper end may pass through a door or window, as may be most convenient; the lower case should be a little crooked to prevent the buckets from rubbing in their descent.

ARTICLE 94.

OF AN ELEVATOR APPLIED TO ELEVATE GRAIN, &c.,
WROUGHT BY A MAN.

In Plate VII., fig. 18, A B, are two ratchet wheels, with two deep grooves in each of them, for ropes to run in; they are fixed close together, on the same shaft with the upper pulley of the elevator, so that they will turn easily on the shaft the backward way, whilst a click falls into the ratchet, and prevents them from turning forwards. Fig. 19 is a side view of the wheel, ratchet, and click. C D are two levers, like weavers' treadles, and from lever C there is a light staff passes to the foreside of the groove wheel B, and is made fast by a rope half way round the wheel; and from said lever C there is a rope passing to the backside of the wheel A; and from lever D there is a light staff passing to the foreside of the groove wheel A, and a rope to the backside of the groove wheel B.

The man who is to work this machine stands on the treadles, and holds by the staffs with his hands: and as he treads on D it descends, and the staff pulls the wheel A forward, and the rope pulls the wheel B backward, and as he treads on C, the staff pulls forward the wheel B, and the rope pulls backward the wheel A: but the click falls into the ratchet, so that the wheels cannot move forward without turning the elevator pulley, it is thus moved one way by both treadles; and in order to keep up a regular motion, a heavy flying wheel F, is added, which should be of cast iron, to prevent much obstruction from the air.

To calculate what quantity a man can raise to any height, let us suppose his weight to be 150 lbs., which is the power to be applied; and suppose he be able to walk about 70 feet up stairs in a minute, by the strength of both his legs and arms, or, which is the same thing, to move his weight on the treadles 70 steps in a minute; then suppose we allow, as by Art. 29—42, to lose 1-3d of the power to gain velocity and overcome friction, (which will be a large allowance in this case, because in the experiment in the table, in Art. 37, when 7lbs. were charged with 6 lbs. they moved with the velocity of 2 feet in half a second,) then there will remain 100 lbs. raised 70 feet in a minute, equal to 200 lbs. raised 35 feet, to the top of the third story, per minute, equal to 200 bushels per hour, 2400 bushels in twelve hours.

The great advantages of this application of the elevator, and of this mode of applying man's strength, will appear from these considerations; namely: he uses the strength of both his legs and arms, to move his weight only from one treadle to the other, which weight does the work; whereas, in carrying bags on his back, he uses the strength of his legs only, to raise both the weight of his body and the burden; add to this, that he generally takes a very circuitous route to the place where he is to empty the bag, and returns empty; whereas, the elevator takes the shortest direction to the place of emptying, and is always steadily at work.

The man must sit on a high bench, as a weaver does, on which he can rest part of his weight, and rest himself occasionally, when the machine moves lightly, and have a beam above, that he may push his head against, to overcome extraordinary resistances. This is probably the best means of applying man's strength to produce rotary motion.

DESCRIPTION OF PLATE IX.

The grain is emptied into the spout A, by which it descends into the garner B; whence, by drawing the gate at C, it passes into the elevator C D, which raises it to D, and empties it into the crane spout E, which is so fixed on gudgeons that it may be turned to any of the surrounding garners; into the screen hopper F, for instance, (which has two parts F and G,) out of which it is let into the rolling screen at H, by drawing the small gate a. It passes through the fan I, and falls into the little sliding-hopper K, which may be moved, so as to guide it into either of the hanging garners, over the stones, L or M, and it is let into the stone-hoppers by the little bags b b, as fast as it can be ground. When ground, it falls into the conveyer N N, which carries it into the elevator at O O, this raises and empties it into the hopper-boy at P, which is so constructed as to carry it round in a ring, gathering it gradually towards the centre, till it sweeps into the bolting hoppers Q Q.

The tail flour, as it falls, is guided into the elevator to ascend with the meal, and, that a proper quantity may be elevated, there is a regulating board R, set under the superfine cloths, on a joint x, so that it will turn towards the head or tail of the reel, and send more or less into the elevator, as may be required.

There may be a piece of coarse cloth, or wire, put on the tails of the superfine reels, that will let all pass through except the bran which falls out at the tail, and a part of which is guided into the elevator with the tail flour, to assist the bolting in warm weather; the quantity is regulated by a small board r, set on a joint under the ends of the reels. Beans may be used to keep the cloths open, and still be returned into the elevator to ascend again. What passes through the coarse cloth or wire, and the remainder of the bran, are guided into the reel S, to be bolted.

To clean Wheat several times.

Suppose the grain to be in the screen hopper E; draw the gate a; shut the gate e; move the sliding hopper K, over the spout K c d; and let it run into the elevator to be raised again. Turn the crane spout over the empty hopper G, and the wheat will be all deposited there nearly as soon as it is out of the hopper F. Then draw the gate e, shut the gate a, and turn the crane spout over F; and so on, alternately, as often as necessary. When the grain is sufficiently cleaned, slide the hopper K over the hole that leads into the stones.

The screenings fall into a garner, hopperwise; to clean them, draw the gate f, and let them run into the elevator, to be elevated into the screen hopper F. Then proceed with them as with the wheat, till sufficiently clean. To clean the fannings, draw the little gate h, and let them into the elevator, &c., as before.

Fig. II. is a perspective view of the conveyer, as it lies in its troughs, at work; and shows the manner in which it is joined to the pulleys, at each side of the elevator.

Fig. III. exhibits a view of the pulley of the meal elevator, as it is supported on each side, with the strap and buckets descending to be filled.

Fig. IV. is a perspective view of the under side of the arms of the hopper-boy, with flights complete. The dotted lines show the track of the flights of one arm; those of the other following, and tracking between them. A A are the sweepers. These carry the meal round in a ring, trailing it regularly all the way, the flights drawing it to the centre, as already mentioned. B B are the sweepers that drive it into the bolting hoppers.

Fig. V. is a perspective view of the bucket of the wheat-elevator; and shows the manner in which it is fastened, by a broad piece of leather, which passes through and under the elevator-strap, and is nailed to the sides with little tacks.

CHAPTER XI.

OF THE CONSTRUCTION OF THE SEVERAL MACHINES.

ARTICLE 95.

OF THE WHEAT ELEVATOR.

To construct a wheat-elevator, first determine how many bushels it should hoist in an hour, and where it shall be set, so as, if possible, to answer all the following purposes:—

1. To elevate the grain from a wagon or ship.
2. From the different garners into which it may be stored.
3. If it be a two-story mill, to hoist the wheat from the tail of the fan, as it is cleaned, to a garner over the stones.
4. To hoist the screenings, to clean them several times.

5. To hoist the wheat from a shelling-mill, if there be one.

One elevator may effect all these objects in a mill rightly planned, and most of them can be accomplished in mills ready built.

Suppose it be wished to hoist about 300 bushels in an hour, make the strap $4\frac{1}{2}$ inches wide, of good, strong, white harness leather, in one thickness. It must be cut and joined together in a straight line, with the thickest, and, consequently, the thinnest ends together, so that if they be too thin, they may be lapped over and doubled, until they are thick enough singly. Then, to make wooden buckets, take the but of a willow or water birch, that will split freely; cut it in bolts, 15 inches long, and rive and shave it into staves, $5\frac{1}{2}$ inches wide, and three-eighths of an inch thick; these will make one bucket, each. Set a pair of compasses to the width of the strap, and make the sides and middle of the bucket equal thereto at the mouth, but let the sides be only two-thirds of that width at the bottom, which will make it of the form of fig. 9, Plate VI.; the ends being cut a little circular, to make the buckets lie more closely to the strap and wheel, as it passes over. Make a pattern of the form of fig. 9, by which to describe all the rest. This makes a bucket of a neat form, to hold about 75 solid inches, or somewhat more than a quart. To make them bend to a square at the corners e c, cut a mitre square across where they are to bend, about 2-8ths through; boil them and bend them hot, tacking a strip of leather across them, to hold them in that form until they get cold, and then put bottoms to them of the thin skirts of the harness leather. These bottoms are to extend from the lower end to the strap that binds it on. To fasten them on well, and with despatch, prepare a number of straps, $1\frac{3}{4}$ inches wide, of the best cuttings of the harness leather; wet them and stretch them as hard as possible, which reduces their width to about $1\frac{1}{2}$ inches. Nail one of these straps to the side of a bucket, with 5 or 6 strong tacks that will reach through the bucket, and clinch inside. Then take a $1\frac{1}{2}$ inch chisel, and strike it through the main strap about a quarter of an inch from each edge, and put one end of the binding-strap through the slits, draw the bucket very closely to the strap, and nail it on the other side of the bucket, which will finish it. See B in fig. 2, Plate VI. C is a meal-bucket fastened in the same manner, but is bottomed only with leather at the lower end, the main strap making the bottom side of it. This is the best way I have yet discovered to make wooden buckets. The scraps of the harness leather, out of which the elevator-straps are cut, are generally about enough to complete the buckets.

To make Sheet-Iron Buckets.

Cut the sheet in the form of fig. 8, Plate VI., making the middle part c, and the sides, a and b, nearly equal to the width of the strap, and nearly $5\frac{1}{2}$ inches long, as before. Bend them to a right angle at every dotted line, and the bucket will be formed :—c will be the bottom side next to the strap; and the little holes a a and b b will meet, and must be riveted to hold it together. The two holes c are for fastening it to the straps by rivets. The part a b is the part that dips up the wheat, and the point, being doubled back, strengthens it, and tends to make it wear well. The bucket being completely formed, and the rivet holes made, spread one out again, as fig. 8, to describe all the rest by, and to mark for the holes, which will meet again when folded up. They are fastened to the strap by two rivets with thin heads put inside the bucket, and a double burr of sheet iron put on the under side of the strap, which fastens them on very tightly. See A, fig. 2, Plate VI. These buckets will hold about 1,3 quarts, or 88 cubic inches. This is the best way I have found to make sheet-iron buckets. D is a meal-bucket of sheet-iron, riveted on by two rivets, with their heads inside the strap; the sides of the buckets are turned a little out, and holes made in them for the rivets to pass through. Fig. 11 is the form of one spread out, and the dotted lines show where they are to be bent at right angles to form them. The strap forms the bottom side of these buckets.

Make the pulleys 24 inches in diameter, as thick as strap is wide, and half an inch higher in the middle than at the sides, to make the strap keep on; give them a motion of 25 revolutions in a minute, and put on a sheet-iron bucket for every 15 inches; then 125 buckets will pass per minute, which will carry 162 quarts, and hoist 300 bushels in an hour, and 3600 bushels in 12 hours. If you wish to hoist faster, make the strap wider, the buckets larger in proportion, and increase the velocity of the pulley, but not to above 35 revolutions in a minute, nor place more buckets than one for every 12 inches; otherwise, they will not empty well. A strap of 5 inches, with buckets 6 inches long, and of a width and proportion suiting the strap, (4½ inches wide,) will hold 1,8 quarts each; and 35 revolutions of the pulley will pass 175 buckets, which will carry 315 quarts in a minute, and 590 bushels in an hour. If the strap be 4 inches wide, and the wooden buckets 5 inches deep, and in proportion to the strap, they will hold ,8 of a quart: then, if there be one for every 15 inches, and the pulley makes 27 revolutions in a minute, it will hoist 200 bushels in an hour. Where there is a good garner to empty the wheat into, this is the size they are commonly made, and is sufficient for unloading wagons.

Plate VI., Fig. 6, represents the gudgeon of the lower pulley; fig. 7, the gudgeon for the shaft on which the upper pulley is fixed. Fix both the pulleys in their places, but not firmly, so that a line, stretched from one pulley to the other, will cross the shafts or gudgeons at right angles. This must always be the case to make the straps work fairly. Put on the strap with the buckets; draw it tightly, and buckle it; put it in motion, and if it do not keep fairly on the pulleys, their position may be altered a little. Observe how much the descending strap swags by the weight of the buckets, and make the case round it so curved, that the points of their buckets will not rub in their descent, which will cause them to wear long and work easily. The side boards need not be made crooked in dressing out, but may be bent sufficiently by sawing them half way, or two-thirds, through, beginning at the upper edge, holding the saw very much aslant, the point downwards and inwards, so that in bending, the parts will slip past each other. The upper case must be nearly straight; for if it be made much crooked, the buckets will incline to turn under the strap. Make the cases 3-4ths of an inch wider inside, than the strap and buckets, and $1\frac{1}{2}$ inches deeper, that they may play freely; but do not give them room to turn upside down. If the strap and buckets be 4 inches, then make the side boards $5\frac{1}{2}$, and the top and bottom boards $6\frac{3}{4}$ inches wide, of inch boards. Be careful that no shoulders nor nail-points be left inside of the cases, for the buckets to catch in. Make the ends of each case, where the buckets enter as they pass over the pulleys, a little wider than the rest of the case. Both the pulleys are to be nicely cased round to prevent waste, not leaving room for a grain to escape, continuing the case of the same width round the top of the upper, and bottom of the lower pulley; then, if any of the buckets should ever get loose, and stand askew, they will be kept right by the case; whereas, if there were any ends of boards or shoulders, they would catch against them. See A B, fig. 1, Plate VI. The bottom of the case of the upper pulley must be descending, so that what grain may fall out of the buckets in passing over the pulleys, may be guided into the descending case. The shaft passing through this pulley is made round where the case fits to it: half circles are cut out of two boards, so that they meet and embrace it closely. The undermost board, where it meets the shaft, is ciphered off inside next the pulley, to guide the grain inward. But it is full as good a way to have a strong gudgeon to pass through the upper pulley, with a tenon at one end, to enter a socket, which may be in the shaft, that is to give it motion. This will suit best where the shaft is short, and has to be moved to put the elevator out of, and into gear.

The way that I have generally cased the pulleys is as follows; namely: The top board of the upper strap-case, and the bottom board of the lower strap-case, are extended past the lower pulley to rest on the floor; and the lower ends of these boards are made two inches narrower, as far as the pulley-case extends; the side-board of the pulley is nailed, or rather screwed, to them, with wooden screws. The rest of the case boards join to the top of the pulley-case, both being of one width. The block, which the gudgeons of this pulley run in, is screwed fast to the outside of the case boards; the gudgeons do not pass quite through, but reach to the bottom of the hole, which keeps the pulley in its place.

The top and bottom boards, and, also, the side-boards of the strap-cases, are extended past the upper pulley, and the side-boards of the pulley-case are screwed to them; but this leaves a vacancy between the top of the side-boards of the strap-cases, and shoulders for the buckets to catch against, and this vacancy is to be filled up by a short board, guiding the buckets safely over the upper pulley. The case must be as close to the points of the buckets, where they empty, as is safe, that as little as possible may fall down again. There is to be a long hole cut into the case at B, for the wheat to fall out at, and a short spout guiding it into the crane spout. The top of the short spout next B, should be loosely fastened in with a button, that it may be taken off, to examine if the buckets empty well, &c. Some neat workmen have a much better way of casing the pulleys, which is not easily described; what I have described is the cheapest, and answers very well.

The wheat should be let in at the bottom, to meet the buckets; and a gate should shut as near to the point of them as possible, as at A, fig. 1, Plate VI. Then, if the gate be drawn sufficiently to fill the buckets, and the elevator be stopped, the wheat will stop running in, and the elevator will be free to start again; but if it had been let in any distance up, then, when the elevator stopped, it would fill from the gate to the bottom of the pulley, and the elevator could not start again. If it be, in any case, let in at a greater distance up, the gate should be so fixed that it cannot be drawn so far as to let in the wheat faster than the buckets can take it, else the case will fill and stop the buckets. If it be let in faster at the hindmost side of the pulley than the buckets will carry it, the same evil will occur; because the buckets will push the wheat before them, being more than they can hold, and give room for too much to come in; therefore, there should be a relief gate at the bottom, to let the wheat out, should too much happen to get in.

The motion is to be given to the upper pulley of all elevators, if it can be done, because the weight in the buckets causes the strap to hang tightly on the upper, and slackly on the lower pulley; therefore, the upper pulley will carry the greatest quantity without slipping. All elevators should stand a little slanting, because they will discharge the better. The boards for the cases should be of unequal lengths, so that two joints may never come close together; this greatly strengthens the case. Some have joined the cases at every floor, which is a great error. There must be a door in the ascending case, at the place most convenient for buckling the strap, &c. &c.

Of the Crane Spout.

To make a crane spout, fix a board 18 or 20 inches broad, truly horizontal, or level, as a, under B, in fig. 1, Plate VI. Through the middle of this board the wheat is conveyed, by a short spout, from the elevator. Then make the spout of 4 boards, 12 inches wide at the upper, and about 4 or 5 inches at the lower end. Cut the upper end off aslant, so as to fit nicely to the bottom of the board; hang it to a strong pin, passing through the broad board near the hole through which the wheat passes, so that the spout may be turned in any direction, and still cover the hole, at the same time it is receiving the wheat, and guiding it into any garner, at pleasure. In order that the pin may have a strong hold of the board and spout, there must be a piece of scantling, 4 inches thick, nailed on the top of the board, for the pin to pass through; and another to the bottom, for the head of the pin to rest on. But if the spout be long and heavy, it is best to hang it on a shaft, that may extend down to the floor, or below the collar-beams, with a pin through it, as x, to turn the spout by. In crane spouts for meal, it is sometimes best to let the lower board reach to, and rest on the floor. If the elevator-cases and crane spout be well fixed, there can neither grain nor meal escape, or be wasted, that enters the elevator, until it comes out at the end of the crane-spout again.

*Of an Elevator to elevate Wheat from a Ship's Hold.**

Make the elevator complete (as it appears 35—39, Plate VIII.) on the ground, and raise it to its place afterwards. The pulleys are to be both fixed in their places and cased; and the blocks that the gudgeon of the upper pulley is to run in, are to be riveted fast to the case-boards of the pulley, and these case-boards screwed to the strap-cases by long screws, reaching through the case-boards edgeways. Both sides of the pulley-case are fastened by one set of screws. On the outside of these blocks, round the centre of the gudgeons, are circular knobs, 6 inches diameter, and 3 inches long, strongly riveted, to keep them from splitting off, because, by these knobs the whole weight of the elevator is to hang. In the moveable frame 40, o o, o o, are these blocks with their knobs, which are let into the pieces of the frame $B C r s$. The gudgeons of the upper pulley p pass through these knobs and play in them. Their use is to bear the weight of the elevator that hangs by them; the gudgeons, by this means, bear only the weight of the strap and its load, as is the case with other elevators. Their being circular gives the elevator liberty to swing out from the wall to the hold of the ship.

The frame 40 is made as follows: the top piece $A B$ is 9 by 8 inches, strongly tenoned into the side pieces $A D$ and $B C$ with double tenons, which side pieces are 8 by 6. The piece r s is put in with a tenon, 3 inches thick, which is dove-tailed, keyed, and draw-pinned, with an iron pin, so that it can easily be taken out. In each side piece $A D$ and $B C$ there is a row of cogs, set in a circle, that are to play in circular rabbets in the posts

---

* See the description of this elevator in Art. 90.

p. 41. These circles are to be described with a radius, whose length is from the centre of the joint gudgeons *G*, to the centre of the pulley 39; and the posts must be set up, so that the centre of the circle will be the centre of the gudgeon *G*; then the gears will be always right, although the elevator rises and falls to suit the ship or tide. The top of those circular rabbets ought to be so fixed, that the lower end of the elevator may hang near the wall. This may be regulated by fixing the centre of the gudgeon *G*. The length of these rabbets is regulated by the distance the vessel is to rise and fall, to allow the elevator to swing clear of the vessel when light, at high water. The best way to make the circular rabbets is, to dress two pieces of 2 inch plank for each rabbet, of the right circle, and to pin them to the posts, at such a distance, leaving the rabbet between them.

When the gate and elevator are completed, and tried together, the gate hung in its rabbets, and played up and down, then the elevator may be raised by the same power that is to raise and lower it, as described, Art. 94.

ARTICLE 96.

OF THE MEAL ELEVATOR.

Little need be said of the manner of constructing the meal elevator, after what has been said in Art. 90, except giving the dimensions. Make the pulleys $3\frac{1}{2}$ inches thick, and 18 inches diameter. Give them no more than 20 revolutions in a minute. Make the strap $3\frac{1}{2}$ inches wide, of good, pliant, white harness leather; make buckets either of wood or sheet-iron, to hold about half a pint each; put one for every foot of the strap; make the cases tight, especially round the upper pulley, slanting much t bottom, so that the meal which falls out of the buckets, may be guided into the descending case. Let it lean a little, that it may discharge the better. The spout that conveys the meal from the elevator to the hopper-boy, should not have much more than 45 degrees descent, that the meal may run easily down, and not cause a dust; fix it that the meal will spread thinly over its bottom in its descent, and it will cool the better. Cover the top of the spout half-way down, and hang a thin, light cloth, at the end of this cover, to check all the dust that may rise, by the fall of the meal from the buckets. Remember to take a large cipher off the inside of the board, where it fits to the undermost side of the shaft of the upper pulley: the meal will otherwise work out along the shaft. Make all tight, as directed, and it will effectually prevent waste.

In letting meal into an elevator, it must be let in some distance above the centre of the pulley, that it may fall clear from the spout that conveys it in; otherwise, it will clog and choke. Fig. 4, Plate VI., is the double socket gudgeon of the lower pulley, to which the conveyer joins. Fig. 3, a b c d, is a top view of the case that the pulley runs in, which is constructed thus; a b is a strong plank, 14 by 3 inches, stepped in the sill, dove-tailed and keyed in the meal-beam, and is called the main bearer. In this, at the determined height, are framed the gudgeon bearers a c b d, which are planks 15 by 1½ inches, set 7½ inches apart, the pulley running between, and resting on them. The end piece c d, 7 inches wide and 2 thick, is set in the direction of the strap-case, and extends 5 inches above the top of the pulley; to this the bearers are nailed. On the top of the bearers, above the gudgeons, are set two other planks, 13 by 1½ inches, rabbetted into the main bearer, and screwed fast to the end piece c d: these are 4 inches above the pulley. The bottom piece of this case slides in between the bearers, resting on two cleets, so that it can be drawn out to empty the case, if it should ever, by any means, be overcharged with meal; this completes the case. In the gudgeon bearer, under the gudgeons, are mortises, made about 12 by 2 inches, for the meal to pass from the conveyer into the elevator; the bottom board of the conveyer trough rests on the bearer in these mortises. The strap-case, joins to the top of the pulley-case, but is not made fast, but the back board of the descending case is stepped into the inside of the top of the end piece c d. The bottom of the ascending case is to be supported steadily to its place, and the board at the bottom must be ciphered off at the inside, with long and large ciphers, making them, at the point, only $\frac{1}{4}$ of an inch thick; this is to make the bottom of the case wide for the buckets to enter, if any of them should be a little askew; the pulley-case is wider than the strap-cases, to give room for the meal from the conveyer to fall into the buckets; and, in order to keep the passage open, there is a piece 3 inches wide, and $1\frac{3}{4}$ inches thick, put on each side of the pulley, to stand at right angles with each other, extending $3\frac{1}{2}$ inches at each end, past the pulley; these are ciphered off, so as to clear the strap, and draw the meal under the buckets: they are called bangers.

ARTICLE 97.

OF THE MEAL CONVEYER.

Fig. 3, Plate VI., is a conveyer joined to the pulley of the elevator. (See it described, Art. 88.) Fig. 4 is the gudgeon that is put through the lower pulley, to which the conveyer is joined by a socket, as represented. Fig. 5 is a view of the said socket and the band, as it appears on the end of the shaft. The tenon of the gudgeon is square, that the socket may fit it every way alike. Make the shaft $5\frac{1}{2}$ inches diameter, of eight equal sides, and put on the socket and the gudgeon; then, to lay it out for the flights, begin at the pulley, mark as near the end as possible, on the one side, and turning the shaft the way it is to work, at the distance of $1\frac{1}{2}$ inches towards the other end, set a flight on the next side, and thus go on to mark for a flight on every side, still advancing $1\frac{1}{2}$ inches to the other end, which will form the dotted spiral line, which would drive the meal the wrong way; but the flights are to be set across this spiral line, at an angle of about 30 degrees, with a line square across the shaft; and then they will drive the meal the right way, the flights operating like ploughs.

To make the flights, take good maple, or other smooth, hard wood; saw it into 6 inch lengths, split it always from the sap to the heart; make pieces $2\frac{1}{2}$ inches wide, and $\frac{3}{4}$ of an inch thick; plane them smooth on one side, and make a pattern to describe them by, and make a tenon $2\frac{1}{2}$ inches long, to suit a $\frac{3}{4}$ inch auger. When they are perfectly dry, having the shaft bored, and the inclination of the flights marked by a scribe, drive them in and cut them off $2\frac{1}{2}$ inches from the shaft; dress them with their foremost edge sharp, taking all off from the back side, leaving the face smooth and straight, to push forward the meal; make their ends nearly circular. If the conveyer be short, put in lifting flights, with their broad side foremost, half the number of the others, between the spires of them; they cool the meal by lifting and letting it fall over the shaft.

To make the trough for it to run in, take 3 boards, the bottom one 11, back 15, and front 13 inches. Fix the block for the gudgeon to run in at one end, and fill the corners with cleets, to make the bottom nearly circular, that but little meal may lie in it; join it neatly to the pulley-case, resting the bottom on the bottom of the hole cut for the meal to enter, and the other end on a supporter, that it can be removed and put to its place again with ease, without stopping the elevator.

A meal elevator and conveyer thus made, of good materials, will last 50 years, with very little repair, and save an immense quantity of meal from waste. The top of the trough must be left open, to let the steam of the meal out; and a door, about 4 feet long, may be made in the ascending case of the elevator, to buckle the strap, &c. The strap of the elevator turns the conveyer, so that it can be easily stopped if any thing should be caught in it; it is dangerous to turn it by cogs. This machine is often applied to cool the meal, without the hopper-boy, and to attend the bolting-hopper, by extending it to a great length, and conveying the meal immediately into the hopper, which answers very well; but where there is room a hopper-boy is preferable.

ARTICLE 98.

OF A GRAIN CONVEYER.

This machine has been constructed in a variety of ways; the following appears to be the best; namely: First, make a round shaft, 9 inches diameter; and then, to make the spire, take strong sheet-iron, make a pattern 3 inches broad, and of the true arch of a circle; the diameter of which (being the inside of the pattern) is to be 12 inches; this will give it room to stretch along a 9 inch shaft, so as to make a rapid spiral, that will advance about 21 inches along the shaft every revolution. By this pattern cut the sheet-iron into circular pieces, and join the ends together by riveting and lapping them, so as to let the grain run freely over the joints; when they are joined together they will form several circles, one above the other, slip it on the shaft, and stretch it along as far as you can, till it comes tight to the shaft, and fasten it to its place by pins, set in the shaft at the back side of the spire, and nail it to the pins: it will now form a beautiful spiral, with returns, 21 inches apart, which distance is too great; there should, therefore, be two or three of these spirals made, and wound into each other, and all put on together, because, if one be put on first, the others cannot be got on so well afterwards; if there be three, they will then be 7 inches apart, and will convey wheat very fast. If these spirals be punched full of holes like a grater, and the trough be lined with sheet-iron, punched full of small holes, it will become an excellent rubber; will clean the wheat of the dust and down, that adhere to it, and supersede the necessity of any other rubbing machine.

The spirals may also be formed with either wooden or

iron flights, set so near to each other in the spiral lines, as to convey the wheat from one to another.

ARTICLE 99.

OF THE HOPPER-BOY.

This machine, also, has appeared under various constructions, the best of which is represented by fig. 12, Plate VII.—(See the description Art. 88.)

To make the flight-arms C D, take a piece of dry poplar or other soft scantling 14 feet long, 8 by $2\frac{1}{2}$ inches in the middle, 5 by $1\frac{1}{2}$ inches at the end, and straight at the bottom; on this strike the middle line a b, fig. 13. Consider which way it is to revolve, and cipher off the under side of the foremost edge from the middle line, leaving the edge $\frac{3}{4}$ of an inch thick, as appears by the shaded part. Then, to lay out the flights, take the following

RULE.

Set your compasses at $4\frac{1}{2}$ inches distance, and, beginning with one foot in the centre c, step towards the end b, observing to lessen the distance one sixteenth part of an inch every step; this will set the flights closer together at the end than at the centre. Then, to set the flights of one arm to track truly between those of the other, and to find their inclination, with one point in the centre c, sweep the dotted circles across every point in one arm; then, without altering the centre or distance, make the little dotted marks on the other arm, and between them the circles are to be swept for the flights in it. To vary their inclination regularly, from the end to the centre, strike the dotted line c d half an inch from the centre c, and $2\frac{1}{2}$ inches from the middle line at d, and then, with the compasses set to half an inch, set off the inclination from the dotted circles, on the line c d; the line c d then approaches the middle line, the inclination is greater near the centre than at the end, and varies regularly. Dove-tail the flights into the arm, observing to put the side that is to drive the meal, to the line of inclination. The bottoms of them should not extend past the middle line, the ends being all rounded and dressed off at the back side, to make the point sharp, leaving the driving side quite straight, like the flight r. (See them complete in the end c a.) The sweepers should be 5 or 6 inches long, screwed on behind the flights, at the back side of the arms, one at each end of the arm, and one at the part that passes over the hopper: their use is described in Art. 88.

The upright shaft should be 4 by 4 inches, and made round for about $4\frac{1}{2}$ feet at the lower end, to pass lightly through the centre of the arm. To keep the arm steady, there is a stay-iron 15 inches high, its legs $\frac{1}{2}$ inch by $\frac{3}{4}$, to stride 2 feet. The ring at the top should fit the shaft neatly, and be smooth and rounded inside, that it may slide easily up and down; by this the arm hangs to the rope that passes over a pulley at the top of the shaft, 8 inches diameter, with a deep groove for the rope or cord to run in. Make the leading arm 6 by $1\frac{1}{4}$ inches in the middle, 2 by 1 inch at the end, and 8 feet long. This arm must be braced to the cog-wheel above, to keep it from splitting the shaft by an extra stress.

The weight of the balance w, must be so nearly equal to the weight of the arm, that when it is raised to the top it will descend quietly.

In the bottom of the upright shaft is the step-gudgeon (fig. 15,) which passes through the square plate 4 by 4 inches (fig. 14;) on this plate the arm rests, before the flights touch the floor. The ring on the lower end of the shaft is less than the shaft, that it may pass through the arm: this gudgeon comes out every time the shaft is taken out of the arm.

If the machine is to attend but one bolting-hopper, it need not be above 12 or 13 feet long. Set the upright shaft close to the hopper, and the flights all gather as the end c b, **fig. 13.** But, if it is to attend for the grinding of two pair of stones, and two hoppers, make it 15 feet long, and set it between them a little to one side of both, so that the two ends may not both be over the hoppers at the same time, which would make it run unsteadily; then the flights between the hoppers and the centre must drive the meal outwards to the sweepers, at the end c a, fig. 13.

If it be to attend two hoppers, and cannot be set between them for want of room, then set the shaft near to one of them; make the flights so that they all gather to the centre, and put sweepers over the outer hopper, which will be first supplied, and the surplus carried to the other. The machine will regulate itself to attend both, although one should feed three times as fast as the other.

If it be to attend three hoppers, set the shaft near the middle one, and put sweepers to fill the other two, the surplus will come to the centre one, and it will regulate to feed all three; but should the centre hopper ever stand while the others are going, (of either of these last applications,) the flights next the centre must be moveable, that they may be turned, and set to drive the meal out from the centre. Hopper-boys should be driven by a strap in some part of their movement, that they may easily stop if any thing catch in them; but many millwrights prefer cogs: they should not revolve more than 4 times in a minute.

ARTICLE 100.

OF THE DRILL.

(See the description, Art. 88.) The pulleys should not be less than 10 inches diameter for meal, and for wheat, more. The case they run in is a deep, narrow trough, say 16 inches deep, and 4 wide, pulleys and strap 3 inches. The rakes are little, square blocks of willow or poplar, or any soft wood, they will not split by driving the nails; they should all be of one size, that each may take an equal quantity; they are nailed to the strap with long, small nails, with broad heads, which are inside the strap; the meal should always be let into them above the centre of the pulley, or at the top of it, to prevent its choking, which it is apt to do, if let in low. The motion should be slow for meal, but may be more lively for wheat.

Directions for using a Hopper-boy.

1. When the meal elevator is set in motion to elevate the meal, the hopper-boy must be set in motion also, to spread and cool it; and as soon as the circle is full, the bolts may be started; the grinding and bolting may likewise be carried on regularly together; which is the best way of working.

2. But if you do not choose to bolt as you grind, turn up the feeding sweepers and let the hopper-boy spread and cool the meal, and rise over it; and when you begin to bolt, turn them down again.

3. If you choose to keep the warm meal separate from the cool, shovel about 18 inches of the outside of the circle, in towards the centre, and turn the end flights, to drive the meal outwards; it will then spread the warm meal outwards, and gather the cool meal into the bolting hopper. As soon as the ring is full with warm meal, rake it out of the reach of the hopper-boy, and let it fill again.

4. To mix tail-flour or bran, &c., with a quantity of meal that is under the hopper-boy, make a hole for it in the meal quite to the floor, and put it in; and the hopper-boy will mix it regularly with the whole.

5. If it do not keep the hopper full, turn the feeding sweeper a little lower, and throw a little meal on the top of the arm, to make it sink deeper into the meal. If the spreading sweepers discharge their loads too soon, and do not trail the meal all around the circle, turn them a little lower; if they do not discharge, but keep too full, raise them a little.

ARTICLE 101.

OF THE UTILITY OF THESE INVENTIONS AND IMPROVEMENTS.

In order to dry the meal in the most rapid and effectual manner, it is evident, that it should be spread as thinly as possible, and be kept in motion from the moment it leaves the stones, until it be cold, that it may have a fair opportunity of discharging its moisture, which will be done more effectually at that time, than after it has grown cold in a heap, and has retained its moisture; this immediate drying does not allow time for insects to deposite their eggs, which, in time, breed the worms that are often found in the heart of barrels of flour well packed; and, by the moisture being expelled more effectually, it will not be so apt to sour. The first great advantage, therefore, is, that *the meal is better prepared for bolting, for packing, and for keeping, in much less time than usual.*

2. *They do the work to much greater perfection,* by cleaning the grain and screenings more effectually, hoisting and bolting over great part of the flour, and grinding and bolting over the middlings, all at one operation, mixing those parts that are to be mixed, and separating such as are to be separated.

3. *They save much meal from being wasted,* if they be well constructed, because there is no necessity for trampling in it, which trails it wherever we walk, nor shoveling it about to raise a dust that flies away, &c. This article of saving will soon pay the cost of making the machinery, and of keeping it in repair afterwards.

4. *They afford more room than they take up,* because the whole of the meal loft that heretofore was little enough to cool the meal on, may be spared for other uses, excepting the circle described by the hopper-boy; and the wheat garners may be filled from one story to another, up to the crane-spout, above the collar-beams; so that a small part of the house will hold an unusual

quantity of wheat, and it may be drawn from the bottom into the elevator, as wanted.

5. *They tend to despatch business*, by finishing as they go; so that there is not as much time expended in grinding over middlings, which will not employ the power of the mill, nor in cleaning and grinding the screenings, they being cleaned every few days, and mixed with the wheat; and as the labour is easier, the miller can keep the stones in better order, and more regularly and steadily at work, especially in the night time, when they frequently stop for want of help; whereas, one man would be sufficient to attend six pair of stones, running (in one house) with well constructed machinery.

6. *They last a long time, with but little expense of repair,* because their motions are slow and easy.

7. *They hoist the grain and meal with less power, and disturb the motion of the mill much less than the old way,* because the descending strap balances the ascending one, so that there is no more power used, than to hoist the grain or meal itself; whereas, in the old way, for every 3 bushels of wheat, which fill a 4 bushel tub with meal, the tub has to be hoisted, the weight of which is equal to a bushel of wheat; consequently, the power used is as 3 for the elevator to 4 for the tubs, which is one fourth less with elevators than tubs; besides, the weight of 4 bushels of wheat, thrown at once on the wheel, always checks the motion; before the tub is up, the stone sinks a little, and the mill is put out of tune every tub-full, which makes a great difference in a year's grinding; this is worthy of notice when water is scarce.

8. *They save a great expense of attendance.* One-half of the hands that were formerly required are now sufficient, and their labour is easier. Formerly, one hand was required for every 10 barrels of flour that the mill made daily; now, one for every 20 barrels is sufficient. A mill that made 40 barrels a day, required four men and a boy; two men are now sufficient.

CHAPTER XII.

BILLS OF MATERIALS TO BE PROVIDED FOR BUILDING AND CONSTRUCTING THE MACHINERY.

ARTICLE 102.

For a Wheat Elevator 43 feet high, with a Strap 4 inches wide.

Three sides of good, firm, white harness-leather.

220 feet of inch pine, or other boards that are dry, of about $12\frac{1}{2}$ inches width, for the cases; these are to be dressed as follows:

86 feet in length, 7 inches wide, for the top and bottom.

86 feet in length, 5 inches wide, with the edges truly squared, for the side boards.

A quantity of inch boards for the garners, as they may be wanted.

Sheet-iron, or a good but of willow wood, for the buckets.

2000 tacks, 14 and 16 ounce size, the largest about half an inch long, for the buckets.

3 lbs. of 8 penny, and 1 lb. of 10 penny nails, for the cases.

2 dozen of large, wood screws, (but nails will do,) for pulley cases.

16 feet of 2 inch plank, for pulleys.

16 feet of ditto, for cog-wheels, and dry pine scantling, $4\frac{1}{2}$ by $4\frac{1}{2}$, or 5 by 5 inches, to give it motion.

Smith's Bill of Iron.

1 double gudgeon $\frac{3}{4}$ inch, (such as fig. 6, Plate VI.,) 5 inches between the shoulders, $3\frac{3}{4}$ inches between the holes, the necks, or gudgeon part, 3 inches.

1 small gudgeon, of the common size, $\frac{3}{4}$ inch thick.

1 gudgeon, an inch thick, (fig. 7,) neck $3\frac{1}{4}$, tang. 10 inches, to be next the upper pulley.

BILLS OF MATERIALS.

2 small bands, 4¼ inches from the outsides.
1 harness-buckle, 4 inches from the outsides, with 2 tongues, of the form of fig. 12.
Add whatever more may be wanting for the gears, that are for giving it motion.

For a Meal Elevator 43 Feet high, Strap 3½ Inches wide, and a Conveyer for two pair of Stones.

270 feet of dry pine, or other inch boards, most of them 11½ or 12 inches wide, of any length, that they may suit to be dressed for the case boards, as follows:
86 feet in length, 6½ inches wide, for tops and bottoms of the cases.
86 feet in length, 4½ inches wide, for the side boards, truly squared at the edges.
The back board of the conveyer trough 15 inches, bottom do. 11 inches, and front 13 inches wide.
Some 2 inch plank for the pulleys and cog-wheels.
Scantling for conveyers 6 by 6, or 5½ by 5½ inches, of dry pine or yellow poplar, (prefer light wood;) pine for shafts, 4½ by 4½, or 5 by 5 inches.
2½ sides of good, pliant, harness leather.
1500 of 14 ounce tacks.
A good, clean butt of willow for buckets, unless the pieces that are left, which are too small for the wheat-buckets, will make the meal-buckets.
4 lbs. of 8 penny, and 1 lb. of 10 penny nails.
2 dozen of large wood screws, (nails will do,) for the pulley cases.

Smith's Bill of Iron.

1 double gudgeon, (such as fig. 4, Plate VI.,) 1½ inches thick, 7½ inches between the necks, 3¼ between the key-holes, the necks 1½ inches long, and the tenons at each end of the same length, exactly square, that the socket may fit every way alike.
2 sockets, one for each tenon, such as appears on one end of fig. 4. The distance between the outside of the straps, with the nails in, must be 5¼ inches; fig. 5 is an end view of it, and the band that drives over

it at the end of the shaft, as they appear on the end of the conveyer.

2 small ¾ inch gudgeons for the other ends of the conveyers.

4 thin bands 5½ inches from the outsides, for the conveyers.

1 gudgeon an inch thick, neck 3¼ inches, and tang. 10 inches, for the shaft in the upper pulley; but if a gudgeon be put through the pulley, let it be of the form of fig. 6, with a tenon and socket at one end, like fig. 4.

1 harness-buckle, 3½ inches from the outsides, with two tongues; such as fig. 12, Plate VI.

Add whatever more small gudgeons and bands may be necessary for giving motion.

For a Hopper-Boy.

1 piece of dry, hard, clean, pine scantling, 4½ by 4½ inches, and 10 feet long, for the upright shaft.

1 piece of dry poplar, soft pine, or other soft, light wood, not subject to crack and split in working, 8 by 2½ inches, 15 or 16 feet long, for the flight arms.

Some 2 inch plank for wheels, to give it motion, and scantling 4½ by 4½ inches for the shafts.

60 flights, 6 inches long, 3 inches wide, and ½ inch at one, and ¼ at the other edge, thinner at the fore than hind end, that they may drive in tight like a dove-tail wedge. These may be made out of green, hard maple, split from sap to heart, and set to dry.

Half a common bed-cord, for a leading line, and balance rope.

Smith's Bill of Iron.

1 stay-iron, C F E, Plate VII. fig. 12. The height from the top of the ring F, to the bottom of the feet C E, is 15 inches; distance of the points of the feet C E 24 inches; size of the legs ½ by ¾ inch; size of the ring F, 1 by ¼ inch, round and smooth inside; 4 inches diameter, the inside corners rounded off, to keep it from cutting the shaft; there must be two little loops, or

eyes, one in each quarter, that the balance rope may be hung to either.

2 screws with thumb-nuts, (that are turned by the thumb and fingers) $\frac{1}{4}$ of an inch thick, and 3 inches long, for the feet of the stay-iron.

2 do. for the end flights, $3\frac{1}{2}$ inches long, rounded $1\frac{1}{2}$ inches next the head, and square $1\frac{1}{4}$ inches next the screw, the round part thickest.

2 do. for the end sweepers, $6\frac{1}{4}$ inches long, rounded 1 inch next the head, $\frac{1}{4}$ inch thick.

2 do. for the hopper sweepers, $8\frac{1}{2}$ inches long, and $\frac{1}{4}$ inch thick, or long nails, with rivet heads, will answer the purpose.

1 step-gudgeon, (fig. 15,) $2\frac{1}{4}$ inches long below the ring, and tang 9 inches, $\frac{3}{4}$ inches thick.

1 plate, 4 by 4, and $\frac{1}{8}$ inch thick, for the step-gudgeon to pass through, (fig. 14.)

1 band for the step-gudgeon, $3\frac{3}{4}$ inches diameter; from the outsides it has to pass through the stay-iron.

1 gudgeon and band, for the top of the shaft, gudgeon $\frac{3}{4}$ inch, band 4 inches diameter, measuring the outside.

The smith can, by the book, easily understand how to make these irons; and the reader may, from these bills of materials, make a rough estimate of the whole expense, which he will find trifling, compared with their utility.

ARTICLE 103.

A MILL FOR CLEANING AND HULLING RICE.

Fig. 2, Plate X. The rice brought to the mill in boats, is to be emptied into the hopper 1, out of which it is conveyed by the conveyer into the elevator at 2, which elevates it into the garner 3, on the third floor; thence it descends into the garner 4, which hangs over the stones 5, and supplies them regularly. The stones are to be dressed with a few deep furrows, with but little draught, and picked full of large holes; they must be set more than the length of the grain apart. The hoop should be lined inside with strong sheet-iron, and this, if punched full of holes, will be thereby improved. The grain is to be kept under the stone as long as necessary; this is effected by forcing it to rise some distance up the hoop, to be discharged through a hole, which is to be raised, or lowered, by a gate sliding in the bottom of it.

The principle by which the grains are hulled, is that of rubbing them against one another, between the stones, with great force; by which means they hull one another without being much broken by the stones. As the grain passes through the stones 5, it should fall into a rolling-screen or shaking-sieve 6, made of wire, with such meshes as will let out all the sand and dust, which may, if convenient, run through the floor into the water; the rice, and most of the heavy chaff, should fall through into the conveyer, which will convey it into the elevator at 2. The light chaff, &c., that does not pass through the sieve, will fall out at the tail, and, if useless, may also run into the water and float away. There may be a fan put on the spindle, above the trundle, to make a light blast, to blow out the chaff and dust, which should be conveyed out through the wall; and this fan may supersede the necessity of the shaking-sieve. The grain and heavy chaff are to be elevated into garner 7; thence they are to descend into garner 8, and pass through the stones 9, which are to be fixed and dressed in the same way as the others, but are to rub the grain harder. The outside of the chaff, from its sharpness, will cut off all the inside hull from the grain, and leave it perfectly clean: as it falls from these last stones, it passes through the wind of the fan 10, fixed on the spindle of the stones 9, which will blow out the chaff and dust, and they then drop into the room 21; the wind should escape through the wall. There is a regulating board that moves on a joint at 21, so as to take all the grain into the conveyer, which will convey it into the elevator at 11, which elevates it into the garner 12, to pass through the rolling screen 13; this should have meshes of three different sizes; first, to take out the dust, which falls into part 17, by itself; secondly, to pass the small rice into apartment 16; the whole grains then fall into garner 14, perfectly clean, and are drawn into barrels at 15. The fan 18 blows out the dust, and lodges it in the room 19, and the wind passes out at 20; the head rice falls at the tail of the screen, and runs into the hopper of the stones 5, to go through the whole operation again. Thus, the whole work is completely performed by the water, with the help of the machinery, taking it from the boat, and operating upon it until it be put into the barrel, without the least manual labour.

Perhaps it may be advantageous to make a few furrows in the edge of the stone, slanting, at an angle of about 30 degrees with a perpendicular line; these furrows will throw up the grain next the stone, on the top of that in the hoop, which will change its position continually; but this, probably, may not be found necessary.

PART THE FOURTH.

On the Process of manufacturing Grain into Flour, as practised by the most skilful Millers in the United States.

CHAPTER XIII.

ARTICLE 104.

THE PRINCIPLES OF GRINDING EXPLAINED, TOGETHER WITH SOME OBSERVATIONS ON LAYING OUT THE FURROWS IN THE STONES WITH A PROPER DRAUGHT.

The end we have in view in grinding the grain, is to reduce it to such a degree of fineness, as is found by experience to fit it to make the best bread; and to put it in such a state, that the flour may be most effectually separated from the bran, or skin of the grain, by means of sifting or bolting. It has been proved by experience, that to grind grain fine with dull mill-stones, will not answer said purpose, because it kills or destroys that quality of the grain, which causes it to ferment and raise in the baking; it also makes the meal so clammy, that it sticks to the cloth, and chokes up the meshes in bolting; hence it appears, that it should be made fine with as little pressure as possible; and it is evident that this cannot be done without sharp instruments. Let us suppose we undertake to operate on one single grain, it seems to accord with reason, that we should first cut it into several pieces, with a sharp instrument, to put it into a state suitable for being passed between two planes, in order to its being reduced to one regular degree of fineness. The planes should have on their faces a number of little sharp edges, to scrape off the meal from the bran, and should be set at such a distance apart as to reduce the meal to the required fineness, and no finer; so that no part can escape unground. The same rules or principles will apply to any quantity that will serve for one grain.

To prepare the stones for grinding to the greatest perfection, we may conclude, therefore, that their faces must be put into such order, that they will first cut the grain into several pieces, and then pass it between them in such a manner, that none can escape without being ground to a certain degree of fineness, whilst, at the same time, it scrapes the meal off cleanly from the bran or skin.

The best way that I have yet found to effect this, is (after the stones are faced with the staff, and the pick,) to grind between them a few quarts of fine, sharp sand; this will face them to fit each other so exactly, that no meal can pass them without being ground; this is also the best way of sharpening all the little edges on the face, that are formed by the pores of the stone; instead of sand, water may be used, the stones then face each other; they will then scrape the meal off of the bran, without too much pressure being applied. But as the meal will not pass from the centre to the periphery or verge of the stones, with sufficient rapidity, without some assistance, there must be a number of furrows, to aid it in its egress; and these furrows must be set with such a draught that the meal will not pass too far along them at once, without passing over the land, or plane, lest it should get out unground. They should also be of sufficient depth, to permit air enough to pass through the stones to carry out the heat generated by the friction of grinding; but if they have too much draught, they will not bear to be deep, or the meal will escape along them unground. These furrows ought to be made sharp at the feather edge, which is the hinder edge of the furrow, and the foremost edge of the land; this serves the purpose of cutting down the grain; they should be more numerous near the centre, because there the office of the stone is to cut the grain, and near the periphery the office of the two planes is to reduce the flour to the required fineness, and scrape the bran clean, which is effected by the edges, formed by the numerous little pores with which the burr stone abounds. We must consider, however, that it is not best to have the stones too sharp near the eye, because they then cut the bran too fine. The stones incline to keep open near the eye, unless they be too close. If they be porous, (near the eye,) and will keep open without picking, they will remain a little dull, which will flatten the bran, without cutting it too much: but if they be soft next the eye, they will keep too open, and that part of the stone will be nearly useless; they, therefore, should be very hard and porous.

It is also necessary, that the face of the stone be dressed in such a form, as to allow room for the grain, or meal, in every stage of its passage between the stones. In order to understand this, let us conceive the stream of wheat entering the eye of the stone, to be about the thickness of a man's finger, but instantly spreading every way over the whole face of the stone; this stream must, therefore, get thinner, as it approaches the periphery, where it would be thinner than a fine hair, if it did not pass slower as it becomes finer, and if the stones were not kept apart by the bran; for this reason, the stones must be so dressed, that they will not touch at the centre, within about a 16th or 20th part of an inch, but get closer gradually, till within about 10 or 12 inches from the verge of the stone, proportioned to the diameter, and from that part out they must fit nicely together. This close part is called the flouring of the stone. The furrows should be deep near the centre, to admit wheat in its chopped state, and the air, which tends to keep the stones cool.

ARTICLE 105.

OF THE DRAUGHT NECESSARY TO BE GIVEN TO THE FURROWS OF MILL-STONES.

From these principles and ideas, and the laws of central forces, explained at Art. 13, I form my judgment of the proper draught of the furrows, and the manner of dress; points in which I find but few of the best millers to agree; some prefer one kind, and some another, which shows that this necessary part of the miller's art is not yet well understood. In order to illustrate this matter, I have constructed fig. 3, Plate XI. A B represents the eight quarter, C D the twelve quarter, and E A the central dress. Now, we observe that in the eight quarter dress, the short furrows at F have about five times as much draught as the long ones, and cross one another like a pair of shears opened so wide that they will drive all before them, and cut nothing; and if these furrows be deep they will drive out the meal as soon as it gets into them, and thereby make much coarse meal, such as middlings and ship stuff or carnel; the twelve quarter dress appears to be better; but the short furrows at G have about four times as much draught as the long ones, the advantage of which I cannot perceive, because if we have once found the draught that is right for one furrow, so as to cause the meal to pass through the stone in a proper time, it appears reasonable that the draught of every other furrow should be equal to it.

In the central dress E A, the furrows have all one draught, and if we could once determine exactly how much is necessary, I have no doubt we should find this to be the correct plan; and I apprehend that we shall find the best draught to be in a certain proportion to the size and velocity of the stone; because the centrifugal force that the circular motion of the stones gives the meal, has a tendency to move it outward, and this force will be in inverse proportion to the diameter of the stones, their velocities being the same, by the fourth law of circular motion. E e is a furrow of the running stone, and we may see by the figure, that the furrows cross one another at the centre at a much greater angle than near the periphery, which I conceive to be right, because the centrifugal force is much less towards the centre than near the periphery. But we must also consider, that the grain, whole or but little broken, requires less draught and centrifugal force to send it out, than it does when ground fine; which shows that we must not, in practice, follow the theory laid down in Art. 13, respecting the laws of circular motion and central forces; because the grain, as it is ground into meal, is less affected by the central force to drive it out; the angles, therefore, with which the furrows cross each other, must be greater near the verge or skirt of the stone, and less near its centre than would be assigned by that theory; and what ought to be the amount of this variation is a question which practice has not yet determined.

From the whole of my speculations on this difficult subject, added to observations on my own and others' practice and experience, I propose the following rule for laying out a five foot mill-stone. (See fig. 1, Plate XI.)

1. Describe a circle with 3 inches, and another with 6 inches radius, round the centre of the stone.
2. Divide the 3 inches space between these two circles into 4 spaces, by 3 circles equi-distant; call these five circles draught circles.
3. Divide the stone into 5 parts, by describing 4 circles equi-distant between the eye and the verge.
4. Divide the circumference of the stone into 18 equal parts, called quarters.
5. Then take a straight-edged rule, lay one end at one of the quarters at 6, at the verge of the stone, and the other end at the outside draught circle, 6 inches from the centre of the stone, and draw a line for the furrow from the verge of the stone to the circle 5 : then shift the rule from draught circle 6, to the draught circle 5, and continue the furrow line towards the centre, from circle 5 to 4 : then shift in the rule to draught circle 4, and continue to 3; shift to 3, and continue to

2; shift to 2, and continue to 1, and the curve of the furrow is formed, as 1—6 in the figure.

6. To this curve form a pattern, by which to lay out all the remainder.

The furrows with this curve will cross each other with the following angles, shown fig. 1,

at circle 1, which is the eye of the stone, at 75 degrees angle,
— 2 - - 45 ———
— 3 - - 35 ———
— 4 - - 31 ———
— 5 - - 27 ———
— 6 - - 23 ———

These angles, as shown by the lines G r, H r, G s, H s, &c. &c., will I think, do well in practice, will grind smooth, and make but little coarse meal, &c.

Supposing the greatest draught circle to be 6 inches radius, then, by theory, the angles would have been
at circle 1 - - - 138 degrees angle.
— 2 - - - 69 ———
— 3 - - - 46 ———
— 4 - - - 35,5 ———
— 5 - - - 27,5 ———
— 6 - - - 23 ———

If the draught circle had been 5 inches radius, and the furrows straight, the angles would then have been at

circle. degrees.
1 about 180

And 6 inches from centre, as shown by lines G 1, H 1, — 110

2 — 60
3 — 38
4 — 29
5 — 23
6 — 18

Here, the angles near the centre are much too great to grind, and they will push the grain before them; to remedy this disadvantage, take the aforesaid rule, which forms the furrows, as shown at 6—7, fig. 1, which is 4 of 18 qrs. H 8 represents a furrow of the runner, showing the angles where they cross those of the bed-stone, in every part. Here I have supposed the extremes of the draught of 6 inches for the verge, and 3 inches for the eye of the stone, to be right for a stone 5 feet diameter, revolving 100 times in a minute; but of this I am, by no means, certain. Yet, by experience the extremes may be ascertained for stones of all sizes, with different velocities; no kind of dress of which I can conceive, appears to me likely to be brought to perfection excepting this, and it certainly appears, both by reason and by inspecting the figure, that it will grind the smoothest of all the different kinds exhibited in the plate.

The principle of grinding is partly that of shears, clipping; the planes of the face of the stones serving as guides to keep the grain in the edge of the shears, the furrows and pores forming the edges; if the shears cross one another at too great an angle, they cannot cut; it follows, therefore, that all the strokes of the pick should be parallel to the furrows.

To give two stones of different diameters the same draught, we must make their draught circles in direct proportion to their diameters; then the furrows of the upper and lower stones of each size, will cross each other with equal angles in all proportional distances, from their centres to their peripheries. But when we come to consider that the mean circles of all stones are to have nearly equal velocities, and that their centrifugal forces will be in inverse proportion to their diameters, we must perceive that small stones must have much less draught than large ones, in proportion to their diameters. (See the proportion for determining the draught, Art. 13.)

It is very necessary that the true draught of the furrows should be determined to suit the velocity of the stone, because the centrifugal force of the meal will vary, as the squares of the velocity of the stone, by the 5th law of circular motion. But the error of the draught may be corrected, in some measure, by the depth of the furrows. The less the draught, the deeper must be the furrow; and the greater the draught, the shallower the furrow, to prevent the meal from escaping unground;

but if the furrows be too shallow, there will not a sufficient quantity of air pass through the stones to keep them cool. But in the central dress the furrows meet so near together, that they cut the stone too much away at the centre, unless they be made too narrow; I, therefore, prefer what is called the quarter dress, but divided into so many quarters, that there will be little difference between the draught of the furrows; suppose 18 quarters in a 5 foot stone, then each quarter takes up about $10\frac{1}{2}$ inches of the circumference of the stone, which suits for a division into about 4 furrows and 4 lands, if the stone be close; but, if it be open, 2 or 3 furrows to each quarter will be enough. This rûle will give 4 feet 6 inch stones, 16; and 5 feet 6 inch stones, 21; and 6 feet stones, 23 quarters. But the number of quarters is not very important; it is better, however, to have too many than too few. If the quarters be few, the disadvantage of the short furrows crossing at too great an angle, and throwing out the meal too coarse, may be remedied, by making the land widest next the verge, thereby turning the furrows towards the centre, when they will have less draught, as in the quarter H I, fig. 3.

ARTICLE 106.

OF FACING MILL-STONES.

The burr mill-stones are generally left in such face by the maker, that the miller need not spend much labour and time on them with picks, before he may hang them, and grind them together with water or dry sand. After they have been ground together for a sufficient length of time, they must be taken up, and the red staff tried over their faces,* and if it touch in circles, the project-

---

* The red staff is made longer than the diameter of the stones, and three inches thick on the edge, which is made perfectly straight; on this is rubbed red clay, mixed with water, which shows the highest parts of the faces of the stones, when rubbed over them, by leaving the red on those high parts.

ing parts should be well cracked with picks, and again ground with a small quantity of water or sand; after this, take them up, and try the staff on them, picking off the red parts as before, and repeat this operation, until the staff will touch nearly alike all the way across, and until the stone comes to a face in every part, that the quality thereof may plainly appear; then, with a red or black line, proceed to lay out the furrows, in the manner determined upon, from the observations already laid down in the last article. After having a fair view of the face and quality of the stone, we can judge of the number of furrows most suitable, observing that where the stone is most open and porous, fewer furrows will be wanted; but where it is close and smooth, the furrows ought to be more numerous, and both they and the lands narrow, (about 1⅛ inches wide,) that they may form a greater number of edges, to perform the grinding. The furrows, at the back, should be made nearly the depth of the thickness of a grain of wheat, but sloped up to a feather edge, not deeper than the thickness of a finger-nail;* this edge is to be made as sharp as possible, which cannot be done without a very sharp, hard pick. When the furrows are all made, try the red staff over them, and if it touch near the centre, the marks must be quite taken off about a foot next to it, but observing to crack lighter the farther from it, so that when the stones are laid together, they will not touch at the centre, by about one-twentieth part of an inch, and close gradually, so as to touch and fit exactly, for about 10 or 12 inches from the verge. If the stones be now well hung, having the facing

* For the form of the bottom of the furrow, see fig. 3, Plate XI. The curve line e b shows the bottom, b the feather edge, and e the back part. If the bottom had been made square at the back, as at e, the grain would lie in the corner, and by the centrifugal force, would work out along the furrows without passing over the lands, and a part would escape unground. The back edge must be sloped for two reasons: 1st, that the meal may be pushed on to the feather edge; 2dly, that the furrow may grow narrower, as the faces of the stones wear away, to give liberty to sharpen the feather edge, without making the furrows too wide. Fig. 5 represents the face of two stones, working together, the runner moving from a to d. When the furrows are just over each other, as at a, there is room for a grain of wheat; when they move to the position of b, it is flattened, and at c, is clipped in two by the feather edges, and the lands or planes operate on it, as at d.

and furrowing neatly done, they will be found in the most excellent order that they can possibly be put, for grinding wheat, because they are in good face, fitting so neatly together that the wheat cannot escape unground, and all the edges being perfectly sharp, so that the grain can be ground into flour, with the least pressure possible.

ARTICLE 107.

OF HANGING MILL-STONES.

If the stone have a balance-ryne it is an easy matter to hang it, for we have only to set the spindle perpendicular to the face of the bed-stone; which is done by fastening a staff on the cock-head of the spindle, so that the end may reach to the edge of the stone, and be near the face. In this end we put a piece of elastic material, such as of whalebone or quill, so as to touch the stone, that, on turning the trundle head, the quill may move round the edge of the stone, and when it is made to touch alike all the way round, by altering the wedges of the bridge, the stone may be laid down and it will be ready hung;*

---

* But here we must observe, whether the stone be of a true balance, as it hangs on the cock-head, and, if not, it must be truly balanced, by running lead into the lightest side. This ought to be carefully attended to by the maker, because the stone may be made to balance truly when at rest; yet, if every opposite part do not balance each other truly, the stone may be greatly out of balance when in motion; and this is the reason why the bush of some stones can be kept tight but a few hours, while others will keep so for several months, the spindles being good, and the stones balanced when at rest. The reason why a stone that is balanced at rest, will sometimes not be balanced in motion, is, that if the upper part be heaviest on one side, and the lower part be heaviest on the other side of the centre, the stone may balance at rest, yet, when set in motion, the heaviest parts draw outwards most, by the centrifugal force, which will put the stone out of balance while in motion: and if the stone be not round, the parts farthest from the centre will have the greatest centrifugal force, because the centrifugal force is as the square of the distance from the centre. The neck of the spindle will wear next the longest side, and the bush get loose: and this argues in favour of a stiff ryne. The best method that I have seen for hanging stones with stiff horned rynes, is as follows: Fix a screw to each horn to regulate by, which is done thus—after the horns are bedded, sink under each horn a strong burr, through which the screw is to pass from the neck of the stone, and fasten them in with lead; then, after the stone is laid down, put in the screws from the top of the stone, screwing them

but if we have a stiff-ryne, it will be much more difficult, because we have not only to fix the spindle perpendicular to the face of the bed-stone, but we must set the face of the runner perpendicular to the spindle, and all this must be done with the greatest exactness, because the ryne, being stiff, will not give way to suffer the runner to form itself to the bed-stone, as will the balance-ryne.

The bed of the ryne being first carefully cleaned out, the ryne is put into it and tied, until the stone be laid down on the cock-head; then we find the part that hangs lowest, and, by putting the hand thereon, we press the stone down a little, turning it about at the same time, and observing whether the lowest part touches the bed-stone equally all the way round; if it do not, it is adjusted by altering the wedges of the bridge-tree, until it touches equally, and then the spindle will stand perpendicular to the face of the bed-stone. Then, to set the face of the runner perpendicular, or square, to the spindle, we remain in the same place, turning the stone, and pressing on it at every horn of the ryne, as it passes, and observing whether the runner will touch the bed-stone equally at every horn, which, if it do not, we strike with an iron bar on the horn that bears the stone highest, which, by its jarring, will settle itself better into its bed, and thereby let the stone down a little in that part; but, if this be not sufficient, there must be paper put on the top of the horn that lets the stone too low; observing to mark the high horns, that when the stone is taken up, a little may be taken off the bed, and the ryne will soon become so neatly bedded, that the stone will hang very easily. But I have always found that every time the stone is taken up, the bridge is moved a little out of place; or, in other words, the spindle moved a little from its true perpendicular position, with respect to the face of the bed-stone, which is a great objection to the stiff horn ryne; for if the spindle be thrown but very little out of place, the stones cannot come together equally;

till the points bear tightly on the horn: then proceed to hang the stone, which is very easily done, by turning the screws.

whilst, with a balance ryne, if it be considerably out of place, it will do but little or no injury in the grinding, because the running stone has liberty to form itself to the bed-stone.

ARTICLE 108.

OF REGULATING THE FEED AND WATER IN GRINDING.

The stone being well hung, proceed to grind, and, when all things are ready, draw as much water as is judged to be sufficient; then observe the motion of the stone, by the noise of the damsel, and feel the meal; if it be too coarse, and the motion too slow, give less feed, and it will grind finer, and the motion will be quicker; if it yet grind too coarse, lower the stone; then, if the motion be too slow, draw a little more water; but if the meal feel to be too low ground, and the motion right, raise the stone a little, and give a little more feed. If the motion and feed be too great, and the meal be ground too low, shut off part of the water. But if the motion be too slow, and the feed be too small, draw more water.

The miller must here remember, that there is a certain portion of feed that the stones will bear and grind well, which will be in proportion to their size, velocity, and sharpness, and, if these be exceeded, there will be a loss, as the grinding will not then be well done. But no rule can be laid down, to ascertain the proper portion of feed, a knowledge depending upon that skill which is only to be attained by practice; as is also the art of judging of the right fineness. I will, however, lay down such rules and directions as may be of some assistance to the beginner.

ARTICLE 109.

RULE FOR JUDGING OF GOOD GRINDING.

Catch your hand full of meal as it falls from the stones, and feel it lightly between your fingers and thumb; and if it feel smooth, and not oily or clammy, and will not stick much to the hand, it shows it to be fine enough, and the stones to be sharp. If there be no lumps to be felt larger than the rest, but all is of one fineness, it shows the stones to be well faced, and the furrows not to have too much draught, as none has escaped unground.

If the meal feel very smooth and oily, and stick much to the hand, it shows it to be too low ground, hard pressed, and the stones dull.

If it feel in part oily, and in part coarse and lumpy, and will stick much to the hand, it shows that the stones have too much feed; or, that they are dull, and badly faced; or have some furrows that have too much draught; or are too deep, or, perhaps, too steep at the back edge, as part has escaped unground, and part is too much pressed, and low.

Catch your hand full, and holding the palm up, shut it briskly; if the greatest quantity of the meal fly out and escape between your fingers, it shows it to be in a fine and lively state, the stones sharp, the bran thin, and that it will bolt well: But the greater the quantity that stays in the hand, the more faulty is the flour.

Catch a handful of meal in a sieve, and sift the meal clean out of the bran; then feel it, and if it be soft and springy, or elastic, and, also, feel thin, with but little sticking to the inside of the bran, and no pieces found much thicker than the rest, the stones are shown to be sharp, and the grinding well done.*

* Instead of a sieve, you may take a shovel and hold the point near the stream of meal, and it will catch part of the bran, with but little meal mixed with it; which may be separated by tossing it from one hand to the other, wiping the hand at each toss.

But if it be broad and stiff, and the inside white, it is a sure sign that the stones are dull, or over-fed. If you find some parts that are much thicker and harder than the rest, such as half or quarter grains, it shows that there are some furrows that have too much draught, or are too deep, or steep, at the back edge; else, that you are grinding with less feed than the depth of the furrows, and velocity of the stone will bear.

ARTICLE 110.

OF DRESSING AND SHARPENING THE STONES WHEN DULL.

When the stones get dull they must be taken up, that they may be sharpened; to do this in the best manner, we must be provided with sharp, hard picks, with which the feather edge of the furrows are to be dressed as sharp as possible; which cannot be done with soft or dull picks. The bottoms of the furrows are likewise to be dressed, to keep them of the proper depth; but here the dull picks may be used.* The straight staff must now, also, be run over the face carefully, and if there be any parts harder or higher than the rest, the red will be left on them; they must be cracked lightly, with many cracks, to make them wear as fast as the softer parts, in order to keep the face good. These cracks also form the edges that help to clean the bran; and the harder and closer the stone, the more numerous are they to be. They are to be made with a very sharp pick, parallel to the furrows; the damper the grain, the more the stone is to be cracked; and the drier and harder it is, the smoother must the face be. The hard smooth places which glaze, may be made to wear more evenly, by striking them, either

---

* To prevent the steel from striking your fingers, take a piece of leather about 5 by 6 inches square; make a hole through the middle, and put the handle of the pick through it, keeping it between your hands and the pick, making a loop in the lower edge, through which put one of your fingers, to keep up the lower part from the stone.

with a smooth or rough-faced hammer, many light strokes, until a dust begins to appear, which frets the flinty part, and makes it softer and sharper. The stone will never be in the best order for cleaning the bran, without first grinding a little sand, to sharpen all the little edges formed by the pores of the stone; the same sand may be used several times. The stones may be sharpened without being taken up, or even stopped; to do this, take half a pint of sand, and hold the shoe from knocking, to let them run empty; then pour in the sand, and this will take the glaze off the face, and whet up the edges so that they will grind considerably better: this ought to be often done.*

Some are in the practice of letting stones run for months without being dressed; but I am well convinced that those who dress them well twice a week, are fully paid for their trouble.

ARTICLE 111.

OF THE MOST PROPER DEGREE OF FINENESS FOR FLOUR.

As to the most proper degree of fineness for flour, millers differ in their opinion; but a great majority, and many of the most experienced, and of the best judgment, agree in this; that if the flour be made very fine, it will be killed, (as it is termed,) so that it will not rise or ferment so well in baking; but I have heard many good millers give it as their opinion, that flour cannot be made too fine, if ground with sharp, clean stones, provided they be not suffered to rub against each other; and some of those millers do actually reduce almost all the meal they get out of the wheat into superfine flour; by which means they have but two kinds; namely, superfine flour, and horse-feed; which is what is left after the

---

* Care should be taken to prevent the sand from getting mixed with the meal it should be caught in some vessel, the stone being suffered to run quite empty; the small quantity that will remain in the stone will not injure the flour.

flour is made, and is not fit to make even the coarsest kind of ship-bread.

To test the properties of the finest flour, I contrived to catch so much of the dust of that which was floating about in the mill, as made a large loaf of bread, which was raised with the same yeast, and baked in the same oven, with other loaves, that were made out of the most lively meal; the loaf made of the dust of the flour was equally light, and as good, if not better, than any of the others; it was more moist, and pleasant to the taste, though made of flour that, from its fineness, felt like oil.

I conclude, therefore, that it is not the degree of fineness that destroys the life of the flour, but the degree of heat, produced by the too great pressure applied in grinding; and that flour may be reduced to the greatest degree of fineness, without injuring the quality, provided it be done with sharp, clean stones, and with little pressure.

CHAPTER XIV.

ARTICLE 112.

OF GARLIC, WITH DIRECTIONS FOR GRINDING WHEAT MIXED THEREWITH; AND FOR DRESSING THE STONES SUITABLE THERETO.

In many parts of America there is a species of onion, called garlic, that grows spontaneously with the wheat. It bears a head resembling a seed onion, which contains a number of grains about the size of a grain of wheat, but somewhat lighter.* It is of a glutinous texture, and ad-

---

* The complete separation of this garlic from the wheat, is so difficult, that it has hitherto baffled all our art. Those grains that are larger, and those that are smaller than the wheat, can be separated by screens; and those that are much lighter, may be blown out by fans; but those that are of the same size, and nearly of the same weight, cannot be separated without putting the wheat in water, where the wheat will sink, and the garlic swim. But this method is too tedious for the miller to practise, except it be once a year, to clean up the headings, rather than lose the wheat that is mixed with the garlic, which cannot be otherwise sufficiently separated. Great care should be taken by the farmers to prevent

heres to the stone, in such a manner as apparently to blunt the edges, so that they will not grind to any degree of perfection. We are, therefore, obliged to take the runner up, and wash the glaze off with water, scrubbing the faces with stiff brushes, and drying up the water with cloths or sponges; this laborious operation must be repeated twice, or perhaps four times in 24 hours, if there be about 10 grains of garlic in a handful of wheat.

To put the stones in the best order to grind garlicky wheat, they must be cracked roughly all over the face, and dressed more open about the eye; they then break the grains of garlic less suddenly, giving the glutinous substance of the garlic more time to incorporate itself with the meal, and preventing its adherence to the stone. The rougher the face, the longer will the stones grind, because the more time will the garlic be in filling all the edges.

The best method that I have yet discovered for manufacturing garlicky wheat is as follows; namely :—

First, clean it over several times, in order to take out all the garlic that can be separated by the machinery, (which is easily done if you have a wheat elevator well fixed, as directed in Art. 94, Plate IX.) then chop or half grind it; which will break the garlic (it being softer than the wheat;) the moisture will then diffuse itself through the chopped wheat, so that it will not injure the stones so much, in the second grinding. By this means a considerable quantity can be ground, without taking up the stones. The chopping may be done at the rate of 15 or 20 bushels in an hour, and with but little trouble or loss of time, provided there be a meal elevator that will hoist it up to the meal loft, from whence it may descend to the hopper by spouts, to be ground a second time, when it will grind faster than if it had not been chopped. Great care should be taken, that it be not chopped so fine that it will not feed by the knocking of the shoe; as, likewise, that it be not too coarse, lest the

---

this troublesome thing from getting root in their farms; which if it does, it will be almost impossible ever to root it out again, as it propagates both by seed and root, and is very hardy.

garlic be not sufficiently broken. If the chopped grain could lie a considerable time, the garlic would dry, and it would grind much better.

But, although every precaution be taken, if there be much garlic in the wheat, the bran will not be well cleaned; besides which, there will be much coarse meal made, such as middlings, and stuff, which will require to be ground over again, in order to make the most profit of the grain; this I shall treat of in the next chapter.

CHAPTER XV.

ARTICLE 113.

OF GRINDING THE MIDDLINGS OVER, AND, IF NECESSARY, THE STUFF, AND BRAN OR SHORTS, TO MAKE THE MOST OF THEM.

ALTHOUGH we may grind the grain in the best manner we possibly can, so as to make any reasonable despatch, there yet will appear in the bolting, a species of coarse meal, called middlings, and stuff, a quality between superfine and shorts, which will contain a portion of the best part of the grain: but in this state they will make very coarse bread, and, consequently, will command but a low price. For this reason, it is oftentimes profitable to the miller to grind and bolt them over again, and to make them into superfine flour, and fine middlings; this may easily be done by proper management.

The middlings are generally hoisted by tubs, and laid in a convenient place on the floor, in the meal-loft, near the hopper-boy, until there is a large quantity gathered: when the first good opportunity offers, it is bolted over without any bran or shorts mixed with it, in order to take out all that is already fine enough to pass through the superfine cloth. The middlings will pass through the middlings' cloth, and will then be round and lively, and in a state fit for grinding, being freed from the fine part that would have prevented it from feeding freely. The small specks of bran that were before mixed with it, being lighter than the rich round part, will not pass through the middlings' cloth, but will pass on to the stuff's cloth. The middlings will, by this means, be richer than before, and when made fine, may be mixed with the ground meal, and bolted into superfine flour.

The middlings may now be put into the hanging garner, over the hopper of the stones, out of which it will run into the hopper, and keep it full, as does the wheat, provided the garner be rightly constructed, and a hole about 6 by 6 inches be made for it to issue out at. There must be a rod put through the bar that supports the upper end of the damsel, the lower end of which must reach into the eye of the stone, near to the bottom, and on one side thereof, to prevent the meal from sticking in the eye, which, if it do, it will not feed. The hole in the bottom of the hopper must not be less than four inches square. Things being thus prepared, and the stones being sharp and clean, and nicely hung, draw a small quantity of water, (for meal does not require above one-tenth part that grain does) taking great care to avoid pressure, because the bran is not now between the stones, to prevent their coming too closely together. If you lay on as much weight as when grinding grain, the flour will be killed; but if the stones be well hung, and it be pressed lightly, the flour will be lively, and will make much better bread, without being bolted, than it would before it was ground. As fast as it is ground, it may be elevated and bolted; but a little bran will now be necessary to keep the cloth open; and all that passes through the superfine cloth in this operation, may be mixed with what passed through in the first bolting of the middlings, and be hoisted up, and mixed (by the hopper-boy) regularly with the ground meal, and bolted into superfine flour, as directed Art. 89.*

---

* All this trouble and loss of time may be saved by a little simple machinery; namely: As the middlings fall by the first bolting, let them be conveyed into the eye of the stone, and ground with the wheat, as directed Art. 89, Plate VIII; by which means, the whole thereof may be made into superfine flour, without any

The stuff, which is a degree coarser than middlings, if it be too poor for ship bread, and too rich to feed cattle on, is to be ground over in the same manner as the middlings. But if it be mixed with fine flour, (as it sometimes is,) so that it will not feed freely, it must be bolted over first; this will take out the fine flour, and, also the fine specks of bran, which, being lightest, will come through the cloth last. When it is bolted, the part that passes through the middlings' and stuff's parts of the cloth, are to be mixed and ground together; by this means the rich particles will be reduced to flour, and, when bolted, will pass through the finer cloths, and will make tolerably good bread. What passes through the middlings' cloth will make but indifferent ship-bread, and what passes through the ship-stuff's cloth, will be what is called brown-stuff, roughings, or horse-feed.

The bran and shorts seldom are worth the trouble of grinding over, unless the stones have been very dull, or the grinding been but slightly performed, or the wheat very garlicky. When it is done, the stones must be very sharp, and more water and pressure are required, than in grinding grain. The flour, thus obtained, is generally of an indifferent quality, being made of that part of the grain that lies next the skin, and a great part of it is the skin itself, cut fine.*

---

loss of time, or danger of being too hard pressed for want of the bran, to keep the stones apart. This mode I first introduced, and several others have since adopted it.

*The merchant miller is to consider, that there is a certain degree of closeness or perfection that he is to aim at in manufacturing, which will yield him the greatest profit possible, in a given time. And this degree of care and perfection will vary with the prices of wheat and flour, so that what would yield the greatest profit at one time, would sink money at another; because, if the difference in the price of wheat and of flour be but little, then we must make the grain yield as much as possible, to obtain any profit. But if the price of flour be much above that of the wheat, then we had best make the greatest despatch, even if we should not do it so well, in order that the greater quantity may be done while these prices last; whereas, if we were to make such a despatch when the price of flour was but little above that of wheat, we should sink money.

CHAPTER XVI.

ARTICLE 114.

OF THE QUALITY OF MILL-STONES, TO SUIT THE QUALITY OF THE WHEAT.

It has been found, by experience, that different qualities of wheat require different qualities of stones, to grind to the highest perfection.

Although there be several species of wheat, of different qualities; yet, with respect to the grinding, we may divide them into three kinds only, namely :—

A TABLE

Showing the Product of a Bushel of Wheat of different weights and qualities, ascertained by Experiments in grinding parcels.

Weight per bushel.	Superfine flour.	Tail flour and middlings.	Ship stuff.	Bread stuff, shorts and bran.	Screenings and loss in grinding.	Proof.	Quality of the grain.
lbs.	lbs.	lbs.	lbs.	lbs.	lbs.	lbs.	
59.5	38.5	3.68	2.5	13.1	1.72	59.5	White wheat, clean.
59	40.23	3.65	2.12	12	1	59	Do. do. well cleaned.
60	38.7	3.6	1.61	8.52	7.57	60	Red do. not well cleaned.
61	39.7	5.68	2.4	9.54	3.68	61	White do. mixed with green garlic.
56	35.81	5	1.85	7.86	5.48	56	White do. very clean.
59.25	35.26	4.4	1.47	11.33	6.79	59.25	Red do. with some cockle and light grains.

If the screenings had been accurately weighed, and the loss in weight occasioned by the grinding ascertained, this table would have been more interesting. A loss of weight does take place by the evaporation of the moisture by the heat of the stones in the operation.

The author conceived that if a complete separation of the skin of the wheat from the flour could be effected, and the flour be reduced to a sufficient degree of fineness, it might all pass for superfine; and having made the experiments in the table, he effected such improvements in the manufacture, by dressing the mill-stones to grind smooth; and, by means of the machinery which he invented, returning the middlings into the eye of the stone, to be ground over with the wheat, and elevating the tail-flour to the hopper-boy, to be bolted over again, &c., that in making his last 2000 barrels of superfine flour, he left no middlings or ship-stuff, which was not too poor for any kind of bread, excepting some small quantities which were retained in the mill; and the flour passed the inspection with credit. Others have since pursued the same principles, and put them more fully and completely into operation.

1. The dry and hard.
2. The damp and soft.
3. Wheat that is mixed with garlic.

When the grain that is to be ground is dry and hard, such as is raised on high and clayey lands, threshed in barns, and kept dry,* the stones for grinding it should be of that quality of the burr, that is called close and hard, with few large pores, in order that they may have more face. The grain being brittle and easily broken into pieces, requires more face, or plane parts, (spoken of in Art. 104,) to reduce it to the required fineness, without cutting the skin too much.

When the grain that is to be ground is somewhat damp and soft, such as is raised on a light sandy soil, is trodden out on the ground, and is carried in the holds of ships to market, which tends to increase the dampness, the stones should be more open and porous, because the grain is tough, difficult to be broken into pieces, and requires more sharp edges, and less face (or plane surface,) to reduce it to the required fineness.† (See Art. 104.)

When there is garlic, or wild onion, (mentioned Art. 111,) mixed with the wheat, the stones require to be open, porous, and sharp; because the glutinous substance of the garlic adheres to the face of the stones, and blunts the edges; by which means little can be ground before the stones get so dull that they will require to be taken up and sharpened; and the more porous and sharp the stones are, the longer they will run, and grind a larger quantity without getting dull. There is a quality of the burr stone which may be denominated mellow or soft,

* Such wheat as is produced by the mountainous and clay lands of the country, distant from the sea and tide waters, is generally of a brownish colour, the grain appearing flinty, and sometimes the inside a little transparent, when cut by a sharp knife. This transparent kind of wheat is generally heavy, and of a thin skin, and will make as white flour, and as much of it, as the whitest grain.

† Such is the wheat that is raised in all the low, level, and sandy lands, of countries near the sea and tide waters of America, where it is customary to tread out their wheat on the ground by horses; and where it sometimes gets wet by rain and dew, and the dampness of the ground. This grain is naturally of a fairer colour, and softer; and, when broken, the inside is white, which shows it to be nearer to a state of pulverization; it is more easily reduced to flour, and will not bear so much pressure as the grain that is raised on high and clay lands; or such that, when broken, appears solid and transparent.

to distinguish it from the kind which is hard and flinty; these are not so subject to glaze on the face; and it is found by experience that stones of this texture will grind at one dressing three or four times as much grain mixed with garlic, as those of a hard quality.* (See Art. 111.)

CHAPTER XVII.

ARTICLE 115.

OF BOLTING REELS AND CLOTHS, WITH DIRECTIONS FOR BOLTING AND INSPECTING THE FLOUR.

THE effect we wish to produce by sifting, or bolting, is to separate the different qualities of flour from each other, and from the skin, shorts, or bran; let us now inquire which are the most proper means of attaining this end.

* It is very difficult to convey my ideas of the quality of the stones to the reader, for want of something with which to measure or compare their degrees of porosity or closeness, hardness or softness. The knowledge of these different qualities is only to be attained by practice and experience; but I may observe, that pores in the stone, larger in diameter than the length of a grain of wheat, are injurious; for how much soever they are larger, is so much loss of the face, because it is the edges that do the grinding; therefore, all large pores in stones are a disadvantage. The greater the number of pores in the stone, (so as to leave a sufficient quantity of touching surface, to reduce the flour to a sufficient degree of fineness,) the better.

Mill-stone makers ought to be acquainted with the true principles on which grinding is performed, and with the art of manufacturing grain into flour, that they may be judges of the quality of the stones suitable to the quality of the wheat, of different parts of the country; also, of the best manner of disposing of the different pieces of stone, of different qualities, in the same mill-stone, according to the office of the several parts, from the centre to the verge of the stone. (See Art. 104.)

Mill-stones are generally but very carelessly and slightly made; whereas, they should be made with the utmost care, and to the greatest nicety. The runner must be balanced exactly on its centre, and every corresponding opposite part of it should be of equal weight, or else the spindle will not keep tight in the bush; (see Art. 107.)—and if it is to be hung on a balance ryne, it should be put in at the formation of the stone, which should be nicely balanced thereon.

But, above all, the kind of stone should be most attended to, that no piece of an unsuitable quality for the rest be put in; it being known to all experienced millers, that they had better give a high price for an extraordinary good pair, than to have an indifferent pair for nothing.

Observations concerning Bolting.

1. Suppose that we try a sieve, the meshes of which are so large, as to let all the bran and meal through: it is evident that we could never thus attain the end proposed by the use thereof.

2. Suppose we try a finer sieve, that will let all the meal through, but none of the bran; by this we cannot separate the different qualities of the flour.

3. We provide as many sieves of the different degrees of fineness, as we intend to make different qualities of flour; and which, for distinction, we name—Superfine, Middlings, and Carnel.

The superfine sieve we make of meshes, so fine as to let through the superfine flour, but none of the middlings: the middlings' sieve, so fine as to let the middlings pass through, but none of the carnel: the carnel sieve, so fine as to let none of the shorts or bran pass through.

Now it is evident, that if we would continue the operation long enough, with each sieve, beginning with the superfine, that we might effect a complete separation. But if we do not continue the operation a sufficient length of time, with each sieve, the separation will not be complete; for part of the superfine will be left, and will pass through with the middlings, and part of the middlings with the carnel, and a part of the carnel with the shorts; and this would be a laborious and tedious work, if performed by hand.

Many inventions have been made to facilitate this business, amongst which the circular sieve, or bolting reel, is one of the foremost; this was, at first, turned and fed by hand; and afterwards it was so contrived as to be turned by water. But many have been the errors in the application of this machine, either from having the cloths too coarse, by which means the middlings and small pieces of bran passed through with the superfine flour, and part of the carnel with the middlings: or by having the cloths too short when they are fine enough, so that the operation could not be continued a sufficient time to take all the superfine out, before it reach the middlings' cloth, and all the middlings before it reach the carnel cloth.

The late improvements made on bolting seem to be principally as follows; namely:—

1. The using finer cloths—but they were found to clog, or choke up, when put on small reels of 22 inches diameter.
2. The enlarging the diameter of the reels to $27\frac{1}{2}$ inches, which gives the meal greater distance to fall, and causes it to strike harder against the cloth, which keeps it open.
3. The lengthening the cloths, that the operation may be continued a sufficient length of time.
4. The bolting a much larger portion of the flour over again, than was done formerly.

The meal, as it is ground, must be hoisted to the meal-loft, where it should be spread thin, and often stirred, that it may cool and dry, to prepare it for bolting. After it is bolted, the tail flour, or that part of the superfine that falls last, and which is too full of specks of bran to pass for superfine, is to be hoisted up again and mixed with the ground meal, to be bolted over again. This hoisting, spreading, mixing, and attending the bolting hoppers, in merchant mills, creates a great deal of hard labour, if performed by hand; and is never completely done at last: but all this, and much more of the labour of mills, can now be accomplished by machinery, moved by water. (See Part III.)

Of Inspecting Flour.

The miller must attain a knowledge of the standard quality passable in the market: to examine it whilst bolting, hold a clean piece of board under the bolt, moving it from head to tail, so as to catch a proportional quantity all the way, as far as is taken for superfine; then, smoothing it well by pressing an even surface on it, will make the specks and colour more plainly appear; if it be not good enough, turn a little more of the tail to be bolted over.

If the flour appear darker than was expected from the quality of the grain, it shows the grinding to be too high, and the bolting too near; because the finer the flour, the whiter its colour.

This proceeding requires a good light; therefore, the best way is for the miller to observe to what degree of poorness he may reduce his tail flour, of middlings, so as to be safe; by which means he may judge with much more safety in the night. But the quality of the tail flour, middlings, &c., will greatly vary in different mills; for those that have the late improvements for bolting over the tail flour, grinding over the middlings, &c., can make nearly all into superfine: whereas, in those that have them not—the quality that remains next to superfine, is common, or fine flour; then rich middlings, ship stuff, &c. Those who have experience will perceive the difference in the profits. If the flour feel soft, dead and oily, yet is white, it shows the stones to have been dull, and too much pressure used. If it appear lively, yet dark-coloured, and too full of very fine specks, it shows the stones to have been too rough and sharp, and that it was ground high and bolted too closely.

CHAPTER XVIII.

Directions for keeping the Mill, and the business of it, in good order.

ARTICLE 116.

THE DUTY OF THE MILLER.

The mill is supposed to be completely finished for merchant work, on the new plan; supplied with a stock of grain, flour casks, nails, brushes, picks, shovels, scales, weights, &c., when the millers enter on their duty.

If there be two of them capable of standing watch, or taking charge of the mill, the time is generally divided as follows. In the day-time they both attend to business, but one of them has the chief direction. The night is divided into two watches, the first of which ends at one o'clock in the morning, when the master miller should enter on his watch, and continue till day-light, that he may be ready to direct other hands to their business early. The first thing he should do, when his watch begins, is to see whether the stones are grinding, and the cloths bolting well. And, secondly, he should review all the moving gudgeons of the mill, to see whether any of them want grease, &c.; for want of this, the gudgeons often run dry, and heat, which bring on heavy losses in time and repairs; for when they heat, they get a little loose, and the stones they run on crack, after which they cannot be kept cool. He should also see what quantity of grain is over the stones, and if there be not enough to supply them till morning, set the cleaning machines in motion.

All things being set right, his duty is very easy—he has only to see to the machinery, the grinding, and bolting, once in an hour; he has, therefore, plenty of time to amuse himself by reading, or otherwise.

Early in the morning all the floors should be swept, and the flour dust collected; the casks nailed, weighed, marked, and branded, and the packing begun, that it may be completed in the fore part of the day; by this means, should any unforeseen thing occur, there will be spare time. Besides, to leave the packing till the afternoon, is a lazy practice, and keeps the business out of order.

When the stones are to be sharpened, every thing necessary should be prepared before the mill is stopped, (especially if there be but one pair of stones to a waterwheel) that as little time as possible may be lost: the picks should be made quite sharp, and not be less than 12 in number. Things being ready, the miller is then to take up the stone; set one hand to each, and dress them as soon as possible, that they may be set to work again; not forgetting to grease the gears and spindle foot.

In the after part of the day, a sufficient quantity of

grain is to be cleaned down, to supply the stones the whole night; because it is best to have nothing more to do in the night, than to attend to the grinding, bolting, gudgeons, &c.

ARTICLE 117.

PECULIAR ACCIDENTS BY WHICH MILLS ARE SUBJECT TO CATCH FIRE.

1. There being many moving parts in a mill, if any piece of timber fall, and lie on any moving wheel, or shaft, and the velocity and pressure be great, it will generate fire, and perhaps consume the mill.

2. Many people use wooden candlesticks, that may be set on a cask, bench, or the floor, and forgetting them, the candle burns down, sets the stick, cask, &c., on fire, which, perhaps, may not be discovered until the mill is in a flame.

3. Careless millers sometimes stick a candle to a cask, or post, and forget it, until it has burnt a hole in the post, or set the cask on fire.

4. Great quantities of grain sometimes bend the floor so as to press the head blocks against the top of the upright shafts, and generate fire, (unless the head blocks have room to rise as the floor settles:) mill-wrights should consider this, and be careful to guard against it as they build.

5. Branding irons, carelessly laid down, when hot, and left, might set a mill on fire.

6. The foot of the mill-stone spindle, and gudgeons, frequently heat, and sometimes set the bridge-tree or shaft on fire. It is probable, that, from such causes, mills have taken fire, when no person could discover how.

ARTICLE 118.

OBSERVATIONS ON IMPROVING MILL-SEATS.

I may end this part with a few observations on improving mill-seats. The improvement of a mill-seat at a great expense, is an undertaking worthy of mature deliberation, as wrong steps may increase it 10 per centum, and the improvement be incomplete: whereas, right steps may reduce it 10 per centum, and render them perfect.

Strange as it may appear, it is yet a real fact, that those who have least experience in the milling business, frequently build the best and most complete mills. The reasons are evident—

The professional man is bound to old systems, and relies on his own judgment in laying all his plans; whereas, the inexperienced man, being conscious of his deficiency, is perfectly free from all prejudice, and is disposed to call on all his experienced friends, and to collect all the improvements that are extant.

A merchant who knows but little of the miller's art, or of the structure or mechanism of mills, is naturally led to the following steps; namely:

He calls several of the most experienced millers and mill-wrights, to view the seat separately, and point out the spot for the mill-house, dam, &c., and notes their reasonings. The first, perhaps, fixes on a pretty level spot for the mill-house, and a certain rock, that nature seems to have prepared to support the breast of the dam, and an easy place to dig the race, mill-seat, &c.

The second passes by these places without noticing them; explores the stream to the boundary line; fixes on another place, the only one he thinks appointed by nature for building a lasting dam, the foundation a solid rock, that cannot be undermined by the tumbling water; fixing on a rugged spot for the seat of the house: assigning for his reasons, that the whole fall must be taken in, that all may be right at a future day. He is then informed of the opinion of the other, against which he gives substantial reasons.

The mill-wright, carpenter, and mason, who are to undertake the building, are now called together, to view the seat, fix on the spot for the house, dam, &c. After their opinion and reasons are heard, they are informed of the opinions and reasons of the others; all are joined together, and the places are fixed on. They are then desired to make out a complete draught of the plan for the house, &c., and to spare no pains; but to alter and improve on paper, till all appear to meet right, in the simplest and most convenient manner, (a week may be thus well spent,) making out complete bills of every piece of timber, the quantity of boards, stone, lime, &c.; a bill of iron work, the number of wheels, their diameters, number of cogs, &c. &c., and every thing else required in the whole work. Each person can then make out his charge, and the costs can be very nearly counted. Every species of materials may be contracted for, to be delivered in due time: the work then goes on regularly without disappointment; and when done, the improvements are complete, and a considerable sum of money saved.

PART THE FIFTH.

CHAPTER XIX.

Practical Instructions for building Mills, with all their proportions, suitable to all falls, of from three to thirty-six feet. Received from Thomas Ellicott, Mill-Wright.

PREFATORY REMARKS.

This part, as appears from the heading, was written by Mr. Thomas Ellicott; a part of his preface, published in the early editions of this work, it has been thought best to omit. After some remarks upon the defective operation of mills upon the old construction, he proceeds to say—

In the new way, all these inconveniences and disadvantages are completely provided against: (See Plate XXII;) which is a representation of the machinery, applied in the whole process of the manufacture; taking the grain from the ship or wagon, and passing it through the whole process by water, until it is completely manufactured into superfine flour. This is a mill of my planning and draughting, now in actual practice, built on Occoquam river, in Virginia, with 3 water-wheels, and 6 pair of stones.

If the wheat come by water to the mill, in the ship Z, it is measured and poured into the hopper A, and thence conveyed into the elevator at B, which elevates it, and drops it into the conveyer C D, which conveys it along under the joists of the second floor, and drops it into the hopper garner at D, out of which it is conveyed into the main wheat elevator at E, which carries it up into the peak of the roof, and delivers it into the rolling-screen at F, which (in this plan) is above the collar beams, out of which it falls into the hopper G, thence into the short elevator at H, which conveys it up into the fan I, from whence it runs down slanting, into the middle of the long conveyer at J, that runs towards both ends of the mill, and conveys the grain, as cleaned, into any garner K K K K K K, over all the stones, which is done by shifting a board under the fan, to guide the grain to either side of the cog-wheel j; and although each of these garners should contain 2000 bushels of wheat, over each pair of stones 12000 bushels in 6 garners, yet nearly all may be ground out without handling it, and the feed of the stones will be more even and regular than is possible in the old way. As it is ground by the several pairs of stones, the meal falls into the conveyer at M M M, and is conveyed into the common meal elevator at N, which raises it to O; from thence it runs down the hopper-boy at P, which spreads and cools it over a circle of 10 or 15 feet diameter, and (if thought best) will rise over it, and form a heap two or three feet high, perhaps thirty barrels of flour, or more, at a time, which may be bolted down at pleasure. When it is bolting, the hopper-boy gathers it into the bolting hoppers at Q, and attends them more regularly than is ever done by hand. As it is bolted, the conveyer R, in the bottom of the superfine chest, conveys the superfine flour to a hole through the floor at S, into the packing chest, which mixes it completely. Out of the packing chest it is filled into the barrel at T, weighed in the scale U, packed at W by water, headed at X, and rolled to the door Y, then lowered down by a rope and windlass, into the ship again at Z.

If the wheat come to the mill by land, in the wagon 7, it is emptied from the bags into a spout that is in the wall, and it runs into the scale 8, which is large enough to hold a wagon load; and as it is weighed it is (by drawing a gate at bottom) let run into the garner D, out of which it is conveyed into the elevator at E, and so through the same process as before.

As much of the tail of the superfine reels 37 as we think will not pass inspection, we suffer to pass on into the short elevator, (by shutting the gates at the bottom of the conveyer next the elevator, and opening one farther towards the other end.) The rubblings, which fall at the tail of said reels, are also hoisted into the bolting hoppers of the sifting reel 39, which is covered with a fine cloth, to take out all the fine flour dust, which will stick to the bran in warm, damp weather; and all that passes through it is conveyed by the conveyer 40, into the elevator 41, which elevates it so high that it will run freely into the hopper-boy at O; and is bolted over again with the ground meal. The rubblings, that fall at the tail of the sifting reel 39, fall into the hopper of the middlings' reel 42; and the bran falls at the tail into the lower story. Thus, you have it in your power, either by day or night, without any hand labour, except to shift the sliders, or some such trifle, to make your flour to suit the standard quality; and the greatest possible quantity of superfine is made out of the grain, and finished completely at one operation.

Agreeably to request, I shall now attempt to show the method of making and putting water on the several kinds of water-wheels commonly used, with their dimensions, &c., suited to falls and heads of from 3 to 36 feet. I have also calculated tables for gearing them to millstones; and made draughts* of several water-wheels with their forebays, and manner of putting on the water, &c.

THOMAS ELLICOTT.

*All my draughts are taken from a scale of eight feet to an inch, except Plate XVII., which is four feet to an inch.

ARTICLE 119.

OF UNDERSHOT MILLS.

Fig. 1, Plate XIII., represents an undershot wheel, 18 feet diameter, with 3 feet total head and fall. It should be 2 feet wide for every foot the mill-stones are in diameter; that is, 8 feet between the shrouds for a 4 feet, and 10 feet wide for a 5 feet stone. It should have three sets of arms and shrouds, on account of its great width. Its shaft should be at least 26 inches in diameter. It requires 12 arms, 18 feet long, $3\frac{1}{2}$ inches thick, by 9 wide; and 24 shrouds, $7\frac{1}{2}$ feet long, 10 inches deep, by 3 thick, and 32 floats, 15 inches wide. Note—It may be geared the same as an overshot wheel, of equal diameter. Fig. 2 represents the forebay, with its sills, posts, sluice, and fall: I have, in this case, allowed 1 foot fall and 2 feet head.

Fig. 3 represents an undershot wheel, 18 feet diameter, with 7 feet head and fall. It should be as wide between the shrouds as the stone is in diameter; its shaft should be 2 feet in diameter; requires 8 arms, 18 feet long, $3\frac{1}{4}$ inches thick, by 9 wide; and 16 shrouds, $7\frac{1}{2}$ feet long, 10 inches deep, by 3 thick. It may be geared the same as an overshot wheel 13 feet in diameter, because their revolutions per minute will be nearly equal.

Fig. 4 represents the forebay, sluice, and fall: the head and fall about equal.

Fig. 5 represents an undershot wheel, 12 feet diameter, with 15 feet total head and fall. It should be 6 inches wide for every foot the stone is in diameter. Its shaft 20 inches in diameter; requires 6 arms, 12 feet long, 3 by 8 inches; and 12 shrouds, $6\frac{1}{2}$ feet long, $2\frac{1}{2}$ inches thick, and 8 deep. It suits well to be geared to a 5 feet stone with single gears, 60 cogs in the cogwheel, and 16 rounds in the trundle; to a $4\frac{1}{2}$ feet stone, with 62 cogs and 15 rounds; and, to a 4 feet stone, with 64 cogs and 14 rounds. These gears will do well till the fall is reduced to 12 feet, only the wheel must be less, as the falls are less, so as to make the same number of revolutions in a minute; but this wheel requires more water than a breast-mill, with the same fall.

Fig. 6 is the forebay, gate, shute, and fall. Forebays should be wide, in proportion to the quantity of water they are to convey to the wheels, and should stand 8 or 10 feet in the bank, and be firmly joined, to prevent the water from breaking through; which it will certainly do, unless they be well secured.

ARTICLE 120.

DIRECTIONS FOR MAKING FOREBAYS.

The best way with which I am acquainted, for making this kind of forebays, is shown in Plate XVII., fig. 7. Make a number of solid frames, each consisting of a sill, two posts, and a cap; set them cross-wise, (as shown in the figure,) $2\frac{1}{2}$ or 3 feet apart; to these the planks are to be spiked, for there should be no sills lengthwise, as the water is apt to find its way along them. The frame at the head next the water, and one 6 or 8 feet downwards in the bank, should extend 4 or 5 feet on each side of the forebay into the bank, and be planked in front, to prevent the water, and vermin, from working round. Both of the sills of these long frames should be well secured, by driving down plank edge to edge, like piles, along the upper side, from end to end.

The sills being settled on good foundations, the earth or gravel must be rammed well on all sides, full to the top of the sills. Then lay the bottom with good, sound plank, well jointed and spiked to the sills. Lay your shute, extending the upper end a little above the point of the gate when full drawn, to guide the water in a right direction to the wheel. Plank the head to its proper height, minding to leave a suitable sluice, to guide the water smoothly down. Fix the gate in an upright position—hang the wheel, and finish it off, ready for letting on the water.

A rack must be made across the stream, to keep off the floating matter that would break the floats and buckets of undershot, breast, and pitch-back wheels, and injure the gates. (See it at the head of the forebay, fig. 7, Plate XVII.) This is done by setting a frame 3 feet in front of the forebay, and laying a sill 2 feet in front of it, for the bottom of the rack; in it the staves are put, made of laths, set edgewise with the stream, 2 inches apart, their upper ends nailed to the cap of the last frame; which causes them to lean down stream. The bottom of the race must be planked between the forebay and rack, to prevent the water from making a hole by tumbling through the rack when choked; and the sides must be planked outside of the posts to keep up the banks. This rack must be twice as long as the forebay is wide, or else the water will not come fast enough through it to keep the head up; for the head is the spring of motion of an undershot mill.

ARTICLE 121.

OF THE PRINCIPLE OF UNDERSHOT MILLS.

They differ from all others in principle, because the water loses all its force by the first stroke against the floats; and the time this force is spending, is in proportion to the difference of the velocities of the wheel and the water, and the distance of the floats. Other mills have the weight of the water after the force of the head is spent, and will continue to move: but an undershot will stop as soon as the head is spent, as they depend not on the weight. They should be geared so, that when the stone goes with a proper motion, the water-wheel will not run too fast, as they will not, then, receive the full force of the water; nor too slow, so as to lose its power by its rebounding and dashing over the buckets. This matter requires very close attention, and to find it out by theory, has puzzled our mechanical philosophers. They give us for a rule, that the wheel must move just one-third the velocity of the water: perhaps this may suit where the head is not much higher than the float-boards, but I am fully convinced that it will not suit high heads.

Experiments for determining the proper Motion for Undershot Wheels.

I drew a full sluice of water on an undershot wheel with 15 feet head and fall, and counted its revolutions per minute; then geared it to a mill-stone, set it to work properly, and again counted its revolutions, and the difference was not more than one-fourth slower. I believe, that if I had checked the motion of the wheel to be equal to one-third the motion of the water, the water would have rebounded and flown up to the shaft. Hence, I conclude, that the motion of the water must not be checked by the wheel more than one-third, nor less than one-fourth, else it will lose in power; for, although the wheel will carry a greater load with a slow, than with a swift motion, yet it will not produce so great an effect, its motion being too slow. And again, if the motion be too swift, the load or resistance it will overcome will be so much less, that its effect will be lessened also. I conclude, that about two-thirds the velocity of the water is the proper motion for undershot wheels; the water will then spend all its force in the distance of two float-boards. It will be seen that I differ greatly with those learned authors who have concluded that the velocity of the wheel ought to be but one-third of that of the water. To confute them, suppose the floats 12 inches, and the column of water striking them, 8 inches deep; then, if two-thirds of the motion of this column be checked, it must instantly become 24 inches deep, and rebound against the backs of the floats, and the wheel would be wallowing in this dead water; whereas, when only one-third of its motion is checked, it becomes 12 inches deep, and runs off from the wheel in a smooth and lively manner.

Directions for gearing Undershot Wheels, 18 *feet in diameter, where the head is above* 3 *and under* 8 *feet, with double gears: counting the head from the point where the water strikes the floats.*

1. For 3 feet head and 18 feet wheel, see 18 feet wheel in the overshot table.
2. For 3 feet 8 inches head, see 17 feet wheel in do.
3. For 4 feet 4 inches head, see 16 feet wheel in do.
4. For 5 feet head, see 15 feet wheel in do.
5. For 5 feet 8 inches head, see 14 feet wheel in do.
6. For 6 feet 4 inches head, see 13 feet wheel in do.
7. For 7 feet head, see 12 feet wheel in do.

The revolutions of the wheels will be nearly equal; therefore the gears may be the same.

The following table is calculated to suit for any sized stone, from 4 to 6 feet diameter, different sized water-wheels from 12 to 18 feet diameter, and different heads from 8 to 20 feet above the point it strikes the floats; and to make 5 feet stones revolve 88 times; 4 feet 6 inch stones 97 times; and 4 feet stones 106 times, in a minute, when the water-wheel moves two-thirds the velocity of the striking water.

MILL-WRIGHT'S TABLE FOR UNDERSHOT MILLS, SINGLE GEARED.

Height of the head of water in feet.	Diameter of the water-wheel in feet.	Velocity of the water per minute in feet.	Velocity of the water wheel per minute in feet.	Revolutions of the water-wheel per minute.	Revolutions of the stone per minute.	Number of cogs in the cog-wheel.	Number of rounds in the trundle head.	Revolutions of the mill-stone for one of the water-whs.	Diameter of the stones in feet.
8	12	1360	906	24	88	56	15	3¾	5
9	13	1448	965	23½	88	58	15	3⅞	5
10	14	1521	1014	23 1/7	88	58	15	3 6/7	5
11	15	1595	1061	22¾	88	58	15	3¾	5
12	16	1666	1111	22¼	88	58	15	3⅞	5
13	16	1735	1157	23 1/7	88	60	16	3¾	5
14	16	1800	1200	24	88	59	16	2⅔	5
15	16	1863	1242	24⅘	88	60	17	3½	5
16	16	1924	1283	25⅔	88	59	17	3⅜	5
17	17	1983	1322	25	88	62	17	3¾	5
18	17	2041	1361	25⅔	88	62	17	3⅜	5
19	18	2097	1398	25	88	62	17	3¾	5
20	18	2152	1435	25½	88	60	17	3⅜	5
1	2	3	4	5	6	7	8	9	10

Note that there are nearly 60 cogs in the cog-wheel, in the foregoing table, and 60 inches is the diameter of a 5 feet stone: therefore, it will do, without sensible error, to put 1 cog more in the wheel for every inch that the stone is less than 60 inches diameter, down to 4 feet; the trundle head and water-wheel remaining the same; and for every three inches that the stone is larger than 60 inches in diameter, put 1 round more in the trundle, and the motion of the stone will be nearly right, up to 6 feet diameter.

ARTICLE 122.

OF BREAST WHEELS.

Breast wheels differ but little in their structure or motion from overshots, excepting only, that the water passes under, instead of over them, and they must be wider in proportion as their fall is less.

Fig. 1, Plate XIV., represents a low breast with 8 feet head and fall. It should be 9 inches wide for every foot of the diameter of the stone. Such wheels are generally 18 feet diameter; the number and dimensions of their parts being as follows: 8 arms 18 feet long, $3\frac{1}{4}$ by 9 inches; 16 shrouds 8 feet long, $2\frac{1}{2}$ by 9 inches; 56 buckets; and shaft, 2 feet diameter.

Fig. 2 shows the forebay, water-gate, and fall, and manner of striking on the water.

Fig. 3 is a middling breast wheel, 18 feet diameter, with 12 feet head and fall. It should be 8 inches wide for every foot the stone is in diameter.

Fig. 4 shows the forebay, gate, and fall, and manner of striking on the water.

Fig. 5 and 6, is a high breast wheel, 16 feet diameter, with 3 feet head in the forebay, and 10 feet fall. It should be 7 inches wide for every foot the stone is in diameter. The number and dimensions of its parts are 6 arms, 16 feet long, $3\frac{1}{4}$ by 9 inches; 12 shrouds, 8 feet 6 inches long, $2\frac{1}{2}$ by 8 or 9 inches deep, and 48 buckets.

ARTICLE 123.

OF PITCH-BACK WHEELS.

Pitch-back wheels are constructed exactly similar to breast wheels, only the water is struck on them at a greater height. Fig. 1, Plate XV, is a wheel 18 feet diameter, with 3 feet head in the penstock, and 16 feet fall below it. It should be 6 inches wide for every foot of the diameter of the stone.

Fig. 2 shows the trunk, penstock, gate, and fall; the gate sliding on the bottom of the penstock, and drawn by the lever A, turning on a roller. This wheel is much recommended by some mechanical philosophers, for the saving of water; but I do not join them in opinion, but think that an overshot with an equal head and fall, is fully equal to it in power; besides the saving of the expense in building so high a wheel, and the greater difficulty of keeping it in order.*

ARTICLE 124.

OF OVERSHOT WHEELS.

Overshot wheels receive their water on the top, being moved by its weight; and are much to be recommended where there is fall enough for them. Fig. 3 represents one 18 feet diameter, which should be about 6 inches wide for every foot the stone is in diameter. It should hang 8 or 9 inches clear of the tail water, otherwise the water will be drawn back under it. The head in the penstock should be generally about 3 feet, which will spout the water about one-third faster than the wheel moves. Let the shute have about 3 inches fall, and direct the water into the wheel at the centre of its top.

I have calculated a table for gearing overshot wheels, which will suit equally well any of the others of equal diameter, that have equal heads above the point where the water strikes the wheel.

* On this subject see the Appendix.—EDITOR.

Dimensions of this wheel, 8 arms 18 feet long, 3 by 9 inches; 16 shrouds 7 feet 9 inches long, 2½ by 7 or 8 inches; 56 buckets; and shaft 24 inches diameter.

Fig. 4 represents the penstock and trunk, &c., the water being let on the wheel by drawing the gate G.

Fig. 1 and 2, Plate XVI., represents a low overshot, 12 feet diameter, which should be in width equal to the diameter of the stone. Its parts and dimensions are, 6 arms 12 feet long, 3½ by 9 inches; 12 shrouds 6½ feet long, 2½ by 8 inches; shaft 22 inches diameter, and 30 buckets.

Fig. 3 represents a very high overshot 30 feet diameter, which should be 3½ inches wide for every foot of the diameter of the stone. Its parts and dimensions are, 6 main arms, 30 feet long, 3¼ inches thick, 10 inches wide at the shaft, and 6 at the end; 12 short arms 14 feet long, of equal dimensions; which are framed into the main arms near the shaft, as in the figure, for if they were all put through the shaft, they would make it too weak. The shaft should be 27 inches diameter, the whole being very heavy and bearing a great load. Such high wheels require but little water.

CHAPTER XX.

ARTICLE 125.

OF THE MOTION OF OVERSHOT WHEELS.

After trying many experiments, I concluded that the circumference of overshot wheels geared to mill-stones, grinding to the best advantage, should move 550 feet in a minute; and that of the stones 1375 feet in the same time; that is, while the wheel moves 12, the stone moves 30 inches, or in the proportion of 2 to 5.

Then, to find how often the wheel we propose to make, will revolve in a minute, take the following steps: 1st, Find the circumference of the wheel by multiplying the diameter by 22, and dividing by 7, thus:

Suppose the diameter to be 16 feet, then, 16 multiplied by 22, produces 352; which, divided by 7, gives 50 2-7 for the circumference.

```
    16
    22
    —
    32
    32
    ——
  7)352
    ——
   50 2-7
```

By which we divide 550, the distance the wheel moves in a minute, and it gives 11, for the revolutions of the wheel per minute, casting off the fraction 2-7, it being small.

```
  5|0)55|0
     —
   11 times.
```

To find the revolutions of the stone per minute, 4 feet 6 inches (or 54 inches) diameter, multiply 54 inches by 22, and divide by 7, and it gives 169 5-7 (say 170) inches, the circumference of the stone.

```
    54
    22
   ———
   108
   108
   ———
  7)1188
   ———
   169 5-7
```

By which divide 1375 feet, or 16500 inches, the distance of the skirt of the stone should move in a minute, and it gives 97; the revolutions of a stone per minute, 4½ feet diameter.

```
   1375
     12
  17|0)1650(0 |97
       153
       ———
       120
       119
       ———
         1
```

To find how often the stone revolves for once of the water-wheel, divide 97, the revolutions of the stone, by 11, the revolutions of the wheel, and it gives 8 9-11, (say 9 times.)

```
  11)97
   —
   8 9-11
```

ARTICLE 126.

OF GEARING.

If the mill were to be single geared, 99 cogs and 11 rounds would give the stone the right motion, but the cog-wheel would then be too large, and the trundle too small; it must, therefore, be double geared.

Chap. 20.]   OF GEARING.   289

    Suppose we choose 66 cogs in the big cog-wheel and 48 in the little one, and 25 rounds in the wallower, and 15 in the trundle.

    Then, to find the revolutions of the stone for one of the water-wheel, multiply the cog-wheels together, and the wallower and trundle together, and divide one product by the other, and it will give the answer, $8\frac{168}{375}$, not quite $8\frac{1}{2}$ revolutions, instead of 9.

```
    25
    15
   ---
   125
    25
   ---
   375

    66
    48
   ---
   528
   264
   ---
375)3168(8 168-375
    3000
    ----
     168
```

    We must, therefore, devise another proportion—Considering which of the wheels we had best alter, and wishing not to alter the big cog-wheel or trundle, we put one round less in the wallower, and two cogs more in the little cog-wheel, and multiplying and dividing as before, we find the stone will turn $9\frac{1}{4}$ times for once of the water-wheel, which is as near as we can get. The mill now stands thus, a 16 feet overshot wheel, that will revolve 11 times in a minute, geared to a stone $4\frac{1}{2}$ feet in diameter; the big cog-wheel 66 cogs, $4\frac{1}{2}$ inches from centre to centre of the cogs; (which we call the pitch of the gear) little cog-wheel 50 cogs, $4\frac{1}{4}$ pitch; wallower 24 rounds, $4\frac{1}{2}$ pitch; and trundle 15 rounds, $4\frac{1}{4}$ inches pitch.

ARTICLE 127.

RULES FOR FINDING THE DIAMETER OF THE PITCH CIRCLES.

    To find the diameter of the pitch circle, that the cogs stand in, multiply the number of cogs by the pitch, which gives the circumference; this, multiplied by 7, and divided by 22, gives the diameter in inches; which, divided by 12, reduces it to feet and inches, thus:

```
      66
      4½
     ---
     264
      33
     ---
     297
       7
     ---
22)2079(94½ in.
     198
     ---
      99
      88
     ---
      11
```

37

For the cog-wheel of 66 cogs, and 4½ inches pitch, we find 7 feet 10½ inches to be the diameter of the pitch circle; to which I add 8 inches, for the outside of the cogs, which makes 8 feet 6½ inches, the diameter from out to out.

By the same rules, I find the diameters of the pitch circles of the other wheels to be as follows; namely:—

	feet.	inches.	
Little cog-wheel 50 cogs, 4¼ inches pitch,	5	7½ 10/22	pitch cir.
I add, for the outside of the circle,		7½	
Total diameter from out to out,	6	3	
Wallower 24 rounds, 4½ inches pitch,	2	11¾ 4/22	do.
Add, for outsides,	0	3 13/22	
Total diameter from the outsides,	3	3	
Trundle head 15 rounds, 4¼ inches pitch,	1	8¼ 3/22	do.
Add, for outsides,	0	2½ 19/22	
Total diameter for the outsides,	1	11	

Thus, we have completed the calculations for one mill, with a 16 feet overshot water-wheel, and stones 4½ feet diameter. By the same rules we may calculate for wheels of all sizes from 12 to 30 feet, and stones from 4 to 6 feet diameter, and may form tables that will be of great use even to master workmen, in despatching of business, in laying out work for their apprentices and other hands, getting out timber, &c.; but more especially to those who are not sufficiently skilled in arithmetic to make the calculations. I have from long experience been sensible of the need of such tables, and have therefore undertaken the task of preparing them.

ARTICLE 128.

MILL-WRIGHTS' TABLES,

Calculated to suit overshot water-wheels with suitable heads above them, of all sizes, from 12 to 30 feet diameter, the velocity of their circumferences being about 550 feet per minute, showing the number of cogs and rounds in all the wheels, double gear, to give the circumference of the stone a velocity of 1375 feet per minute, also the diameter of their pitch circles, the diameter of the outsides, and revolutions of the water-wheel, and stones, per minute.

For particulars, see what is written over the head of each table. Table I. is to suit a 4 feet stone, Table II. a 4½, Table III. a 5 feet, and Table IV. a 5½ feet stone.

N. B. If the stones should be an inch or two larger or less than those above described, make use of the table that comes nearest to it, and likewise for the water-wheels. For farther particulars see " Draughting Mills."

Use of the following Tables.

Having levelled your mill-seat, and found the total fall, after making due allowances for the fall in the races, and below the wheel, suppose there be 21 feet 9 inches, and the mill-stones be 4 feet in diameter, then look in Table I, (which is for 4 feet stones,) column 2, for the fall that is nearest yours, and you find it in the 7th example; and against it in column 8, is 3 feet, the head proper to be above the wheel; in column 4 is 18 feet, for the diameter of the wheel, &c., for all the proportions of the gears to make a steady moving mill; the stones to revolve 106 times in a minute.*

---

* The following tables are calculated to give the stones the revolution per minute mentioned in them, as near as any suitable number of cogs and rounds would permit, which motion I find is 8 or 10 revolutions per minute slower than proposed by Evans, in his table;—his motion may do best in cases where there is plenty of power, and steady work on one kind of grain; but, in country mills, where they are continually changing from one kind to another, and often starting and stopping, I presume a slow motion will work most regularly. His table being calculated for only one size of mill-stones, and mine for four, any one choosing his motion, may look for the width of the water-wheel, number of cogs, and rounds and size of the wheels to suit them, in the next example following, keeping to my table in other respects, and you will have his motion nearly.

TABLE I. For Overshot Mills, with Stones 4 feet diameter, to revolve 106 times in a minute, pitch of the gear of the great cog-wheel and wallowers 4½ inches, and of lesser cog-wheel and trundle 4¼ inches.

No. of Examples.	Total falls of water from the top of that in the penstock to that in the tail-race.	Different heads of water above the water-wheel.	Diameters of water-wheels from out to out.	Widths of water-wheels in the clear.	No. of cogs in the great and lesser cog-wheels.	Diameters of pitch circles of great and lesser wheels.	Diameters of cog-wheels, from out to out.	No. of rounds in the wallowers and trundles.	Diameters of pitch circles in wallowers and trundles.	Total diameters of wallowers and trundles.	Revolutions of great wheel per minute, nearly.
	ft. in.	ft. in.	feet.	ft. in.		ft. in.	ft. in.		ft. in.	ft. in.	
1	15 3	2 6	12	3 0	66 / 48	7 10. 5 / 5 4.87	8 6. 5 / 6 0. 5	25 / 15	2 11.75 / 1 8.33	3 3 / 1 11.33	13
2	16 4	2 7	13	2 10	69 / 48	8 2.33 / 5 4.87	8 10.33 / 6 0. 5	25 / 15	2 11.73 / 1 8.33	3 3 / 1 11.33	12.5
3	17 5	2 8	14	2 8	69 / 48	8 2.33 / 5 4.87	8 10.33 / 6 0. 5	26 / 15	3 1.25 / 1 8.33	3 5.25 / 1 11.33	12
4	18 6	2 9	15	2 6	69 / 50	8 2.33 / 5 7. 5	8 10.33 / 6 3	25 / 15	2 11.75 / 1 8.33	3 3 / 1 11.33	11.5
5	19 7	2 10	16	2 4	72 / 52	8 7.25 / 5 10.33	9 3 / 6 6	26 / 15	3 1.25 / 1 8.33	3 5.25 / 1 11.33	11
6	20 8	2 11	17	2 3	72 / 52	8 7.25 / 5 10.33	9 3 / 6 6	25 / 14	2 11.75 / 1 7	3 3 / 1 10	10.5
7	21 9	3 0	18	2 2	72 / 52	8 7.25 / 5 10.33	9 3 / 6 6	24 / 14	2 10.33 / 1 7	3 1. 5 / 1 10	10
8	22 10	3 1	19	2 1	75 / 52	8 11.33 / 5 10.33	9 7.33 / 6 6	24 / 14	2 10.33 / 1 7	3 1. 5 / 1 10	9.66
9	23 11	3 2	20	2 0	75 / 52	8 11.33 / 5 10.33	9 7.33 / 6 6	23 / 14	2 9 / 1 7	3 0 / 1 10	9.25
10	25 1	3 4	21	1 11	78 / 52	9 3. 5 / 5 10.33	9 11. 5 / 6 6	24 / 14	2 10.33 / 1 7	3 1. 5 / 1 10	8.87
11	26 3	3 6	22	1 10	78 / 52	9 3. 5 / 5 10.33	9 11. 5 / 6 6	23 / 14	2 9 / 1 7	3 0 / 1 10	8.5
12	27 5	3 8	23	1 9	78 / 54	9 3. 5 / 6 1	9 11. 5 / 6 8. 5	23 / 14	2 9 / 1 7	3 0 / 1 10	8.25
13	28 7	3 10	24	1 8	81 / 54	9 8 / 6 1	10 4 / 6 8. 5	23 / 14	2 9 / 1 7	3 0 / 1 10	8
14	29 9	4 0	25	1 7	81 / 56	9 8 / 6 3.75	10 4 / 6 11.25	23 / 14	2 9 / 1 7	3 0 / 1 10	7.75
15	30 11	4 2	26	1 6	84 / 56	10 0.25 / 6 3.75	10 8.25 / 6 11.25	23 / 14	2 9 / 1 7	3 0 / 1 10	7.5
16	32 1	4 4	27	1 5	84 / 58	10 0.25 / 6 6.25	10 8.25 / 7 1.75	23 / 14	2 9 / 1 7	3 0 / 1 10	6.75
17	33 3	4 6	28	1 4	84 / 56	10 0.25 / 6 3.75	10 8.25 / 6 11.25	23 / 13	2 9 / 1 5.25	3 0 / 1 8.25	6.66
18	34 6	4 9	29	1 3	84 / 56	10 0.25 / 6 3.75	10 8.25 / 6 11.25	22 / 13	2 7. 5 / 1 5.25	2 10. 5 / 1 8.25	6. 5
19	35 9	5 0	30	1 2	87 / 56	10 5 / 6 3.75	11 1 / 6 11.25	22 / 13	2 7. 5 / 1 6.25	2 10. 5 / 1 8.25	6.25

TABLE II. For Overshot Mills, with Stones 4 feet 6 inches Diameter, to revolve 99 times in a minute, pitch of the gears 4½ and 4¼ inches.

No. of Examples.	Total falls of water from the top of that in the penstock to that in the tail-race.	Different heads of water above the water-wheel.	Diameters of water-wheels from out to out.	Widths of water-wheels in the clear.	No. of cogs in the great and lesser cog-wheel.	Diameters of pitch circles of great and lesser cog-wheels.	Diameters of cog-wheels from out to out.	No. of rounds in wallowers and trundles.	Diameters of pitch circles in wallowers and trundles.	Total diameters of wallowers and trundles.	Revolutions of the great wheel per minute, nearly.
	ft. in.	ft. in.	feet	ft. in.		ft. in.	ft. in.		ft. in.	ft. in.	
1	15 3	2 6	12	3 6	66 / 48	7 10.5 / 5 4.87	8 6.5 / 6 0.5	26 / 15	3 1.25 / 1 8.33	3 4.25 / 1 11.33	13
2	16 4	2 7	13	3 4	66 / 48	7 10.5 / 5 4.87	8 6.5 / 6 0.5	25 / 15	2 11.75 / 1 8.33	3 3 / 1 11.33	12.5
3	17 5	2 8	14	3 2	69 / 48	8 2.33 / 5 4.87	8 10.33 / 6 0.5	26 / 15	3 1.25 / 1 8.33	3 4.25 / 1 11.33	12
4	18 6	2 9	15	3 0	69 / 48	8 2.33 / 5 4.87	8 10.33 / 6 0.5	25 / 15	2 11.75 / 1 8.33	3 3 / 1 11.5	11.5
5	19 7	2 10	16	2 10	69 / 50	8 2.33 / 5 7.87	8 10.33 / 6 3	25 / 15	2 11.75 / 1 8.33	3 3 / 1 11.5	11
6	20 8	2 11	17	2 8	72 / 52	8 7.12 / 5 10.33	9 3 / 6 6	26 / 15	3 1.25 / 1 8.33	3 4.25 / 1 11.5	10.5
7	21 9	3 0	18	2 6	72 / 52	8 7.25 / 5 10.33	9 3 / 6 6	25 / 14	2 11.75 / 1 7	3 3 / 1 11.5	10
8	22 10	3 1	19	2 4	72 / 52	8 7.25 / 5 10.33	9 3 / 6 6	24 / 14	2 10.33 / 1 7	3 2.5 / 1 11.5	9.5
9	23 11	3 2	20	2 3	75 / 52	8 11.33 / 5 10.33	9 7.33 / 6 6	24 / 14	2 10.33 / 1 7	3 2.5 / 1 11.5	9
10	25 1	3 4	21	2 2	75 / 52	8 11.33 / 5 10.33	9 7.33 / 6 6	23 / 14	2 9 / 1 7	3 0 / 1 11.5	8.75
11	26 3	3 6	22	2 1	78 / 52	9 3.5 / 5 10.33	9 11.5 / 6 6	24 / 14	2 10.33 / 1 7	3 2.5 / 1 11.5	8.5
12	27 5	3 8	23	2 0	78 / 52	9 3.5 / 5 10.33	9 11.5 / 6 6	23 / 14	2 9 / 1 7	3 0 / 1 11.5	8.25
13	28 7	3 10	24	1 11	78 / 54	9 3.5 / 6 1	9 11.5 / 6 8.5	23 / 14	2 9 / 1 7	3 0 / 1 11.5	8
14	29 9	4 0	25	1 10	81 / 54	9 8 / 6 1	10 4 / 6 8.5	23 / 14	2 9 / 1 7	3 0 / 1 11.5	7.75
15	30 11	4 2	26	1 9	81 / 56	9 8 / 6 3.25	10 4 / 6 11.25	23 / 14	2 9 / 1 7	3 0 / 1 11.5	7.5
16	32 1	4 4	27	1 8	84 / 56	10 0.25 / 6 3.25	10 8.25 / 6 11.25	23 / 14	2 9 / 1 7	3 0 / 1 11.5	6.75
17	33 3	4 6	28	1 6	84 / 58	10 0.25 / 6 6.25	10 8.25 / 7 1.25	23 / 14	2 9 / 1 7	3 0 / 1 11.5	6.66
18	34 6	4 9	29	1 5	84 / 56	10 0.75 / 6 3.25	10 8.25 / 6 11.25	23 / 13	2 9 / 1 5.25	3 0 / 1 8.25	6.5
19	35 9	5 0	30	1 4	84 / 56	10 0.75 / 6 3.25	10 8.25 / 6 11.25	22 / 13	2 7.5 / 1 5.25	2 10.5 / 1 8.25	6.25

294   TABLES, &c.

TABLE III. Stones 5 feet Diameter, to revolve 86 times in a minute, the pitch of the gears 4½ and 4¼ inches.

No. of Examples.	Total falls of water from the top of that in the penstock to that in the tail-race.	Different heads of water above the water-wheel.	Diameters of water-wheels from out to out.	Widths of water-wheels in the clear.	No. of cogs in the great and lesser cog wheel.	Diameter of pitch circles of great and lesser cog-wheels.	Diameters of cog-wheels from out to out.	No. of rounds in the wallowers and trundles.	Diameters of pitch circles in wallowers and trundles.	Total diameters of wallowers and trundles.	Revolutions of great wheel per minute, nearly.
	ft. in.	ft. in.	feet	ft. in.		ft. in.	ft. in.		ft. in.	ft. in.	
1	15 3	2 6	12	4 0	63 / 48	7 6.12 / 5 4.87	8 2.12 / 6 0. 5	26 / 16	3 1.25 / 1 9.66	3 4.25 / 2 4.25	13
2	16 4	2 7	13	3 10	66 / 48	7 10. 5 / 5 4.87	8 6. 5 / 6 0. 5	26 / 16	3 1.25 / 1 9.66	3 4.25 / 2 4.25	12.5
3	17 5	2 8	14	3 8	66 / 48	7 10. 5 / 5 4.87	8 6. 5 / 6 0. 5	25 / 15	2 11.75 / 1 8.33	3 3 / 1 11.33	12
4	18 6	2 9	15	3 6	69 / 48	8 2.33 / 5 4.87	8 10.33 / 6 0. 5	26 / 15	3 1.25 / 1 8.33	3 4.25 / 1 11.33	11.5
5	19 7	2 10	16	3 4	69 / 48	8 2.33 / 5 4.87	8 10.33 / 6 0. 5	25 / 15	2 11.75 / 1 8.33	3 3 / 1 11. 5	11
6	20 8	2 11	17	3 2	69 / 50	8 2.33 / 5 7. 5	8 10.33 / 6 3	25 / 15	2 11.75 / 1 8.33	3 3 / 1 11. 5	10.5
7	21 9	3 0	18	3 0	72 / 52	8 7.25 / 5 10.33	9 3 / 6 6	26 / 15	3 1.25 / 1 8.33	3 4.25 / 1 11.33	10
8	22 10	3 1	19	2 10	72 / 52	8 7.25 / 5 10.33	9 3 / 6 6	25 / 14	2 11.75 / 1 7	3 3 / 1 11. 5	9.66
9	23 11	3 2	20	2 8	72 / 52	8 7.25 / 5 10.33	9 3 / 6 6	24 / 14	2 10.33 / 1 7	3 2. 5 / 1 11. 5	9.25
10	25 1	3 4	21	2 6	75 / 52	8 11.33 / 5 10.33	9 7.33 / 6 6	24 / 14	2 10.33 / 1 7	3 2. 5 / 1 11. 5	8.87
11	26 3	3 6	22	2 5	75 / 52	8 11.33 / 5 10.33	9 7.33 / 6 6	23 / 14	2 9 / 1 7	3 0 / 1 11. 5	8.5
12	27 5	3 8	23	2 4	78 / 52	9 3. 5 / 5 10.33	9 11. 5 / 6 6	24 / 14	2 10.33 / 1 7	3 2.33 / 1 11. 5	8.25
13	28 7	3 10	24	2 3	78 / 52	9 3. 5 / 5 10.33	9 11. 5 / 6 6	23 / 14	2 9 / 1 7	3 0 / 1 11. 5	8
14	29 9	4 0	25	2 2	78 / 54	9 3. 5 / 6 1	9 11. 5 / 6 8. 5	23 / 14	2 9 / 1 7	3 0 / 1 11. 5	7.75
15	30 11	4 2	26	2 0	81 / 54	9 8 / 6 1	10 4 / 6 8. 5	23 / 14	2 9 / 1 7	3 0 / 1 11. 5	7.5
16	32 1	4 4	27	1 11	81 / 56	9 8 / 6 3.25	10 4 / 6 11.25	23 / 14	2 9 / 1 7	3 0 / 1 11. 5	6.33
17	33 3	4 6	28	1 9	84 / 56	10 0.25 / 6 3.25	10 8.25 / 6 11.25	23 / 14	2 9 / 1 7	3 0 / 1 11. 5	6.66
18	34 6	4 9	29	1 7	84 / 58	10 0.25 / 6 6.25	10 8.25 / 7 1.25	23 / 14	2 9 / 1 7	3 0 / 1 11. 5	6.25
19	35 9	5 0	30	1 6	84 / 56	10 0.25 / 6 3.25	10 8.25 / 6 11.25	23 / 13	2 9 / 1 5.25	3 0 / 1 8. 5	6.25

TABLE IV. For Overshot Mills with Stones 5 feet 6 inches Diameter, to revolve 80 times in a minute, pitch of the gears 4¾ inches and 4½ inches.

No. of Examples.	Total fall of water from the top of that in the penstock to that in the tail-race.	Different heads of water above the water-wheels.	Diameters of water-wheels from out to out.	Widths of water-wheels in the clear.	No. of cogs in the great and lesser cog-wheels.	Diameters of pitch circles of great and lesser cog-wheels.	Diameters of cog-wheels from out to out.	No. of rounds in wallowers and trundles.	Diameters of pitch circles in wallowers and trundles.	Total diameters of wallowers and trundles.	Revolutions of the great wheel per minute, nearly.
	ft. in.	ft. in.	feet.	ft. in.		ft. in.	ft. in.		ft. in.	ft. in.	
1	15 3	2 6	12	4 6	60 / 48	7 6.75 / 5 8.75	8 2.75 / 6 4.25	26 / 16	3 3.25 / 1 11	3 6.25 / 2 2	13
2	16 4	2 7	13	4 4	63 / 48	7 11.12 / 5 8.75	8 7.12 / 6 4.25	26 / 16	3 3.25 / 1 11	3 6.25 / 2 2	12.5
3	17 5	2 8	14	4 2	66 / 48	8 3.75 / 5 8.75	8 11.75 / 6 4.25	26 / 16	3 3.25 / 1 11	3 6.25 / 2 2	12
4	18 6	2 9	15	4 0	66 / 48	8 3.75 / 5 8.75	8 11.75 / 6 4.25	26 / 15	3 3.25 / 1 9.5	3 6.25 / 2 0.5	11.5
5	19 7	2 10	16	3 10	69 / 48	8 8.33 / 5 8.75	9 4.33 / 6 4.25	26 / 15	3 3.25 / 1 9.5	3 6.25 / 2 0.5	11
6	20 8	2 11	17	3 8	69 / 48	8 8.33 / 5 8.75	9 4.33 / 6 4.25	25 / 15	3 1.75 / 1 9.5	3 4.75 / 2 0.5	10.5
7	21 9	3 0	18	3 6	69 / 50	8 8.33 / 5 11.5	9 4.33 / 6 2.5	25 / 15	3 1.75 / 1 9.5	3 4.75 / 2 0.5	9
8	22 10	3 1	19	3 4	72 / 52	9 0.75 / 6 2.5	9 8.75 / 6 10	26 / 14	3 3.25 / 1 8	3 6.25 / 1 11	9.66
9	23 11	3 2	20	3 2	72 / 72	9 0.75 / 6 2.5	9 8.75 / 6 10	25 / 14	3 1.75 / 1 8	3 4.75 / 1 11	9.25
10	25 1	3 4	21	3 0	52 / 52	9 0.75 / 6 2.5	9 8.75 / 6 10	24 / 14	3 0.75 / 1 8	3 3.75 / 1 11	8.12
11	26 3	3 6	22	2 10	75 / 52	9 5.33 / 6 2.5	10 1.33 / 6 10	24 / 14	3 0.75 / 1 8	3 3.75 / 1 11	8.5
12	27 5	3 8	23	2 8	75 / 52	9 5.33 / 6 2.5	10 1.33 / 6 10	23 / 14	2 10.75 / 1 8	3 1.75 / 1 11	8.25
13	28 7	3 10	24	2 6	78 / 52	9 10.5 / 6 2.5	10 6 / 6 10	24 / 14	3 0.75 / 1 8	3 3.75 / 1 11	8
14	29 9	4 0	25	2 4	78 / 52	9 10.5 / 6 2.5	10 6 / 6 10	23 / 14	2 10.75 / 1 8	3 1.75 / 1 11	7.75
15	30 11	4 2	26	2 2	78 / 54	9 10.5 / 6 5.33	10 6 / 7 1	23 / 14	2 10.75 / 1 8	3 1.75 / 1 11	7.5
16	32 1	4 4	27	2 0	81 / 54	10 2.5 / 6 5.33	10 10.5 / 7 1	23 / 14	2 10.75 / 1 8	3 1.75 / 1 11	6.75
17	33 3	4 6	28	1 11	81 / 56	10 2.5 / 6 8	10 10.5 / 7 3.5	23 / 14	2 10.75 / 1 8	3 1.75 / 1 11	6.66
18	34 6	4 9	29	1 10	84 / 56	10 7 / 6 8	11 3 / 7 3.5	23 / 14	2 10.75 / 1 8	3 1.75 / 1 11	6.5
19	35 9	5 0	30	1 9	84 / 58	10 7 / 6 11	11 3 / 7 6.5	23 / 14	2 10.75 / 1 8	3 1.75 / 1 11	6.25

CHAPTER XXI.

ARTICLE 129.

DIRECTIONS FOR CONSTRUCTING UNDERSHOT WHEELS, SUCH AS SHOWN IN FIGURE 1, PLATE XIII.

1. Dress the arms straight and square on all sides, and find the centre of each; divide each into 4 equal parts on the side; square, centre, scribe, and gauge them from the upper side across each point, on both sides, 6 inches each way from the centre.

2. Set up a truckle or centre-post, for a centre to frame the wheel on, in a level piece of ground, and set a stake to keep up each end of the arms level with the truckle, of convenient height to work on.

3. Lay the first arm with its centre on the centre of the truckle, and take a square notch out of the upper side 3-4ths of its depth, wide enough to receive the 2d arm.

4. Make a square notch in the lower edge of the 2d arm, 1-4th of its depth, and lay it in the other, and they will joint, standing square across each other.

5. Lay the 3d arm just equi-distant between the others, and scribe the lower arms by the side of the upper, and the lower edge of the upper by the sides of the lower arms. Then take the upper arm off and strike the square scribes, taking out the lower half of the 3d arm, and the upper half of the lower arms, and fit and lay them together.

6. Lay the 4th arm on the others, and scribe as directed before; then take 3-4ths of the lower edge of the 4th arm, and 1-4th out of the upper edge of the others, and lay them together, and they will be locked together in the depth of one.

7. Make a sweep-staff with a gimlet hole for the centre at one end, which must be set by a gimlet in the centre of the arms. Measure from this hole half the diameter of the wheel, making a hole there, and another the depth of the shrouds towards the centre, making each edge of this sweep at the end next the shrouds, straight.

towards the centre hole, to scribe the ends of the shrouds by.

8. Circle both edges of the shrouds by the sweep; dress them to the proper width and thickness; lay out the laps 5 inches long; set a gauge to a little more than one-third their thickness; gauge all their ends for the laps from the outsides; cut them all out but the last, that it may be made a little longer, or shorter, as may suit to make the wheel the right diameter; sweep a circle on the arms to lay the shrouds to, while fitting them; put a small draw-pin in the middle of each lap, to draw the joints close; strike true circles both for the inside and outside of the shrouds, and $1\frac{1}{2}$ inches from the inside, where the arms are to be let in.

9. Divide the circle into 8 equal parts, coming as near the middle of each shroud as possible, strike a scribe across each to lay out the notch by, that is to be cut $1\frac{1}{2}$ inches deep, to let in the arm at the bottom, where it is to be forked to take in the remainder of the shroud. Strike a scribe on the arms with the same sweep that the stroke for the notches on the shrouds was struck with.

10. Scribe square down on each side of the arms, at the bottom, where they are to be forked; make a gauge to fit the arms, so wide as just to take in the shrouds, and leave $1\frac{1}{2}$ inches of wood outside of the mortise; bore 1 or 2 holes through each end of the arms to draw-pin the shrouds to the arms when hung; mark all the arms and shrouds to their places, and take them apart.

11. Fork the arms, put them together again, and put the shrouds into the arms; draw-bore them, but not too much, which would be worse than too little; take the shrouds apart again, turn them the other side up, and draw the joints together with the pins, and lay out the notches for 4 floats between each arm, 32 in all, large enough for admitting keys to keep them fast, but allowing them to drive in when any thing gets under the wheel. The ends of the floats must be dove-tailed a little into the shrouds; when one side is framed, frame the other to fellow it. This done, the wheel is ready to hang, but remember to face the shrouds between the arms

with inch boards, nailed on with strong nails, to keep the wheel firmly together.

ARTICLE 130.

DIRECTIONS FOR DRESSING SHAFTS, &c.

The shaft for a water-wheel with 8 arms should be 16 square, or 16 sided, about 2 feet diameter, the tree to make it being 2 feet 3 inches at the top end. When cut down, saw it off square at each end, and roll it on level skids, and if it be not straight, lay the rounding side down and view it, to find the spot for the centre at each end. Set the large compasses to half its diameter, and sweep a circle at each end, plumb a line across each centre, and at each side of the circle, striking chalk lines over the plumb lines at each side from end to end, and dress the sides plumb to these lines; turn it down on one side, setting it level; plumb, line, and dress off the sides to a 4 square; set it exactly on one corner, and plumb, line, and dress off the corner to 8 square. In the same manner dress it to 16 square.

To cut it square off to its exact length, stick a peg in the centre of each end, take a long square, (which may be made of boards,) lay it along the corner, the short end against the end of the peg, mark on the square where the shaft is to be cut, and mark the shaft by it at every corner line, from mark to mark; then cut it off to the lines, and it will be truly square.

ARTICLE 131.

TO LAY OUT THE MORTISES FOR THE ARMS.

Find the centre of the shaft at each end, and strike a circle; plumb a line through the centre at each end to be in the middle of two of the sides; make another scribe square across it; divide the distance equally between them, so as to divide the circle into 8 equal parts, and strike a line from each of them, from end to end, in the middle of the sides; measure from the top end about 3 feet, and mark for the arm of the water-wheel, and the width of the wheel, and make another mark. Take a straight-edged 10 feet pole, and put the end even with the end of the shaft, and mark on it even with the marks on the shaft, and by these marks measure for the arm at every corner, marking and lining all the way round. Then take the uppermost arms of each rim, and by them lay out the mortises, about half an inch longer than they are wide, which is to leave key room; set the compasses a little more than half the thickness of the arms, and set one foot in the centre line at the end of the mortise, striking a scribe each way to lay out the width by; this done, lay out 2 more on the opposite side, to complete the mortises through the shaft. Lay out 2 more, square across the first, one-quarter the width of the arm longer, inwards, towards the middle of the wheel. Take notice which way the locks of the arms wind, whether to right or left, and lay out the third mortises to suit, else it will be a chance whether they suit or not: these must be half the width of the arms longer, inwards.

The 4th set of mortises must be three-fourths longer inwards than the width of the arms; the mortises should be made rather hollowing than rounding, that the arms may slip in easily and stand fair.

If there be 3 (which are called 6) arms to the cogwheel, but one of them can be put through the sides of the shaft fairly; therefore, to lay out the mortises, divide the end of the shaft anew, into but 6 equal parts, by striking a circle on each end; and without altering the compasses, step from one of the old lines, six steps round the circle, and from these points strike chalk lines, and they will be the middle of the mortises, which may be laid out as before, minding which way the arms lock, and making two of the mortises one-third longer than the width of the arm, extending one on one side, and the other on the other side of the middle arm.

If there be but 2 (called 4) arms in the cog-wheel, (which will do where the number of cogs does not exceed 60) they will pass fairly through the sides, whether the shafts be 12 or 16 sided. One of these must be made one-half longer than the width of the arms, to give room to put the arm in.

ARTICLE 132.

TO PUT IN THE GUDGEONS.

Strike a circle on the ends of the shaft to let on the end bands; make a circle all round, 2½ feet from each end, and saw a notch all round, half an inch deep. Lay out a square, round the centres, the size of the gudgeons, near the neck; lay the gudgeons straight on the shaft, and scribe round them for their mortises; let them down within one-eighth of an inch of being in the centre. Dress off the ends to suit the bands; make 3 keys of good, seasoned white oak, to fill each mortise above the gudgeons, to key them in, those next to the gudgeons to be 3¼ inches deep at the inner end, and 1½ inches at their outer end, the wedge or driving key 3 inches at the head, and 6 inches longer than the mortise, that it may be cut off, if it batter in driving; the piece next the band so wide as to rise half an inch above the shaft, when all are laid in. Then take out all the keys and put on the bands, and make 8 or 12 iron wedges about 4 inches long by 2 wide, 1-3d inch thick at the end, not much tapered except half an inch at the small end, on one side next the wood; by means of a set, drive them in on each side the gudgeon extremely hard, at a proper distance apart. Then put in the keys again, and lay a piece of iron under each band, between it and the key, 6 inches long, half an inch thick in the middle, and tapering off at the ends; then grease the keys well with tallow, and drive it well with a heavy sledge: after this, drive an iron wedge, half an inch from the two sides of each gudgeon, 5 inches long, about half an inch thick, and as wide as the gudgeon.

ARTICLE 133.

OF COG-WHEELS.

The great face cog-wheels require 3 (called 6) arms, if the number of cogs exceed 54; if less, 4 will do. We find by the table, example 43, that the cog-wheel must have 69 cogs, with $4\frac{1}{2}$ inches pitch, the diameter of its pitch circle 8 feet $2\frac{1}{3}$ inches, and of its outsides 8 feet $10\frac{1}{3}$ inches. It requires 3 arms, 9 feet long, 14 by $3\frac{3}{4}$ inches; 12 cants, $6\frac{1}{2}$ feet long, 16 by 4 inches. (See it represented, fig. 1, Plate XVII.)

To frame it, dress and lock the arms together, (fig. 6, Pl. XVII.) as directed, Art. 129, only mind to leave one-third of each arm uncut, and to lock them the right way to suit the winding of the mortises in the shaft, which is best found by putting a strip of board in the middle mortise, and supposing it to be the arm, mark which way it should be cut, then apply the board to the arm, and mark it. The arms being laid on a truckle, as directed, Art. 129, make a sweep, the sides directing to the centre, 2 feet from the outer end to scribe by; measure on the sweep, half the diameter of the wheel; and by it circle out the back edges of the cants, all of one width in the middle; dress them, keeping the best faces for the face side of the wheel; make a circle on the arms half an inch larger than the diameter of the wheel, laying 3 of the cants with their ends on the arms, at this circle, at equal distances apart. Lay the other three on the top of them, so as to lap equally; scribe them both under and top, and gauge all for the laps from the face side; dress them out and lay them together, and joint them close: draw-pin them by an inch pin near their inside corners: this makes one-half of the wheel, shown fig. 5. Raise the centre level with that half; strike a circle near the outside, and find the centre of one of the cants; then, with the sweep that described the circle, step on the circle 6 steps, beginning at the middle of the cant, and these steps will show the middle of all the cants, or places for the arms. Make a scribe from the centre across each; strike another circle exactly at the corners, to place the corners of the next half by, and another about $2\frac{1}{2}$ inches farther out than the inside of the widest part of the cant, to let the arms in by; lay on three of the upper cants, the widest part over the narrowest part of the lower half, the inside to be at the point where the corner circle crosses the centre lines. Saw off the ends, at the centre scribes, and fit them down to their places, doing the same with the rest. Lay them all on, and joint their ends together; draw-pin them to the lower half, by inch pins, 2 inches from their inner edges, and 9 inches from their ends. Raise the centre level with the wheel; plane a little of the rough off the face, and strike the pitch circle, and another 4 inches inside, for the width of the face; strike another very near it, in which drive a chisel, half an inch deep, all round, and strike lines, with chalk, in the middle of the edge of the upper cants, and cut out of the solid, half of the upper cants, which raises the face; divide the pitch circle into 69 equal parts, $4\frac{1}{2}$ inches pitch, beginning and ending in a joint; strike two other circles each $2\frac{1}{2}$ inches from the pitch circle, and strike central scribes between the cogs, and where they cross the circles put in pins, as many as there are cogs, half on each circle; find the lowest part on the face, and make the centre level with it; look across in another place, square with the first, and make it level with the centre also; then make the face straight, from these four places, and it will be true.

Strike the pitch circle, and divide it over again, and strike one circle on each side of it, 1 inch distance, for the cog mortises; sweep the outside of the wheel and inside of the face, and two circles $\frac{3}{4}$ths of an inch from them, to dress off the corners; strike a circle of two inches diameter on the centre of each cog, and with the sweep strike central scribes at each side of these circles for the cog mortises; bore and mortise half through; turn the wheel, dress and mortise the back side, leaving the arms from under it; strike a circle on the face edge of the arms, equal in diameter to that struck on the face of the half wheel, to let them in by; saw in square, and take out $4\frac{1}{2}$ inches, and let them into the back of the wheel $1\frac{1}{4}$ inches deep, and bore a hole $1\frac{1}{2}$ inches into each arm, to pin it to the wheel.

Strike a circle on the arms one inch less than the diameter of the shaft; make a key 8 inches long; 1½ thick, 3¼ at the butt, and 2½ inches at the top end, and by it lay out the mortises; two on each side of the shaft, in each arm, to hang the wheel by.

ARTICLE 134.

OF SILLS, SPUR-BLOCKS, AND HEAD-BLOCKS.

See a side view of them in Plates XIII., XIV., XV., and XVI., and a top view of them, with their keys, at the end of the shaft, Plate XVIII. The sills are generally 12 inches square. Lay them on the wall as firmly as possible, and one 3 feet farther out; on these lay the spurs, which are 5 feet long, 7 by 7 inches, 3 feet apart, notched and pinned to the sills: on these are set the head-blocks, 14 by 12 inches, 5 feet long, let down with a dove-tail shoulder between the spurs, to support keys to move it endwise, and let 2 inches into the spurs with room for keys, to move it sidewise, and hold it to its place; see fig. 33 and 34, Plate XVIII. The ends of the shaft are let 2 inches into the head-blocks, to throw the weight more on the centre.

Provide two stones 5 or 6 inches square, very hard and clear of grit, for the gudgeons to run on, let them into the head-blocks, put the cog-wheel into its place, and then put in the shaft on the head-blocks in its place.

Put in the cog-wheel arm, lock them together, and pin the wheel to them; then hang the wheel, first by the keys to make it truly round, and then by side wedges, to make it true in face; turn the wheel, and make two circles, one on each side of the cog mortises, half an inch from them, so that the head of the cogs may stand between them equally.

ARTICLE 135.

OF COGS; THE BEST TIME FOR CUTTING, AND MANNER OF SEASONING THEM.

Cogs should be cut 14 inches long, and $3\frac{1}{4}$ inches square; this should be done when the sap runs at its fullest, at least a year before they are used, that they may dry without cracking. If either hickory or white oak be cut when the bark is set, they will worm-eat, and, if dried hastily, will crack; to prevent which, boil them and dry them slowly, or soak them in water, a year, (20 years in mud and fresh water would not hurt them;) when they are taken out they should be put in a hay-mow, under the hay, where, while foddered away, they will dry without cracking; but this often takes too long a time. I have discovered the following method of drying them, in a few days, without cracking. I have a malt kiln with a floor of laths two inches apart; I shank the cogs, hang them shank downwards, between the laths, cover them with a hair cloth, make a wood fire, and the smoke prevents them from cracking. Some dry them in an oven, which ruins them. Boards, planks, or scantling, are best dried in a kiln, covered so as to keep the smoke amongst them. Instead of a malt kiln, dig a cave in the side of a hill, 6 feet deep, 5 or 6 feet wide, with a post in each corner with plates on them, on which lay laths on edge, and pile the cogs on end, nearly perpendicular, so that the smoke can pass freely through, or amongst them. Cover them slightly with boards and earth, make a slow fire, and close up the sides, and renew the fire once a day, for 12 or 15 days they will then dry without cracking.

ARTICLE 136.

OF SHANKING, PUTTING IN, AND DRESSING OFF COGS.

Straighten one of the heart sides for the shank, make a pattern, the head 4 and shank 10 inches long, and 2 inches wide at the head, $1\frac{3}{4}$ at the point; lay it on the cog, scribe the shank and shoulders, for the head, saw in and dress off the sides; make another pattern of the shank, without the head, to scribe the sides and dress off the backs by, laying it even with the face, which is to have no shoulder; take care in dressing them off, that the axe do not strike the shoulder; if it do, it will crack there in drying, (if they be green;) fit and drive them in the mortises exceedingly tight, with their shoulders foremost, when at work. When the cogs are all in, fix two pieces of scantling, for rests, to scribe the cogs by, one across the cog-pit, near the cogs, another in front of them; fix them firmly. Hold a pointed tool on the rest, and scribe for the length of the cogs, by turning the wheel, and saw them off $3\frac{1}{2}$ inches long; then move the rest close to them, and fix it firmly; find the pitch circle on the end of the cogs, and, by turning the wheel, describe it there.

Describe another line $\frac{1}{4}$th of an inch outside thereof, to set the compasses in to describe the face of the cogs by, and another at each side of the cogs to dress them to their width; then pitch the cogs by dividing them equally, so that, in stepping round, the compasses may end in the point where they began; describe a circle, in some particular place, with the pitch, that it may not be lost; these points must be as nearly as possible of a proper distance for the centre from the back of the cogs; find the cog to the back of which this point comes nearest, and set the compasses from that point to the back of the cog; with this distance set off the backs of all the cogs equally, on the circle, $\frac{1}{4}$th of an inch outside of the pitch circle, and from these points, last made, set off the thickness of the cogs, which should, in this case, be $1\frac{1}{8}$ inches.

Then describe the face and back of the cogs by setting the compasses in the hindmost point of one cog, and sweeping over the foremost point of another, for the face, and in the foremost point of one, sweeping over the hindmost of the other, for the back part; dress them off on all sides, tapering about $\frac{1}{8}$th of an inch, in an inch distance; try them by a gauge, to make them all alike; take a little off the corners, and they are finished.

ARTICLE 137.

OF THE LITTLE COG-WHEEL AND SHAFT.

The process of making this is similar to that of the big cog-wheel. Its dimensions we find by the table, and the same example (43,) to be 52 cogs, 4¼ pitch; diameter of pitch circle 5 feet 10⅓ inches, and from out to out, 6 feet 6 inches.

It requires 2 arms, 6 feet 6 inches long, 11 by 3¼ inches; 8 cants, 5 feet 6 inches, 17 by 3½ inches. (See it, fig. 4, Plate XVII.)

Of the Shaft.

Dress it 8 feet long, 14 by 14 inches square, and describe a circle on each end 14 inches diameter; strike two lines through the centre, parallel to the sides, and divide the quarters into 4 equal parts, each; strike lines across the centre at each part at the end of these lines; strike chalk lines from end to end, to hew off the corners by, and it will be 8 square; lay out the mortises for the arms, put on the bands, and put in the gudgeons, as with the big shaft.

ARTICLE 138.

DIRECTIONS FOR MAKING WALLOWERS AND TRUNDLES.

By example 43, in the table, the wallower is to have 26 rounds 4½ pitch: the diameter of its pitch circle is 3 feet 1¼ inches, and 3 feet 4¼ inches from outsides: (see fig. 3, Plate XVII.) Its head should be 3½ inches thick, doweled truly together, or made with double plank, crossing each other. Make the bands 3 inches wide, $\frac{1}{6}$th of an inch thick, evenly drawn; the heads must be made to suit the bands, by setting the compasses so that they will step round the inside of the band in 6 steps; with this distance sweep the head, allowing about $\frac{1}{10}$th of an inch outside, in dressing, to make such a large band tight. Make them hot alike all around with a chip fire, which swells the iron; put them on the head while hot, and cool them with water, to keep them from burning the wood too much, but not too fast, lest they snap; the same mode serves for hooping all kinds of heads.

Dress the head fair after banding, and strike the pitch circle and divide it by the same pitch with the cogs; bore the holes for the rounds with an auger of at least $1\frac{1}{4}$ inches; make the rounds of the best wood, $2\frac{3}{8}$ inches diameter, and 11 inches between the shoulders, the tenons 4 inches, to fit the holes loosely, until within 1 inch of the shoulder, then drive it tight. Make the mortises for the shaft in the heads, with notches for the keys to hang it by. When the rounds are all driven into the shoulders, observe whether they stand straight; if not, they may be set fair by putting the wedges nearest to one side of the tenon, so that the strongest part may incline to draw them straight: this should be done with both heads.

ARTICLE 139.

OF FIXING THE HEAD-BLOCKS AND HANGING THE WHEELS.

The head-blocks, for the wallower shaft, are shown in Plate XVIII. Number 19 is one called a spur, 6 feet long and 15 inches deep, one end of which, at 19, is let 1 inch into the top of the husk-sill, which sill is $1\frac{1}{2}$ inches above the floor, the other end tenoned strongly into a strong post, 14 by 14 inches, 12 or 14 feet long, standing near the cog-wheel, on a sill in the bottom of the cog-pit; the top is tenoned into the husk-plank; these are called the tomkin posts. The other head-blocks appear at 20 and 28. In these large head-blocks there are small ones let in, that are 2 feet long, and 6 inches square, with a stone in each for the gudgeons to run on. That one in the spur 19 is made to slide, to put the wallower in and out of gear, by a lever screwed to its side.

Lay the centre of the little shaft level with the big one, so as to put the wallower to gear $\frac{2}{3}$ the thickness of the rounds deep, into the cog-wheel; put the shaft into its place, hang the wallower, and gauge the rounds to equal distance where the cogs take. Hang the cog-wheel, put in the cogs, make the trundle as directed for the wallower. (See fig. 4, Plate XVII.)

ARTICLE 140.

DIRECTIONS FOR PUTTING IN THE BALANCE-RYNE.

Lay it in the eye of the stone, and fix it truly in the centre; to do which, make a sweep by putting a long pin through the end, to reach into, and fit, the pivot hole in the balance-ryne; by repeated trials on the opposite side, fix it in the centre; then make a particular mark on the sweep, and others to suit it on the stone, scribe round the horns, and with picks and chisels sink the mortises to their proper depth, trying, by the particular marks made for the purpose, by the sweep, if it be in the centre. Put in the spindle with the foot upwards, and the driver on its place, while one holds it plumb. Set the driver over two of the horns, if it has four, but between them if it has but two. When the neck is exactly in the centre of the stone, scribe round the horns of the driver, and let it into the stone, nearly to the balance, if it has four horns. Put the top of the spindle in the pivot-hole, to try whether the mortises let it down freely on both sides.

Make a tram, to set the spindle square by, as follows: take a piece of board, cut a notch in one side, at one end, and hang it on the top of the spindle, by a little peg in the shoulder of the notch, to go into the hole in the foot, to keep it on; let the other end reach down to the edge of the stone; take another piece, circle out one end to fit the spindle neck, and make the other end fast to the lower end of the hanging piece near the stone, so as to play round level with the face of the stone, resting on the centre-hole in the foot, and against the neck; put a bit of quill through the end of the level piece, that will touch the edge of the stone as it plays round. Make little wedges, and drive them in behind the horns of the driver, to keep both ends, at once, close to the sides of the mortises they bear against when at work, keeping the pivot or cock-head in its hole in the balance; try the tram gently round, and mark where the quill touches the stone first, and dress off the bearing sides of the mortises for the driver, until it will touch equally all round, giving the driver liberty to move endwise, and sidewise, so that the stone may rock an inch either way. The ryne and driver must be sunk ¾ths of an inch below the face of the stone. Then hang the trundle firmly and truly on the spindle; put it in its place, to gear in the little cog-wheel.

ARTICLE 141.

TO BRIDGE THE SPINDLE.

Make a little tram of a piece of lath, 3 inches wide at one end, and 1 inch at the other, make a mortise in the wide end, and put it on the cock-head, and a piece of quill in the small end, to play round the face of the stone: then, while one turns the trundle, another observes where the quill touches first, and alters the keys of the bridge-tree, driving the spindle-foot toward the part the quill touches, until it does so equally all round. Case the stone neatly round, within 2 inches of the face.

ARTICLE 142.

OF THE CRANE AND LIGHTER STAFF.

Make a crane, with a screw and bale, for taking up and putting down the stone. (See it represented in Plate XI., fig. 2 and 3.) Set the post out of the way as much as possible, let it be 9 by 6 inches in the middle, the arm 9 by 6, the brace 6 by 4; make a hole plumb over the spindle, for the screw; put an iron washer on the arm under the female screw, nail it fast; the length of the screw in the worm part should exceed half the diameter of the stone, and it should reach 10 inches below it; the bale must touch only at the ends to give the stone liberty to turn, the pins to be 7 inches long, 1⅛ thick, the bale to be 2½ inches wide in the middle, and 1¾ inches wide at the end; the whole should be made of the best iron, for if either of them break, the danger would be great: the holes in the stone should be nearest to the upper side of it. Raise the runner by the crane, screw, and bale, turn it and lay it down, with the horns of the driving ryne in their right places, as marked, it being down, as it appears in Plate XXI., fig. 9. Make the lighter staff CC, to raise and lower the stone in grinding, about 6 feet long, $3\frac{1}{2}$ by $2\frac{1}{2}$ inches at the large end, and 2 inches square at the small end, with a knob on the upper side. Make a mortise through the but-end, for the bray-iron to pass through, which goes into a mortise 4 inches deep, in the end of the bray at b, and is fastened with a pin; it may be 2 inches wide and half an inch thick, made plain, with 1 hole at the lower, and 5 or 6 at the upper end; it should be set in a staggering position. This lighter is fixed in front of the meal-beam, at such a height as to be handy to raise and lower at pleasure; a weight of 4lbs. is hung to the end of it by a strap, which laps two or three times round, and the other end is fastened to the post below, that keeps it in its place. Play the lighter up and down, and observe whether the stone rises and falls flat on the bed-stone; if it do, draw a little water, and let the stone move gently round; then see that all things be right, and draw a little more water, let the stone run at a moderate rate, and grind the faces a few minutes.

ARTICLE 143.

DIRECTIONS FOR MAKING A HOOP FOR THE MILL-STONE.

Take a white pine or poplar board, 8 inches longer than will go round the stone, and 2 inches wider than the top of the stone is high, dress it smooth, and gauge it one inch thick, run a gauge mark $\frac{1}{8}$th of an inch from the outside, divide the length into 52 parts, and saw as many saw-gates square across the inside to the gauge-line. Take a board of equal width, 1 foot long, nail one-half of it on the outside at one end of the hoop, lay it in water a day or two to soak, or frequently sprinkle the outside with hot water, during an hour or two. Bend it round so that the ends meet, and nail the other end to the short board, put sticks across inside, in various directions, to press out the parts that bend least, and make it truly round. Make a cover for the hoop, (such as is represented in Plate XIX., fig. 23;) 8 square inside, and 1 inch outside the hoop. It consists of 8 pieces lapped one over another, the black lines showing the joints, as they appear when made, the dotted lines the under parts of the laps. Describe it on the floor, and make a pattern to make all the rest by; dress all the laps, fit and nail them together by the circle on the floor, and then nail it on the hoop; put the hoop over the stone and scribe it to fit the floor.

ARTICLE 144.

OF GRINDING SAND TO FACE THE STONES.

Lay boards over the hoop to keep the dust from flying, and take a bushel or two of dry, clean, sharp sand, teem it gently into the eye, while the stone moves at a moderate rate, continuing to grind for an hour or two; then take up the stones, sweep them clean, and pick the smoothest, hardest places, and lay the stone down again, and grind more sand as before, turning off the back, (if it be a burr,) taking great care that the chisel do not catch; take up the stone again, and make a red staff, equal, in length, to the diameter of the stone, and 3 by $2\frac{1}{2}$ inches; paint it with red paint and water, and rub it over the face of the stone in all directions, the red will be left on the highest and hardest parts, which must be picked down, making the bed-stone perfectly plain, and the runner a little concave, about $\frac{1}{6}$th of an inch at the eye, and lessening gradually to about 8 inches from the skirt. If they be close, and have much face, they need not touch, or flour, so far as if they be open, and have but little face; those things are necessarily left to the judgment of the mill-wright and miller.

ARTICLE 145.

DIRECTIONS FOR LAYING OUT THE FURROWS IN THE STONES, &c.

If they be five feet in diameter, divide the skirt into 16 equal parts, called quarters; if 6 feet, into 18; if 7 feet, into 20 quarters. Make two strips of board, one an inch, and the other 2 inches wide; stand with your face to the eye, and if the stone turn to the right when at work, lay the strip at one of the quarter divisions, and the other at the left hand side close to the eye, and mark with a flat pointed spike for a master furrow; they are all to be laid out the same way in both stones, for when their faces are together, the furrows should cross each other, like shears in the best position for cutting cloth. Then, having not fewer than 6 good picks, proceed to pick out all the master furrows, making the edge next the skirt and the end next the eye, the deepest, and the feather edge not half so deep as the back.

When all the master furrows are picked out, lay the broad strip next to the feather edges of all the furrows, and mark the head lands of the short furrows, then lay the same strip next the back edges, and mark for the lands, and lay the narrow strip, and mark for the furrows, and so mark out all the lands and furrows, minding not to cross the head lands, but leaving it between the master furrows and the short ones of each quarter. But if they be close country stones, lay out both furrows and lands with the narrow strip.

The neck of the spindle must not be wedged too tight, else it will burn loose; bridge the spindle again; put a collar round the spindle neck, but under it put a piece of an old stocking, with tallow rolled up in it, about a finger thick; tack it closely round the neck; put a piece of stiff leather about 6 inches diameter on the cock-head under the driver, to turn with the spindle and drive off the grain, &c., from the neck; grease the neck with tallow every time the stone is up.

Lay the stone down and turn off the back smooth, and

grind more sand. Stop the mill, raise the stone a little, and balance it truly with a weight laid on the lightest side. Take lead equal to this weight, melt it, and run it into a hole made in the same place in the plaster; this hole should be largest at bottom to keep it in; fill the hole with the plaster, take up the runner again, try the staff over the stones, and if in good face, give them a nice dressing, and lay them down to grind wheat.

ARTICLE 146.

DIRECTIONS FOR MAKING A HOPPER, SHOE, AND FEEDER.

The dimensions of the hopper of a common mill is 4 feet at the top, and 2 feet deep, the hole in the bottom 3 inches square, with a sliding gate in the bottom of the front to lessen it at pleasure: the shoe 10 inches long, and 5 wide in the bottom, of good sound oak. The side 7 or 8 inches deep at the hinder end, 3 inches at the foremost end, 6 longer than the bottom of the fore end, slanting more than the hopper behind, so that it may have liberty to hang down 3 or 4 inches at the fore end, which is hung by a strap called the feeding-string, passing over the fore end of the hopper-frame, and lapping round a pin in front of the meal-beam, which pin will turn by the hand, and which is called the feeding-screw.

The feeder is a piece of wood turned in a lathe, about 20 inches long, 3 inches diameter in the middle, against the shoe, tapered off to $1\frac{1}{2}$ inches at the top; the lower end is banded, and a forked iron driven in it, that spans over the ryne, fitting into notches made on each side, to receive it, directly above the spindle, with which it turns, the upper end running in a hole in a piece across the hopper-frame. In the large part, next the shoe, 6 iron knockers are set, 7 inches long, half an inch diameter, with a tang at each end, turned square to drive into the wood, these knock against and shake the shoe, and thereby shake in the grain regularly.

You may now put the grain into the hopper, draw water on the mill, and regulate the feed by turning the feed screw, until the stream falling into the eye of the stone, be proportioned to the size thereof, or the power of the mill. Here ends the mill-wright's work, with respect to grinding, and the miller takes the charge thereof.

ARTICLE 147.

OF BOLTING CHESTS AND REELS.

Bolting chests and reels are of different lengths, according to the use for which they are intended. Common country chests (a top view of one of which is shown in Plate XIX., fig. 9,) are usually about 10 feet long, 3 feet wide, and 7 feet 4 inches high, with a post in each corner; the bottom 2 feet from the floor, with a board 18 inches wide, set slanting in the back side, to cast the meal forward in the chest, that it may be easily taken up; the door is of the whole length of the chest, and two feet wide, the bottom board below the door sixteen inches wide.

The shaft of the reel is equal in length with the chest, 4 inches diameter, 6 square, two bands on each end, $3\frac{1}{4}$ and $3\frac{3}{4}$ inches diameter; gudgeons 13 inches long, $\frac{7}{8}$ of an inch diameter, 8 inches in the shaft, rounded at the neck $2\frac{1}{2}$ inches, with a tenon for a socket, or handle; there are six ribs $1\frac{1}{2}$ inch deep, $1\frac{1}{8}$ inch thick, $\frac{1}{2}$ an inch at the tail, and $1\frac{1}{2}$ inch at the head, shorter than the shaft, to leave room for the meal to be spouted in at the head, and the bran to fall out at the tail; there are four sets of arms, that is, 12 of them, $1\frac{1}{2}$ inches wide, and $\frac{5}{8}$ thick. The diameter of the reel from out to out of the ribs, is one-third part of the double width of the cloth. A round wheel, made of inch boards, in diameter equal to the outside of the ribs, and $4\frac{1}{2}$ inches wide, measuring from the outside towards the centre, (which is taken out,) is to be framed to the head of the reel, to keep the meal from falling out at the head, unbolted. Put a hoop $4\frac{1}{4}$ inches wide, and $\frac{1}{4}$ thick, round the tail, to fasten the cloth to. The cloth is sewed, two widths of it together, to reach round the reel, putting a strip of strong linen, 7 inches wide at the head, and 5 inches at the tail of the cloth, by which to fasten it to the reel. Paste on each rib a strip of linen, soft paper, or chamois leather (which is the best) 1½ inches wide, to keep the cloth from fretting. Then put the cloth on the reel tight, sew or nail it to the tail, and stretch it lengthwise as hard as it will bear, nailing it to the head.—Six yards of cloth cover a ten feet reel.

Bolting reels for merchant mills are generally longer than for country work, and every part should be stronger in proportion. They are best when made to suit the wide cloths. The socket gudgeons at the head should be much stronger, they being apt to wear out, and troublesome to repair.

The bolting-hopper is made to pass through the floor above the chest, is 12 inches square at the upper, and 10 inches at the lower end; the foremost side 5 inches, and the back side 7 inches from the top of the chest.

The shoe 2 feet long at the bottom of the side pieces, slanting to suit the hopper at the hinder end, set 4 inches higher at the hinder than the fore end, the bottom 17 inches long, and 10 inches wide. There should be a bow of iron riveted to the fore end, to rest on the top of the knocking wheel, which is fixed on the socket gudgeon at the head of the chest, and is 10 inches diameter, 2 inches thick, with 6 half rounds, cut out of its circumference, forming knockers to strike against the bow, and lift the shoe ¾ of an inch every stroke, to shake in the meal.

ARTICLE 148.

OF SETTING BOLTS TO GO BY WATER.

The bolting reels are set to go by water as follows:—
Make a bridge 6 by 4 inches, and 4 inches longer than the distance of the tomkin post, described Art. 139; set it between them, on rests fastened into them 10 inches below the cogs of the cog-wheel, and the centre of it half the diameter of the spur-wheel in front of them; on this

bridge is set the step gudgeon of an upright shaft, with a spur-wheel of 16 or 18 cogs to gear into the cog-wheel. Fix a head-block to the joists of the 3d floor for the upper end of this shaft; put the wheel 28, (Pl. XIX.) on it; hang another head-block to the joists of the 2d floor, near the corner of the mill at 6, for the step of the short upright shaft that is to be fixed there, to turn the reels 1 and 9. Hang another head-block to the joists of the 3d floor, for the upper end of the said short upright, and fix also head-blocks for the short shaft at the head of the reels, so that the centres of all these shafts will meet. Then fix a hanging post in the corner 5, for the gudgeon of the long horizontal shaft 27—5 to run in. After the head-blocks are all fixed, then measure the length of each shaft, and make them as follows; namely:—

The upright shaft $5\frac{1}{2}$ inches for common mills, but if for merchant-work, with Evans' elevators, &c., added, make it larger, say 6 or 7 inches; the horizontal shaft 27 —5, and all the others 5 inches diameter. Put a socket-gudgeon in the middle of the long shafts, to keep them steady; make them 8, or 16, square, except at the end where the wheels are hung, where they must be 4 square. Band their ends, put in the gudgeons, and put them in their proper places in the head-blocks, to mark where the wheels are to be put on them.

ARTICLE 149.

OF MAKING BOLTING WHEELS.

Make the spur wheel for the first upright, with a $4\frac{1}{2}$ inch plank; the pitch of the cogs, the same as the cog-wheel, into which it is to work; put two bands $\frac{3}{4}$ of an inch wide, one on each side of the cogs, and a rivet between each cog, to keep the wheel from splitting.

To proportion the cogs in the wheels, to give the bolts the right motion, the common way is—

Hang the spur-wheel, and set the stones to grind with a proper motion, and count the revolutions of the upright shaft in a minute; compare its revolutions with the revolutions that a bolt should have, which is about 36 revo-

Chap. 21.] OF MAKING BOLTING WHEELS. 317

lutions in a minute. If the upright go $\frac{1}{6}$ more, put $\frac{1}{6}$ less in the first driving-wheel than in the leader, suppose 15 in the driver, then 18 in the leader: but if their difference be more, (say one-half,) there must be a difference in the next two wheels; observing that if the motion of the upright shaft be greater than that of the bolt should be, the driving-wheel must be proportionably less than the leader; but if it be slower, then the driver must be greater in proportion. The common size of bolting wheels is from 14 to 20 cogs; if less than 14, the head-blocks will be too near the shafts.

Common bolting wheels should be made of plank, at least 3 inches thick, well seasoned, and they are best when as wide as the diameter of the wheel, and banded with bands nearly as wide as the thickness of the wheel; the bands may be made of rolled iron, about $\frac{3}{8}$ of an inch thick. Some make the wheels of 2 inch plank, crossed, and no bands; but this proves no saving, as they are apt to go to pieces in a few years. (For hooping wheels, see Art. 136, and for finding the diameter of the pitch circle, see Art. 126.) The wheels, if banded, are generally two inches more in diameter than the pitch circle; but if not, they should be larger. The pitch or distances of the cogs are different; if to turn 1 or 2 bolts, $2\frac{1}{2}$ inches; but if more $2\frac{3}{4}$; if they are to do much heavy work, they should not be less than 3 inches. Their cogs, in thickness, are half the pitch; the shank must drive tightly in an inch auger hole.

When the mortises are made for the shafts in the head, and notches for the keys to hang them, drive the cogs in and pin their shanks at the back side, and cut them off half an inch from the wheel.

Hang the wheels on the shafts so that they will gear a proper depth, about $\frac{2}{3}$ the thickness of the cogs; dress all the cogs to equal distances by a gauge; then put the shafts in their places, the wheels gearing properly, and the head-blocks all secure; set them in motion by water. Bolting reels should turn so as to drop the meal on the back side of the chest, as it will then hold more, and will not cast out the meal when the door is opened.

ARTICLE 150.

OF ROLLING-SCREENS.

These are circular sieves moved by water, and are particularly useful in cleaning wheat for merchant-work. They are of different constructions.

1st. Those of one coat of wire with a screw in them.

2dly. Those of two coats, the inner one nailed to six ribs, the outer one having a screw between it and the inner one.

3dly. Those of a single coat, and no screw.

The first kind answers well in some, but not in all cases, because they must turn a certain number of times before the wheat can get out, and the grain has not so good an opportunity of separating; there being nothing to change its position, it floats a considerable distance with the same grains uppermost.

The double kinds are better, because they may be shorter, and take up less room; but they are more difficult to keep clean.

The 3d kind has this advantage; we can keep the grain in them a longer or shorter time, at pleasure, by raising or lowering the tail end, and it is also tossed about more; but they must be longer. They are generally 9 or 10 feet long, 2 feet 4 inches diameter, if to clean for two or three pairs of stones; but if for more, they should be larger accordingly: they will clean for, from one, to six pairs of stones. They are made 6 square, with 6 ribs, which lie flatwise, the outer corners taken off to leave the edges $\frac{1}{4}$ of an inch thick; the inner corners are brought nearly to sharp edges; the wire work is nailed on with 14 ounce tacks.

The screens are generally moved by the same upright shaft that moves the bolts, which has a wheel on its upper end, with two sets of cogs: those that strike downwards, gear into a wheel striking upwards, which turns a laying shaft, having two pulleys on the other end, one of 24 inches diameter, to turn a fan with a quick motion, the other of 8 inches, which conducts a strap to a pulley

24 inches diameter, on the gudgeon of the rolling screen, to reduce its motion to about 15 revolutions in a minute. (See fig. 19, Plate XIX.) This strap gearing may do for mills in a small way, but where they are in perfection for merchant-work, with elevators, &c., and have to clean wheat for 2, 3, or 4 pairs of stones, they should be moved by cogs.

ARTICLE 151.

OF FANS.

The Dutch fan is a machine of great use, for blowing the dust and other light stuff from among the wheat; there are various sorts of them; those that are only for blowing the wheat, as it falls from the rolling-screen, are generally about 15 inches long, and 14 inches wide, in the wings, and have no riddle or screen in them.

To give motion to a fan of this kind, put a pulley 7 inches diameter, on its axle, to receive a band from a pulley on the shaft that moves the screen, which pulley may be of 24 inches diameter, to give a swift motion; when the band is slack it slips a little on the small pulley, and the motion is retarded, but when tight the motion is quicker; by this the blast is regulated.

Some use Dutch fans complete, with riddle and screen under the rolling screen, for merchant-work; and again use the fan alone for country-work.

The wings of those which are the common farmers' wind-mills, or fans, are 18 inches long, and 20 inches wide; but in mills they are set in motion with a pulley instead of a cog-wheel and wallower.

ARTICLE 152.

OF THE SHAKING SIEVE.

Shaking sieves are of considerable use in country mills, to sift Indian meal, separating it, if required, into several degrees of fineness; and to take the hulls out of buckwheat meal, which are apt to cut the bolting cloth; also, to take the dust out of the grain, if rubbed before ground; they are sometimes used to clean wheat, or screenings, instead of rolling screens.

If they are for sifting meal, they are 3 feet 6 inches long, 9 inches wide, $3\frac{1}{2}$ inches deep; (see it fig. 16, Plate XVIII.) The wire-work is 3 feet long and 8 inches wide: across the bottom of the tail end is a board 6 inches wide, to the top of which the wire is tacked, and then this board and wire are tacked to the bottom of the frame, leaving an opening at the tail end for the bran to fall into the box 17, the meal falling into the meal-trough 15; the head piece should be strong, to hold the iron bow at 15, through which the lever passes that shakes the sieve, which is effected in the following manner. Take two pieces of hard wood, 15 inches long, and as wide as the spindle, and so thick that when one is put on each side just above the trundle, it will make it $1\frac{1}{2}$ inches thicker than the spindle is wide. The corners of these are taken off to a half round, and they are tied to the spindle with a small, strong cord. These are to strike against the lever that works on a pin near its centre, which is fastened to the sieve, and shakes it as the trundle goes round; (see it represented Plate XVIII.) This lever must always be put to the side of the spindle, contrary to that of the meal-spout; otherwise, it will draw the meal to the upper end of the sieve: there must be a spring fixed to the sieve to draw it forward as often as it is driven back. It must hang on straps and be fixed so as to be easily set to any descent required, by means of a roller in the form of a feeding screw, only longer; round this roller the strap winds.

I have now given directions for making, and putting to work, all the machinery of one of the most complete of the old-fashioned grist-mills, that may do merchant-work in the small way; these are represented by Plates XVIII., XIX., XX., XXI.; but they are far inferior to those with the improvements, which are shown by Plate XXII.

CHAPTER XXII.

ARTICLE 153.

OF THE USE OF DRAUGHTING TO BUILD MILLS BY, &c.

Perhaps some are of opinion that draughts are useless pictures of things, serving only to please the fancy. This is not what is intended by them; but to give true ideas of the machine, &c., described, or to be made. Those represented in the plates are all drawn on a scale of $\frac{1}{8}$th of an inch to a foot, in order to suit the size of the book, except Plate XVII., which is a quarter of an inch to a foot; and this scale I recommend, as most buildings will then come on a sheet of common paper.

N. B. Plate XXIV. was made after the above directions, and has explanations to suit it.

The great use of draughting mills, &c., to build by, is to convey our ideas more plainly, than is possible by writing, or by words alone; these may be misconstrued or forgotten; but a draught, well drawn, speaks for itself, when once understood by the artist; who by applying his dividers to the draught and to the scale, finds the length, breadth, and height of the building; or the dimensions of any piece of timber, and its proper place.

By the draught the bills of scantling, boards, rafters, laths, shingles, &c. &c., are known and made out; it should show every wheel, shaft, and machine, and their places. By it we can find whether the house be sufficient to contain all the works that are necessary to carry on the business; the builder or owner understands what he is about, and proceeds cheerfully and without error: it directs the mason where to put the windows, doors, navel-holes, the inner walls, &c., whereas, if there be no draught, every thing goes on, as it were, in the dark; much time is lost and errors are committed to the loss of many pounds. I have heard a man say, that he believed his mill was 500*l*. better from having employed an experienced artist, to draw him a draught to build it by; and I know, by experience, the great utility of them. Every master builder, at least, ought to understand them.

ARTICLE 154.

DIRECTIONS FOR PLANNING AND DRAUGHTING MILLS.

1st. If it be a new seat, view the ground where the dam is to be, and where the mill-house is to stand, and determine on the height of the top of the water in the head race, where it is taken out of the stream; and level from it for the lower side of the race, down to the seat of the mill-house, and mark the level of the water in the dam there.

2ndly. Begin where the tail-race is to empty into the stream, and level from the top of the water up to the mill seat, noticing the depth thereof, in places, as you pass along, which will be of use in digging it out.

Then find the total fall, allowing one inch to a rod for fall in the races; but if they be very wide and long, less will do.

Then, supposing the fall to be 21 feet 9 inches, which is sufficient for an overshot mill, and the stream too light for an undershot; consider well what size stone will suit; for I do not recommend a large stone to a weak, nor a small one to a strong stream. I have proposed stones 4 feet diameter for light, 4,6 for middling, and 5 or 5 feet 6 inches diameter, for heavy streams. Suppose you determine on stones 4 feet, then look in table I., (which is for stones of that size,) column 2, for the fall that is nearest 21 feet 9 inches, your fall, and you find it in the 7th example. Column 3d contains the head of water over the wheel, 3 feet; 4th, the diameter of the wheel, 18 feet; 5th, its width 2 feet 2 inches, &c., for all the proportions to make the stone revolve 106 times in a minute.

Having determined on the size of the wheels, and also of the house; the heights of the stories, to suit the wheels, and machinery it is to contain, and the business to be carried on therein, proceed to draw a ground plan of the house, such as Plate XVIII., which is 32 by 55 feet. (See the description of the plate.) And for the second story, as plate XIX., &c., and for the 3d, 4th, and 5th floors, if required; taking care to plan every thing, so that one shall not clash with another.

Draw an end view, as Plate XX., and a side view, as Plate XXI. Take the draught to the ground, and stake out the seat of the house. It is, in general, best to set that corner of an overshot mill, at which the water enters, farthest in the bank; but great care should be taken to reconsider and examine every thing, more than once, to see whether it be planned for the best; because, much labour is often lost for want of due consideration, and by setting buildings in, and laying foundations on, wrong places. The arrangements being completed, the bills of scantling and iron work may be made out from the draught.

ARTICLE 155.

BILLS OF SCANTLING FOR A MILL, 32 BY 55 FEET, 3 STORIES HIGH; THE WALLS OF MASON WORK, SUCH AS IS REPRESENTED IN PLATES XVIII., XIX., XX., AND XXI.

For the first Floor.

2 sills, 29 feet long, 8 by 12 inches, to lay on the walls for the joists to lie on.

48 joists, 10 feet long, 4 by 9 inches; all of timber that will last well in damp places.

For the second Floor.

2 posts, 9 feet long, 12 by 12 inches.
2 girders, 30 feet long, 14 by 16 do.
48 joists, 10 feet long, 4 by 9 do.

For the Floor over the Water House.

1 cross girder, 30 feet long, 12 by 14 inches, for one end of the joists to lie on.

2 posts to support the girder, 12 feet long, 12 by 12 inches.

16 joists, 13 feet long, 4 by 9 inches; all of good white-oak, or other timber, that will last in damp places.

For the third Floor.

4 posts, 9 feet long, 12 by 12 inches, to support the girders.
2 girder posts, 7 feet long, 12 by 12 inches, to stand on the water-house.
2 girders, 53 feet long, 14 by 16 inches.
90 joists, 10 feet long, 4 by 9 inches.

For the fourth Floor.

6 posts, 8 feet long, 10 by 10 inches, to support the girders.
2 girders, 53 feet long, 13 by 15 inches.
30 joists, 10 feet long, 4 by 8 do. for the middle tier of the floor.
60 do. 12 feet do. 4 by 8, for the outside tiers, which extend 12 inches over the walls, for the rafters to stand on.
2 plates, 54 feet long, 3 by 10 inches: these lie on the top of the walls, and the joists on them.
2 raising pieces, 55 feet long, 3 by 5 inches; these lie on the ends of the joists for the rafters to stand on.

For the Roof.

54 rafters, 22 feet long, 3 inches thick, $6\frac{1}{2}$ wide at the bottom, and $4\frac{1}{2}$ at the top end.
25 collar beams, 17 feet long, 3 by 7 inches.
2760 feet of laths, running measure.
7000 shingles.

For Doors and Window-Cases.

12 pieces, 12 feet long, 6 by 6 inches, for door-cases.
36 do. 8 feet long, 5 by 5 inches, for window-cases.

For the Water-House.

2 sills, 27 feet long, 12 by 12 inches.
1 do. 14 feet long, 12 by 12 do.
2 spur-blocks, 4 feet 6 inches long, 7 by 7 do.
2 head-blocks, 5 feet long, 12 by 14 do.
4 posts, 10 feet long, 8 by 8, to bear up the penstock.
2 cap-sails, 9 feet long, 8 by 10, for the penstock to stand on.

4 corner posts, 5 feet long, 4 by 6 inches, for the corners of the penstock.

For the Husk of a Mill of one Water-Wheel and two Pair of Stones.

2 sills, 24 feet long, 12 by 12 inches.
4 corner posts, 7 feet long, 12 by 14 inches.
2 front posts, 8 feet long, 8 by 12 do.
2 back posts, 8 feet do. 10 by 12 inches, to support the back ends of the bridge-trees.
2 other back posts, 8 feet long, 8 by 8 inches.
3 tomkin posts, 12 feet long, 12 by 14 do.
2 interties, 9 feet long, 12 by 12 inches, for the outer ends of the little cog-wheel shafts to rest on.
2 top pieces, 10 feet 6 inches long, 10 by 10 inches.
2 beams, 24 feet long, 16 by 16 inches.
2 bray-trees, 8¾ feet long, 6 by 12 inches.
2 bridge-trees, 9 feet long, 10 by 10 inches.
4 planks, 8 feet long, 6 by 14 inches, for the stone-bearers.
20 planks, 9 feet long, 4 by about 15 inches, for the top of the husk.
2 head-blocks, 7 feet long, 12 by 15 inches, for the wallower shafts to run on. They serve as spurs also for the head-block for the water-wheel shaft.

For the Water-Wheel and big Cog-Wheel.

1 shaft, 18 feet long, 2 feet diameter.
8 arms for the water-wheel, 18 feet long, 3 by 9 inches.
16 shrouds, 8½ feet long, 2 inches thick, and 8 deep.
16 face boards, 8 feet long, 1 inch thick, and 9 deep.
56 bucket boards, 2 feet 4 inches long, and 17 inches wide.
140 feet of boards, for soaling the wheel.
3 arms for the cog-wheel, 9 feet long, 4 by 14 inches.
16 cants, 6 feet long, 4 by 17 inches.

For little Cog-Wheels.

2 shafts, 9 feet long, 14 inches diameter.
4 arms, 7 feet long, 3¾ by 10 inches.
16 cants, 5 feet long, 4 by 18 inches.

For Wallowers and Trundles.

60 feet of plank, 3½ inches thick.
40 feet do. 3 inches thick, for bolting gears.

Cogs and Rounds.

200 cogs, to be split, 3 by 3, 14 inches long.
80 rounds, do. 3 by 3, 20 inches long.
160 cogs, for bolting works, 7 inches long, and 1¾ square; but if they be for a mill with machinery complete, there must be more in number, accordingly.

Bolting Shafts.

1 upright shaft, 14 feet long, 5½ by 5½ inches.
2 horizontal shafts, 17 feet long, 5 by 5 inches.
1 upright do. 12 feet long, 5 by 5 inches.
6 shafts, 10 feet long, 4 by 4 do.

ARTICLE 156.

BILL OF THE LARGE IRONS FOR A MILL OF TWO PAIR OF STONES.

2 gudgeons, 2 feet 2 inches long in the shaft; neck 4¼ inches long, 3 inches diameter, well steeled and turned. (See fig. 16, Plate XXIV.)
2 bands, 19 inches diameter inside, ¾ thick; and 3 inches wide, for the ends of the shaft.
2 do. 20½ inches inside, ½ an inch thick, and 2½ inches wide, for do.
2 do. 23 inches do. ½ an inch thick, and 2½ inches wide, for do.
4 gudgeons, 16 inches in the shaft, 3½ inches long, and 2¼ inches diameter in the neck, for wallower shafts: (See fig. 15, Plate XXIV.)
4 bands, 12 inches diameter inside, ½ an inch thick, and 2 wide, for do.
4 do. 12 inches do. ½ an inch thick, and 2 wide, for do.
4 wallower bands, 3 feet 2 inches diameter inside, 3 inches wide, and ¼ of an inch thick.
4 trundle bands, 2 feet diameter inside, 3 inches wide, and ¼ of an inch thick.

2 spindles and rynes; spindles 5 feet 3 inches long from the foot to the top of the necks; cock-heads 7 or 8 inches long above the necks; the body of the spindles $3\frac{1}{4}$ by 2 inches; the neck 3 inches long, and 3 inches diameter : the balance rynes proportional to the spindles, to suit the eye of the stone, which is 9 inches diameter. (See fig. 1, 2, 3, Plate XXIV.)

2 steps for the spindles, fig. 4.

2 sets of damsel-irons, 6 knockers to each set.

2 bray-irons, 3 feet long, $1\frac{3}{4}$ inches wide, $\frac{1}{2}$ an inch thick: being a plain bar, one hole at the lower, and 5 or 6 at the upper end.

Bill of Iron for the Bolting and Hoisting Works, in the common way.

2 spur-wheel bands, 20 inches diameter from outsides, for the bolting spur-wheel, $\frac{3}{4}$ths of an inch wide, and $\frac{1}{4}$th thick.

2 spur-wheel bands, 12 inches diameter from outsides, for the hoisting spur-wheel.

2 step-gudgeons and steps, 10 inches long, $1\frac{3}{8}$ inches thick in the tang, or square part; neck 3 inches long, for the upright shafts. (See fig. 5 and 6, Plate XXIV.)

2 bands for do. 5 inches diameter inside, $1\frac{1}{4}$ wide, and $1\frac{1}{4}$ thick.

2 gudgeons, 9 inches tang; neck 3 inches long, $1\frac{1}{8}$ square, for the top of the uprights.

8 bands, $4\frac{1}{2}$ inches diameter inside.

1 socket-gudgeon, $1\frac{1}{8}$ of an inch thick; tang 12 inches long; neck 4 inches; tenon to go into the socket $1\frac{1}{4}$ inches, with a key-hole at the end. (See fig. 8 and 9.)

14 gudgeons, neck $2\frac{1}{2}$ inches, tangs 8 inches long, and 1 inch square, for small shafts at one end of the bolting-reels.

10 bands for do. 4 inches diameter inside, and 1 inch wide.

4 socket-gudgeons, for the 4 bolting-reels, $1\frac{1}{4}$ square; tangs 8 inches; necks 3 inches, and tenons $1\frac{1}{2}$ inches, with holes in the ends of the tangs for rivets, to keep them from turning; the sockets 1 inch thick at the mortise, and 3 inches between the prongs. (See fig. 8 and 9.) Prongs 8 inches long and 1 wide.

8 bands, 3¼ inches, and 8 do. 4 inches, diameter, for the bolting-reel shafts.

For the Hoisting Wheels.

2 gudgeons, for the jack-wheel, neck 3½ inches, and tang 9 inches long, 1⅛ square.
2 bands for do. 4½ inches diameter.
2 gudgeons, for the hoisting wheel, neck 3½ inches, tang 9 inches long, and 1¼ inches square.
2 bands for do. 7 inches diameter.
6 bands for bolting-heads, 16 inches diameter inside, 2¼ wide, and ⅛th of an inch thick.
6 do. for do. 15 inches do. do.

N. B. All the gudgeons should taper a little, and the sides given are the largest part. The bands for shafts should be widest at the foremost side, to make them drive well; but those for heads should be both sides equal. Six picks for the stones, 8 inches long, and 1¼ wide, will be wanted.

ARTICLE 157.

EXPLANATION OF THE PLATES.

PLATE XVII.

Drawn from a scale of a quarter of an inch for a foot.

Fig. 1—a big cog-wheel, 8 feet 2⅓ inches the diameter of its pitch circle, 8 feet 10⅓ inches from out to out; 69 cogs, 4½ inches pitch.

2—a little cog-wheel, 5 feet 10⅓ inches the diameter of its pitch circle, and 6 feet 6 inches from out to out, to have 52 cogs, 4¼ pitch.

3—a wallower, 3 feet 1¼ inches the diameter of its pitch circle, and 3 feet 4¼ inches from out to out; 26 rounds, 4½ pitch.

4—a trundle, 1 foot 8⅓ inches the diameter of its pitch circle, and 1 foot 11⅓ inches from out to out; 15 rounds, 4¼ inches pitch.

5—the back part of the big cog-wheel.

6—a model of locking 3 arms together.

7—the plan of a forebay, showing the sills, caps, and

where the mortises are made for the posts, with a rack at the upper end to keep off the trash.

PLATE XVIII.—*The Ground-Plan of a Mill.*

Fig. 1 and 8—bolting chests and reels, top view.
2 and 4—cog-wheels that turn the reels.
3—cog-wheel on the lower end of a short upright shaft.
5 and 7—places for the bran to fall into.
6, 6, 6—three garners on the lower floor for bran.
9 and 10—posts to support the girders.
11—the lower door to load wagons, horses, &c., at.
12—the step-ladder, from the lower floor to the husk.
13—the place where the hoisting casks stand when filling.
14 and 15—the two meal-troughs and meal-spouts.
16—meal-shaking sieve for Indian and buckwheat.
17—a box for the bran to fall into from the sieve.
18 and 19—the head-block and long spur-block, for the big shaft.
20—four posts in front of the husks, called bray posts.
21—the water and cog-wheel shaft.
22—the little cog-wheel and shaft, for the lower stones.
23—the trundle for the burr stones.
24—the wallower for do.
25—the spur-wheel that turns the bolts.
26—the cog-wheel.
27—the trundle, head wallower, and bridge-tree, for country stones.
28—the four back posts of the husk.
29—the two posts that support the cross-girder.
30—the two posts that bear up the penstock at one side.
31—the water-wheel, 18 feet diameter.
32—the two posts that bear up the other side of the penstock.
33—the head-blocks and spur-blocks, at water end.
34—a sill to keep up the outer ends.
35—the water-house door.
36—a hole in the wall for the trunk to go through.
37—the four windows of the lower story.

PLATE XIX.—*Second Floor.*

Fig. 1 and 9—a top view of the bolting chests and reels.
2 and 10—places for the bran to fall into.
3 and 8—the shafts that turn the reels.
4 and 7—wheels that turn the reels.
5—a wheel on the long shafts between the uprights.
6—a wheel on the upper end of the upright shaft.
11 and 12—two posts that bear up the girders of the third floor.
13—the long shaft between two uprights.
14—five garners to hold toll, &c.
15—a door in the upper side of the mill-house.
16—a step ladder from 2d to 3d floor.
17—the running burr mill-stone laid off to be dressed.
18—the hatch way.
19—stair way.
20—the running country stone turned up to be dressed.
21—a small step-ladder from the husk to 2d floor.
22—the places where the cranes stand.
24—the pulley-wheel that turns the rolling screen.
25 and 26—the shaft and wheel which turn the rolling screen and fan.
27—the wheel on the horizontal shaft to turn the bolting reels.
28—the wheel on the upper end of the first upright shaft.
29—a large pulley that turns the fan.
30—the pulley at the end of the rolling screen.
31—the fan.
32—the rolling screen.
33—a step ladder from the husk to the floor over the water-house.
34 and 35—two posts that support the girders of the third floor.
36—a small room for the tailings of the rolling screen.
37—a room for the fannings.
38—   do. for the screenings.
39—a small room for the dust.
40—the penstock of water.
41—a room for the miller to keep his books in.

42—a fire-place.
43—the upper end door.
44—ten windows in the second story, twelve lights each.

PLATE XX.

Represents a view of the lower side of a stone millhouse, three stories high, which plan will suit tolerably well for a two story house, if the third story be not wanted. Part of the wall is supposed to be open, so that we have a view of the stones, running gears, &c.

Line 1 represents the lower floor, and is nearly level with the top of the sills, of the husk and water-house.

2, 3, and 4—the second, third, and fourth floors.

5 and 6—windows for admitting air under the lower floor.

7—the lower door, with steps to ascend to it, which commonly suits best to load from.

8—the arch over the tail-race for the water to run from the wheel.

9—the water-house door, which sometimes suits better to be at the end of the house, where it makes room to wedge the gudgeon.

10—the end of the water-wheel shaft.

11—the big cog-wheel shaft.

12—the little cog-wheel and wallower, the trundle being seen through the window.

13—the stones with the hopper, shoe, and feeder, as fixed for grinding.

14—the meal-trough.

There is an end view of the husk frame. There are thirteen windows with twelve lights each.

PLATE XXI.

Represents an outside view of the water-end of a millhouse, and is intended to show to the builders, and millwrights, the height of the walls, floors, and timbers, with the places of the doors and windows, and a view of the position of the stones and husk timbers, supposing the wall open, so that we could see them.

Fig. 1, 2, 3, and 4, shows the joists of the floors.

5—a weather-cock, turning on an iron rod.
6—the end of the shaft, for hoisting outside of the house, which is fixed above the collar-beams over the doors, to hoist into either of them, or either story, at either end of the house, as may suit best.
7—the dark squares, showing the ends of the girders.
8—the joists over the water-house.
9—the mill-stones, with the spindles they run on, and the ends of the bridge-trees as they rest on the brays a a. b b show the ends of the brays, that are raised and lowered by the levers c c, called the lighter-staffs, for raising and lowering the running stone.
10—the water-wheel and big cog-wheel.
11—the wall between the water and cog-wheel.
12—the end view of the two side walls of the house.

Plate XXII. is explained in the Preface.

CHAPTER XXIII.

ARTICLE 158.

OF SAW-MILLS.

Construction of their Water-Wheels.

The wheels for saw-mills have been variously constructed; the most simple, where water is plenty, and the fall above six feet, is the flutter-wheel; but where water is scarce, or the head insufficient to give flutter-wheels the requisite motion, high wheels, double geared, will be found necessary. Flutter-wheels may be adapted to any head above six feet, by making them low and wide, for low heads, and high and narrow for high ones, so as to have about 120 revolutions, or strokes of the saw in a minute: but rather than double gear, I would be satisfied with 100.

A TABLE

OF THE

DIAMETER OF FLUTTER-WHEELS FROM OUT TO OUT, AND THEIR WIDTH IN THE CLEAR, SUITABLE TO ALL HEADS, FROM SIX TO THIRTY FEET.

Head of water.	Diameter.	Width.
feet.	ft. in.	ft. in.
6	2 : 8	5 : 6
7	2 : 10	5 : 0
8	2 : 11	4 : 8
9	3 : 0	4 : 3
10	3 : 1	4 : 0
11	3 : 2	3 : 9
12	3 : 3	3 : 6
13	3 : 4	3 : 3
14	3 : 5	3 : 0
15	3 : 6	2 : 9
16	3 : 7	2 : 6
17	3 : 8	2 : 4
18	3 : 9	2 : 2
19	3 : 10	2 : 0
20	3 : 11	1 : 10
21	4 : 0	1 : 9
22	4 : 1	1 : 8
23	4 : 2	1 : 7
24	4 : 3	1 : 6
25	4 : 4	1 : 5
26	4 : 5	1 : 4
27	4 : 6	1 : 3
28	4 : 7	1 : 2
29	4 : 8	1 : 1
30	4 : 9	1 : 0

N. B.—The above wheels are proposed to be made as narrow as will well do, on account of saving water; but if this be abundant, the wheels may be made wider than directed in the table, and the mill will be the more powerful.

Of Geared Saw-Mills.

Of these I shall say but little, they being expensive and but little used.—They should be geared so as to give the saw 120 strokes in a minute, when at work in a common log. The water-wheel is like that of any other mill, whether of the overshot, undershot, or breast kind; the cog-wheel of the spur kind, and as large as will clear the water. The wallower commonly has 14 or 15 rounds, or such number as will produce the right motion. On the wallower shaft is a balance-wheel, which may be made of stone or wood; this is to regulate the motion. There should be a good head above the water-wheel to give it a lively motion, otherwise the mill will run heavily.

The mechanism of a complete saw-mill is such as to produce the following effects; namely:—

1. To move the saw up and down, with a sufficient motion and power.
2. To move the log to meet the saw.
3. To stop of itself when within 3 inches of being through the log.
4. To draw the carriage with the log back, by the power of the water, so that the log may be ready to enter again.

The mill is stopped as follows; namely:—When the gate is drawn the lever is held by a catch, and there is a trigger, one end of which is within half an inch of the side of the carriage, on which is a piece of wood an inch and a half thick, nailed so that it will catch against the trigger as the carriage moves, which throws the catch off the lever of the gate, and it shuts down at a proper time.

Description of a Saw-Mill.

Plate XXIII. is an elevation and perspective view of a saw-mill, showing the foundation, walls, frame, &c., &c.

Fig. 0, 1—the frame uncovered, 52 feet long, and 12 feet wide.

Fig. 2—The lever for communicating the motion from

the saw-gate to the carriage, to move the log; it is 8 feet long, 3 inches square, tenoned into a roller 6 inches diameter, reaching from plate to plate, and working on gudgeons in them; in its lower side is framed a block, 10 inches long, with a mortise in it two inches wide throughout its whole length, to receive the upper end of the hand pole, having in it several holes for an iron pin, to join the hand pole to it, to regulate the feed; by setting the hand-pole nearer the centre of the roller, less feed is given, and, farther off, gives more.

Fig. 3, the hand-pole or feeder, 12 feet long, and 3 inches square, where it joins the block, (Fig. 4,) and tapering 2 inches at the lower end, on which is the iron hand, 1 foot long, with a socket; the end of this is flattened, steeled, and hardened, and turned down half an inch at each side, to keep it on the rag-wheel.

Fig. 5—the rag wheel. This has four cants, $4\frac{1}{2}$ feet long, 17 by 3 inches in the middle, lapped together to make the wheel 5 feet diameter; is faced between the arms with 2 inch plank, to strengthen the laps. The cramp or ratchet iron is put on as a hoop, nearly 1 inch square, with ratchet notches cut on its outer edge, about 3 to an inch. On one side of the wheel are put 12 strong pins, 9 inches long, to tread the carriage back, when the backing works are out of order. On the other side are the cogs, about 56 in number, 3 inches pitch, to gear into the cog-wheel on the top of the tub-wheel shaft, with 15 or 16 cogs. In the shaft of the rag-wheel are 6 or 7 rounds, 11 inches long in the round part, let in nearly their whole thickness, so as to be of a pitch equal to the pitch of the cogs of the carriage, and gear into them easily: the ends are tapered off outside, and a band is driven on them at each end, to keep them in their places.

Fig. 6 is the carriage; a frame 4 feet wide from outsides, one side 29 feet long, 7 by 7 inches; the other 32 feet long, 8 by 7 inches, very straight and true, the interties at each end 15 by 4 inches, strongly tenoned and braced into the sides to keep the frame from racking. In the under side of the largest piece are set two rows of cogs, 2 inches between the rows, and 9 inches from the foreside of one cog to that of another; the cogs of one row between those of the other, so as to make 4½ inches pitch, to gear into the rounds of the rag-wheel. The cogs are about 66 in number; shank 7 inches long, 1¾ inches square; head 2¾ long, 2 inches thick at the points, and 2¼ inches at the shoulder.

Fig. 7—the ways for the carriage to run on. These are strips of plank 4½ inches wide, 2 inches thick, set on edge, let 1½ inches into the top of the cross sills, of the whole length of the mill, keyed fast on one side, made very straight both side and edge, so that one of them will pass easily between the rows of cogs in the carriage, and leave no room for it to move sideways. They should be of hard wood, well seasoned, and hollowed out between the sills to keep the dust from lodging on them.

Fig. 8—the fender posts. The gate with the saw plays in rabbets 2½ deep and 4 inches wide, in the fender posts, which are 12 feet long, and 12 inches square, hung by hooked tenons, to the front side of the two large cross beams in the middle of the frame, in mortises in their upper sides, so that they can be moved by keys to set them plumb. There are 3 mortises, 2 inches square, through each post, within half an inch of the rabbets, through which pass hooks with large heads, to keep the frame in the rabbets: they are keyed at the back of the posts.

Fig. 9—the saw, which is 6 feet long, 7 or 8 inches wide, when new; hung in a frame 6 feet wide from the outsides, 6 feet 3 inches long between the end pieces, the lowermost of which is 14 by 3 inches, the upper one 12 by 3, the side pieces 5 by 3 inches, 10 feet long, all of the best dry, hard wood. The saw is fastened in the frame by two irons, in form of staples; the lower one with two screw pins passing through the lower end, screwing one leg to each side of the end piece: the legs of the upper one are made into screws, one at each side of the end piece, passing through a broad, flat bar, that rests on the top of the end piece, with strong burrs, 1¾ inches square, to be turned by an iron spanner, made to fit them.

These straps are made of flat bars, 3 feet 9 inches long, 3 inches wide, ¾ths thick before turned; at the turn they are 5 inches wide, square, and split to receive the saw, and tug-pins, then brought near together, so as to fit the gate. The saw is stretched tightly in this frame, by the screws at the top, exactly in the middle, at each end, measuring from the outside; the top end standing about half an inch more forward than the bottom.

Fig. 10—the forebay of water, projecting through the upper foundation wall.

Fig. 11—the flutter-wheel. Its diameter and length according to the head of water, as shown in the table. The floats are fastened in with keys, so that they will drive inward, when any thing gets under them, and not break. These wheels should be very heavy, that they may act as a fly, or balance, to regulate the motion, and work more powerfully.

Fig. 12—the crank, (see it represented by a draught from a scale of 1 foot to an inch, fig. 17, Plate XXIV.) The part in the shaft 2 feet 3 inches long, 3¾ by 2 inches, neck 8 inches long, 3 thick, and 12 inches from the centre of the neck to the centre of the wrist or handle, which is 5 inches long to the key hole, and 2 inches thick.

The gudgeon at the other end of the shaft is 18 inches in the shaft, neck 3½ long, 2¾ diameter.

The crank is fastened in the same way as gudgeons. (See Art. 132.)

Fig. 12, 13—the pitman, which is 3½ inches square at the upper end, 4½ in the middle, and 4 near the lower end; but 20 inches of the lower end is 4½ by 5½, to hold the boxes and key, to keep the handle of the crank tight.

Pitman Irons of an improved Construction.

(See fig. 10, 11, 12, 13, 14, 18, Plate XXIV.) Fig. 10 is a plate or bar, with a hole in each end, through which the upper ends of the lug-pins 11—11 pass, with a strong burr screwed on each; they are 17 inches long,

1⅛ inch square, turned at the lower end to make a round hole 1⅛ diameter, made strong round the hole.

Fig. 12 is a large, flat link, through a mortise near the lower side of the end of the saw frame. The lug-pins pass one through each end of this link, which keeps them close to the gate sides.

Fig. 14 is a bar of iron 2 feet long, 3¼ inches wide, ½ inch thick at the lower, and 1⅛ at the upper end. It is split at the top and turned as in the figure, to pass through the lug-pins. At fig. 13 there is a notch set in the head of the pitman bar 14, 1½ inch long, nearly as deep as to be in a straight line with the lower side of the side-pins, made a little hollow, steeled and made very hard.

Fig. 18 is an iron plate, 1½ inch wide, half an inch thick in the middle, with 2 large nail-holes in each end, and a round piece of steel welded across the middle and hardened, made to fit the notch in the upper end of the pitman, Plate XXVI., and draw close by the lug-pins, to the under side of the saw-frame, and nailed fast. Now, if the bearing part of this joint be in a straight line, the lower end of the pitman may play without friction in the joint, because both the upper and lower parts will roll without sliding, like the centre of a scale beam, and will not wear.

This is the best plan for pitman irons, with which I am acquainted. The first set, so made, has been in my saw-mill 8 years, doing much hard work, and three minutes have not been required to adjust them.

Fig. 14—the tub-wheel, for running the carriage back. This is a very light wheel, 4 feet diameter, and put in motion by means of the foot or hand, at once throwing it in gear with the rag-wheel, lifting off the hand and clicks from the ratchet, and hoisting a little gate to let water on the wheel. The moment the saw stops, the carriage begins to move gently back again with the log.

Fig. 15—the cog-wheel on the top of the tub-wheel shaft, with 15 or 16 cogs.

Fig. 16—the log on the carriage, sawed partly through.

Fig. 17—a crank and windlass, to increase power, by

which one man can draw heavy logs on the mill, and turn them, by a rope passing round the log and windlass.

Fig. 18—a cant hook for rolling logs.

Fig. 19—a double dog, fixed into the hindmost head-block, used by some to hold the log.

Fig. 20 are smaller dogs to use occasionally at either end.

Figs. 21, 22, represent the manner of shuting water on a flutter-wheel by a long, open shute, which should not be nearer to a perpendicular than an angle of 45 degrees, lest the water should rise from the shute and take air, which would cause a great loss of power.

Fig. 23 represents a long, perpendicular tight shute; the gate 23 is always drawn fully, and the quantity of water regulated at the bottom by a little gate r, for the purpose. There must be air let into this shute by a tube entering at a. (See Art. 71.) These shutes are for saving expense where the head is great, and should be much larger at the upper than at the lower end, else there will be a loss of power. They must be very strong, otherwise they will burst. The perpendicular ones suit best where a race passes within 12 feet of the upper side of the mill.

OPERATION.

The sluice drawn from the penstock 10, puts the wheel 11 in motion—the crank 12 moves the saw-gate, and saw 9, up and down; and as they rise they lift up the lever 2, which pushes forward the hand-pole 3, which moves the rag-wheel 5, which gears in the cogs of the carriage 4, and draws forward the log 16 to meet the saw, as much as is proper to cut at a stroke. When it is within 3 inches of being through the log, the cleet C, on the side of the carriage, arrives at a trigger and lets it fly, and the sluice gate shuts down; the miller instantly draws water on the wheel 14, which runs the log gently back, &c.

ARTICLE 159.

DESCRIPTION OF A FULLING-MILL.

Fig. 19, Plate XXIV., is the penstock, water-gate, and spout of an overshot fulling-mill, the whole laid down from a scale of 4 feet to an inch.

Fig. 20—one of the 3 interties, that are framed with one end into the front side of the top of the stock-block; the other ends into the tops of the 3 circular pieces that guide the mallets; they are 6 feet long, 5 inches wide, and 6 deep.

Fig. 21 are two mallets; they are 4 feet 3 inches long, 21 inches wide, and 8 thick, shaped as in the figure.

Fig. 22—their handles, 8 feet long, 20 inches wide, and 3 thick: a roller passes through them, 8 inches from the upper ends, and hangs in the hindermost corner of the stock-post. The other ends go through the mallets, and have each, on their underside, a plate of iron faced with steel and hardened, 2 feet long, 3 inches wide, fastened by screw-bolts, for the tappet-blocks to rub against while lifting the mallets.

Fig. 23—the stock-post, 7 feet long, 2 feet square at the bottom, 15 inches thick at the top, and shaped as in the figure.

Fig. 24—the stock where the cloth is beaten, shaped inside as in the figure, planked inside as high as the dotted line, which planks are put in rabbets in the post, the inside of the stock being 18 inches wide at the bottom, 19 at the top, and 2 feet deep.

Fig. 25—one of the 3 circular guides for the mallets; they are 6 feet long, 7 inches deep, and 5 thick; are framed into a cross sill at bottom, that joins its lower edge to the stock-post. This sill forms a part of the bottom of the stock, and is 4 feet long, 20 inches wide, and 10 thick.

The sill under the stock-post is 6 feet long, 20 inches wide, and 18 thick. The sill before the stock is 6 feet long, and 14 inches square.

Fig. 26—the tappet-arms, 5 feet 6 inches long, 21

inches each side of the shaft, 12 inches wide, and 4 thick. There is a mortise through each of them, 4 inches wide, the length from shaft to tappet, for the ends of the mallet handles to pass through. The tappets are 4 pieces of hard wood, 12 inches long, 5 wide, and 4 thick, made in the form of half circles pinned to the ends of the arms.

Fig. 27—an overshot water-wheel, similar to those in other mills.

Fig. 28—one of the 3 sills, 16 feet long, and 16 inches square, with walls under them, as in the figure.

OPERATION.

The cloth is put in a loose heap in the stock 24; the water being drawn on the wheel, the tappet-arms lift the mallets, alternately, which strike the under part of the heap of cloth, and the upper part is continually falling over, and thereby turning and changing its position under the mallets, which are shaped as in the figure, to produce this effect.

Description of the Drawings of the Iron-work, Plate XXIV.

Fig. 1 is a spindle, 2 the balance ryne, and 3 the driver, for a mill-stone. The length of the spindle from the foot to the top of the neck is about 5 feet 3 inches; cock-head 8 or 9 inches from the top of the neck, which is 3 inches long, and 3 diameter; blade or body $3\frac{1}{2}$ by 2 inches; foot $1\frac{1}{4}$ inch diameter; the neck, foot, and top of the cock-head, steeled, turned, and hardened.

Fig. 2—the balance-ryne is sometimes made with 3 horns, one of which is so short as only to reach to the top of the driver, which is let into the stone directly under it; the other to reach nearly as low as the bottom of the driver: of late, they are mostly made with 2 horns only; this may be made sufficiently fast by making it a little wider than the eye, and letting it into the stone a little on each side, to keep it steady, and prevent its moving sideways. Some choose them with 4 horns, which fill the eye too much.

Fig. 3 is a driver, about 15 inches long.

Fig. 4—the step for the spindle foot to run in. It is a box 6 inches long, 4 inches wide at the top, but less at bottom, and 4 inches deep outside, the sides and bottom half an inch thick. A piece of iron 1 inch thick is fitted to lie tightly in the bottom of this box, but not welded; in the middle of this is welded a plug of steel $1\frac{1}{2}$ inches square, in which is punched a hole a quarter of an inch deep, to fit the spindle-foot. The box must be tight, to hold oil.

Fig. 5—a step-gudgeon for large upright shafts, 16 inches long and 2 square, steeled and turned at the toe.

Fig. 6—the step for it, similar to 4, but proportionably less.

Fig. 7 is a gudgeon for large bolting shafts, 13 inches long, and $1\frac{1}{2}$ square.

Fig. 8—a large joint-gudgeon, tang 14 inches, neck 5, and tenon 2 inches long, $1\frac{1}{2}$ square.

Fig. 9—the socket part to fit the shaft, with 3 rivet-holes in each.

Fig. 10, 14, 18—pitman irons, described Art. 158.

Fig. 15, the wallower gudgeon, tang 16 inches, neck $3\frac{1}{2}$ inches long, and $2\frac{1}{2}$ diameter.

Fig. 16—the water-wheel gudgeon, tang 3 feet 2 inches long, neck $4\frac{1}{2}$ inches long, $3\frac{1}{4}$ square.

Fig. 17—a saw-mill crank, described Art. 158.

N. B.—The spindle-ryne, &c., is drawn from a scale of 2 feet to an inch, and all the other irons 1 foot to an inch.

ARTICLE 160.

To what has been said of Saw-Mills, by **Thomas Ellicott**, I add the following:—

Of hanging the Saw.

First, set the fender posts as nearly plumb every way as possible, and the head-blocks on which the log is to lie, level. Put the saw just in the middle of the gate, measuring from the outsides, set it by the gate and not by a plumb line, with the upper teeth about half an inch farther forward than the lower ones;—this is to give the saw liberty to rise without cutting, and the log room to push forward as it rises. Run the carriage forward, so that the saw may strike the block—stick up a nail, &c., there—run it back again its full length, and standing behind the saw, set it to direct exactly to the mark. Stretch the saw in the frame, rather the most at the edge, that it may be stiffest there. Set it in motion, and hold a tool close to one side of it, and observe whether it touch equal-the whole length of the stroke—try if it be square with the top of the head blocks, else it will not make the scantling square.

Of whetting the Saw.

The edge of the teeth ought to be kept straight, and not suffered to wear hollowing—set the teeth a little out, equal at each side, and the outer corners a little longest—they will then clear their way. Some whet the under side of the teeth nearly level, and others a little drooping down; but it then never saws steadily, but is apt to *wood* too much; the teeth should slope up, although but very little. Try a cut through the log, and if it come out at the mark made to set it by, it is shown to be hung right.

Of springing Logs straight.

Some long small logs will spring so much in sawing as to spoil the scantling, unless they can be held straight; to do which make a clamp to bear with one end against the side of the carriage, the other end under the log, with a post up the side thereof—drive a wedge between the post and log, and spring it straight; this will bend the carriage side—but this is no injury.

Of moving the Logs to the Size of the Scantling, &c.

Make a sliding-block to slide in a rabbet in front of the main head-block; fasten the log to this with a little dog on each side, one end of which being round, is driven into a round hole, in the front side of the sliding-block, the other flattened to drive in the log, cutting across the grain, slanting a little out—it will draw the log tight, and stick in it the better. Set a post of hard wood in the middle of the main block close to the sliding one, and to extend with a shoulder over the sliding one, for a wedge to be driven under this shoulder to keep the block tight. Make a mark on each block to measure from—when the log is moved the key is driven out. The other end next the saw is best held by a sliding dog, part on each side of the saw, pointed like a gouge, with two joint dogs, one on each side of the saw.

Remedy for a long Pitman.

Make it in two parts by a joint 10 feet from the crank, and a mortise through a fixed beam, for the lower end of the upper part to play in, the gate will work more steadily, and all may be made lighter.

The feed of a saw-mill ought to be regulated by a screw fixed to move the hand-pole nearer or farther from the centre of the roller that moves it, which may be done, as the saw arrives at a knot, without stopping the mill.

ARTICLE 161.

The following Observations on Saw-Mills, &c., were communicated by WILLIAM FRENCH, *Mill-Wright, New Jersey.*

SAW-MILLS, with low heads, have been much improved in this state. Mills with two saws, with not more than 7 feet head and fall, have sawed from five to six hundred thousand feet of boards, plank, and scantling, in one year. If the water be put on the wheel in a proper manner, and the wheel of a proper size, (as by the following table,) the saw will strike between 100 and 130 strokes in a minute, (see fig. 1, Plate XXVI.) The lower edge of the breast-beam B to be $\frac{3}{4}$ths the height of the wheel, and one inch to a foot, slanting up stream, fastened to the pen-stock posts with screw-bolts, (see post A) circled out to suit the wheel C; the fall D circled to suit the wheel and extending to F, 2 inches above the lower edge of the breast-beam, or higher, according to the size of the throat or sluice E, with a shuttle, or gate, sliding on F E, shutting against the breast-beam B: then 4 buckets out of 9 will be acted on by the water. The method of fastening the buckets or floats is, to step them in starts mortised in the shaft—see start G—9 buckets in a wheel $4\frac{1}{2}$ inches wide, see them numbered 1, 2, &c.

Fig. 2, the go back, is a tub wheel. Its common size is from $4\frac{1}{2}$ to 6 feet diameter, with 16 buckets. The water is brought on it by the trunk H. The bucket I, is made with a long tenon, so as to fasten it with a pin at the top of the wheel.

TABLE
Of the Dimensions of Flutter-Wheels.

Head	Bucket	Wheel	Throat
12 feet	5 feet	3 feet	$1\frac{3}{4}$ inches
11	$5\frac{1}{2}$	3	2
10	6	3	$2\frac{1}{8}$
9	$6\frac{1}{2}$	2 10 inches	$2\frac{1}{4}$
8	7	2 9	$2\frac{1}{2}$
7	$7\frac{1}{2}$	2 8	$3\frac{1}{4}$
6	8	2 7 p.	$3\frac{1}{2}$
5	9	2 6	$3\frac{3}{4}$

N. B.—The crank about 11 inches, but varies to suit the timber.

The Pile Engine.

Fig. 3, a simple machine for driving piles in soft bottoms for setting mill-walls or dams on. It consists of a frame 6 or 7 feet square, of scantling, 4 by 5 inches, with 2 upright posts 2 inches apart, 10 or 12 feet high, 3 by 3 inches, braced from top to bottom of the frame, with a cap on top 2 feet long, 6 by 8 inches, with a pulley in its middle, for a rope to bend over, fastened to a block I, called a tup, which has 2 pieces, 4 inches wide between the uprights, with a piece of 2 inch plank T, 6 inches wide, framed on the ends, so as to slide up and down the upright posts S. This machine is worked by 4 or 6 men, who draw the tup up by the sticks fastened to the end of the rope K, and let it fall on the pile L; they can thus strike 30 or 40 strokes in a minute, by the swing of their arms.

Of building Dams on Soft Foundations.

The best method is to lay 3 sills across the stream, and frame cross sills into them up and down stream, setting the main mud sills on round piles, and pile them with 2 inch plank, well jointed, and driven closely together, edge to edge, from one end to the other. Taking one corner off the lower end of the plank will cause it to keep a close joint at bottom, and by driving an iron dog in the mud-sill, and a wooden wedge to keep it close at the top end, it will be held to its place when the tup strikes. It is necessary to pile the outside cross sills also in some bottoms, and to have wings to run 10 or 12 feet into the bank at each side; and the wing-posts 2 or 3 feet higher than the posts of the dam, where the water falls over, planked to the top N N, and filled with dirt to the plate O.

Fig. 4 is a front view of the breast of the tumbling dam.

Fig. 5 is a side view of the frame of the tumbling dam, on its piling a b c d e, and f g h is the end of the mud-sills. The posts k, are framed into the main mud-sills with a hook tenon, leaning down stream 6 inches in 7 feet, supported by the braces 11, framed into the cross sills I; the cross sills I to run 25 feet up and down stream, and to be well planked over, and the breast-posts to be planked to the top, (see P, fig. 4,) and filled with dirt on the upper side, within 12 or 18 inches of the plate O, (see Q, fig. 5;) slanting to cover the up-stream ends of the sills 3 or 4 feet deep: R represents the water.

When the heads are high, it is best to plank the braces for the water to run down; but, if low, it may fall perpendicularly on the sheeting.

CHAPTER XXIV.

RULES FOR DISCOVERING NEW IMPROVEMENTS; EXEMPLIFIED IN IMPROVING THE ART OF CLEANING AND HULLING RICE, WARMING ROOMS, VENTING SMOKE BY CHIMNEYS, &c.*

The true Path to Inventions.

NECESSITY is called the mother of invention, but, upon inquiry, we shall find that Reason and Experiment bring it forth; for almost all inventions have resulted from such steps as the following:—

I. To investigate the fundamental principles of the theory, and process of the art, or manufacture, we wish to improve.

II. To consider what is the best plan, in theory, that can be deduced from, or founded on, those principles, to produce the effect we desire.

III. To inquire whether the theory be already put in practice to the best advantage; and what are the imperfections or disadvantages of the common process, and what plans are likely to succeed better.

IV. To make experiments in practice, upon any plans that these speculative reasonings may suggest, or lead to. Any ingenious artist, taking the foregoing steps, will probably be led to improvement on his own art: for we see, by daily experience, that every art may be improved. It will, however, be in vain to attempt improvements, unless the mind be freed from prejudice in favour of established plans.

EXAMPLE I.
On the Art of cleaning Grain by Wind.

I. What are the principles on which the art is founded? When bodies fall through resisting mediums, their velo-

---

* The rules and observations, which formed an appendix to the former editions of this work, contain some suggestions which are worthy of attention. Since they were written, many improvements have been made, in the processes to which they refer; but the path is still open, and perhaps the remarks made by Mr. Evans, may yet lead to useful results; with this hope, they have, with some modifications been retained.

cities are as their specific gravities, and the surface they expose to the medium; consequently, when light and heavy articles are mixed together, the farther they fall, the greater will be their distance apart: on this principle a separation can be effected.

II. What is the best plan in theory? First, make a current of air, as deep as possible, for the grain to fall through; the lightest will then be carried farthest, and the separation be more complete at the end of the fall. Secondly, cause the grain, with the chaff, &c., to fall in a narrow line across the current, that the light parts may meet no obstruction from the heavy, in being carried forward. Thirdly, fix a movable board edgewise to separate between the good clean, and the light grain, &c. Fourthly, cause the same blast to blow the grain several times, and thereby effect a complete separation at one operation.

III. Is this theory in practice already? what are the disadvantages of the common process? We find that the farmers' common fans drop the grain in a line 15 inches wide, to fall through a current of air about 8 inches deep, instead of falling in a line half an inch wide, through a current three feet deep; so that it requires a very strong blast even to blow out the chaff; but garlic, light grains, &c., cannot be thus removed, as they meet so much obstruction from the heavy grains; the grain has, therefore, to undergo two or three such operations, so that the practice appears absurd, when tried by the scale of reason.

IV. The fourth step is to construct a fan to put the theory in practice, by experiment. (See Art. 83.)

EXAMPLE II.

The Art of Distillation.

I. The principles on which this art is founded, are evaporation and condensation. When liquid is heated, the spirit it contains, being more volatile than the watery part, evaporates, before it, into steam, which being condensed again into a liquid, by cold, is obtained in a separate state.

II. The best plan, in theory, for effecting this, appears to be as follows: the fire should be applied to the still, so as to spend the greatest possible part of its heat to heat the liquid. Secondly, the steam should be conveyed into a metallic vessel of any suitable form, and this should be immersed in cold water, to condense the steam; in order to keep the condenser cold, there should be a stream of cold water continually entering the bottom and flowing over the top of the condensing tub; the steam should have no free passage out of the condenser, else the strongest part of the liquor will escape.

III. Is this theory already put in practice, and what are the disadvantages of the common process?—1st, A great part of the heat escapes up the chimney. 2dly, It is almost impossible to keep the grounds from burning in the still. 3dly, The fire cannot be regulated to keep the still from boiling over; we are, therefore, obliged to run the spirit off very slowly; how are we to remedy these disadvantages?—First, to lessen the fuel, apply the fire as much to the surface of the still as possible; enclose the fire by a wall of clay that will not convey the heat away so fast as stone; let in no more air than is necessary to keep the fire burning, for the surplus air carries away the heat of the fire. Secondly, to keep the grounds from burning, immerse the still, with the contained liquor, in a vessel of water, joining their tops together; then, by applying the fire to heat the water in the outside vessel, the grounds will not burn, and by regulating the heat of the outside vessel the still may be kept from boiling over.

IV. A still to be heated through the medium of water, was made, some years ago, by Colonel Alexander Anderson, of Philadelphia, and the experiment tried; but the outside vessel being open, the water in it boiled away, and carried off the heat, and the liquor in the still could not be made to boil—this appeared to defeat the scheme. But, by enclosing the water in a tight vessel, so that the steam could not escape, and that the heat might be increased, it now passed to the liquor in the still, which boiled as well as if the fire had been immediately applied to it. By fixing a valve to be loaded so as to let the steam escape, when it has arrived to such a degree of heat as to require it, all danger of explosion is avoided, and all boiling over prevented.

EXAMPLE III.

The Art of venting Smoke from Rooms by Chimneys.

I. The principles are:—Heat, by repelling the particles of air to a greater distance than when cold, renders it lighter than cold air, and it will rise above it, forming a current upwards, with a velocity proportional to the degree of heat, and the size of the tube or funnel of the chimney, through which it ascends, and with a power proportional to its perpendicular height; which power to ascend will always be equal to the difference of the weight of a column of rarefied air of the size of the smallest part of the chimney, and a column of common air of equal size.

II. What is the best plan, in theory, for venting smoke, that can be founded on these principles?

1st. The size of the chimney must be proportioned to the size and closeness of the room and to the fire; because, if the chimney be immensely large, and the fire small, there will be little current upwards. And again, if the fire be large, and the chimney too small, the smoke cannot be all vented by it: more air being necessary to supply the fire, than can find vent up the chimney, it must spread in the room again, which air, after passing through the fire, is rendered deleterious.

2dly. The narrowest place in the chimney must be next the fire, and in front of it, so that the smoke would have to pass under it to get into the room; the current will there be greatest, and will draw up the smoke briskly.

3dly. The chimney must be perfectly tight, so as to admit no air but at the bottom.

III. The errors in chimneys in common practice, are,

1st. In making them widest at bottom.

2dly. Too large for the size and closeness of the room.

3dly. In not building them high enough, so that the wind, whirling over the tops of houses, blows down them.

4thly. By letting in air any where above the breast, or opening, which destroys the current of it at the bottom.

IV. The cures directed by the principles and theory, are,

1st. If the chimney smoke on account of being too large for the size and closeness of the room, make the chimney less at the bottom—its size at the top may not do much injury, but it will weaken the power of ascent, by giving the smoke time to cool before it leave the chimney; the room may be as tight, and the fire as small as you please, if the chimney be in proportion.

2dly. If it be small at the top and large at the bottom, there is no cure but to lessen it at the bottom.

3dly. If it be too small, which is seldom the case, stop up the chimney and use a stove—it will be large enough to vent all the air that can pass through a two inch hole, which is large enough to sustain the fire in a stove. Chimneys built in accordance with these theories, I believe, are every where found to answer the purpose. (See Franklin's letters on smoky chimneys.)

EXAMPLE IV.

The Art of warming Rooms by Fire.

I. Consider in what way fire operates.

1st. The fire heats and rarefies the air in the room, which gives us the sensation of heat or warmth.

2dly. The warmest part of the air being lightest, rises to the uppermost part of the room, and will ascend through holes (if there be any) to the room above, making it warmer than the one in which the fire is.

3dly. If the chimney be too open, the warm air will fly up it, leaving the room empty; the cold air will then rush in at all crevices to supply its place, which keeps the room cold.

II. Considering these principles, what is the best plan, in theory, for warming rooms?

1st. We must contrive to apply the fire to spend all its heat, to warm the air which comes into the room.

2dly. The warm air must be retained in the room as long as possible.

3dly. Make the fire in a lower room, conducting the heat through the floor into the upper one, and leaving another hole for the cold air to descend to the lower room.

4thly. Make the room so tight as to admit no more cold air, than can be warmed as it comes in.

5thly. By closing the chimney so as to let no warm air escape, but that which is absolutely necessary to sustain the fire—a hole of two square inches will be sufficient for a very large room.

6thly. The fire may be supplied by a current of air brought from without, not using any of the air already warmed. If this theory, which is founded on true principles and reason, be compared with common practice, the errors will appear, and may be avoided.

I had a stove constructed in accordance with these principles, and have found all to answer according to theory.

The operation and effects are as follows; namely:—

1st. It applies the fire to warm the air as it enters the room, and admits a full and fresh supply, rendering the room moderately warm throughout.

2dly. It effectually prevents the cold air from pressing in at the chinks or crevices, but causes a small current to pass outwards.

3dly. It conveys the coldest air out of the room first, consequently,

4thly. It is a complete ventilator, thereby rendering the room healthy.

5thly. The fire may be supplied (in very cold weather) by a current of air from without, that does not communicate with the warm air in the room.

6thly. Warm air may be retained in the room any length of time, at pleasure; circulating through the stove, the coldest entering first, to be warmed over again.

7thly. It will bake, roast, and boil equally well with the common ten plate stove, as it has a capacious oven.

8thly. In consequence of these improvements, it requires not more than half the usual quantity of fuel.

Description of the Philosophical and Ventilating Stove.

It consists of three parts, either cylindrical or square, the greatest surrounding the least. (See fig. 1, Plate X.) S F is a perspective view thereof in a square form, supposed open at one side: the fire is put in at F, into the least part, which communicates with the space next the outside, where the smoke passes to the pipe 1—5. The middle part is about two inches less than the outside part, leaving a large space between it, and above the inner part, for an oven, in which the air is warmed, being brought in by a pipe B D between the joists of the floor, from a hole in the wall at B, it rises under the stove at D, into the space surrounding the oven and the fire, which air is again surrounded by the smoke flue, giving the fire a full action to warm it, whence it ascends into the room by the pipe 2. E brings air from the pipe D B to blow the fire. H is a view of the front end plate, showing the fire and oven doors. I is a view of the back end, the plate being off, the dark square shows the space for the fire, and the light part the air-space surrounding the fire, the dark outside space the smoke surrounding the air; these are drawn on a larger scale. The stove consists of fifteen plates, twelve of which join, by one end, against the front plate H.

To apply this stove to the best advantage, suppose fig. 1, Plate X., to represent a three or four story house, two rooms on a floor—set the stove S F in the partition on the lower floor, half in each room; pass the smoke pipe through all the stories; make the room very close; let no air enter but what comes in by the pipes A B or G C through the wall at A and G, that it may be the more pure, and pass through the stove and be warmed. But to convey it to any room, and take as much heat as possible with it, there must be an air-pipe surrounding the smoke pipe, with a valve to open at every floor. Suppose we wish to warm the rooms No. 3—6, we open the valves, and the warm air enters, ascends to the upper part, depresses the cold air, and if we open the holes a—c, it will descend the pipes, and enter the stove to be warmed again: this may be done in very cold weather. The higher the room above the stove, the more powerfully will the warm air ascend and expel the cold air. But if the room require to be ventilated, the air must be prevented from descending, by shutting the little gate 2 or 5, and drawing 1 or 6, and giving it liberty to ascend and escape at A or G—or up the chimney, letting it in close at the hearth. If the warm air be conveyed under the floor, as between 5—6, and let rise in several places, with a valve at each, it will be extremely convenient and pleasant; if above the floor, as at 4, several persons might set their feet on it to warm. The rooms will be moderately warm throughout—a person will not be sensible of the coldness of the weather.

One large stove of this construction may be made to warm a whole house, ventilate the rooms at pleasure, bake bread, meat, &c.

These principles and improvements ought to be considered and provided for in building.

EXAMPLE V.

Art of Hulling and Cleaning Rice.

STEP I. The principles on which this art may be founded, will appear, by taking a handful of rough rice, and rubbing it hard between the hands—the hulls will be broken off, and, by continuing the operation, the sharp texture of the outside of the hull (which, through a magnifying glass, appears like a sharp, fine file, and, no doubt, is designed by nature for the purpose) will cut off the inside hull, and the chaff being blown out, will leave the rice perfectly clean, without breaking any of the grains.

II. What is the best plan, in theory, for effecting this? (See the plan proposed, represented in Plate X., fig. 2; explained Art. 103.)

EXAMPLE VI.

To save Ships from sinking at Sea.

STEP I. The principle on which ships float, is the difference of their specific gravities from that of the water,—sinking only to displace a quantity of water equal in weight to that of the ship, and its lading; they sink deeper, therefore, in fresh than in salt water. If we can calculate the weight of the cubic feet of water a ship displaces when empty, it will show her weight, and subtracting that from what she displaces when loaded, shows the weight of her load; each cubic foot of fresh water weighing 62,5 lbs. If an empty rum hogshead weigh 62,5 lbs. and measure 15 cubic feet, it will require 875 lbs. to sink it. A vessel of iron, containing air only, and so large as to make its whole bulk lighter than so much water, will float, but if it be filled with water, it will sink. Hence, we may conclude, that a ship loaded with any thing that will float, will not sink if filled with water; but if loaded with any thing specifically heavier than water, it will sink as soon as filled.

II. This appears to be the true theory:—How is it to be applied in case a ship spring a leak, that gains on the pumps?

III. The mariner, who understands well the above principles and theory, will be led to the following steps:

1st. To cast overboard such things as will not float, and carefully to reserve every thing that will float, for by them the ship may at last be buoyed up.

2dly. To empty every cask or thing that can be made water-tight, to put them in the hold, and fasten them down under the water, filling the vacancies between them with billets of wood; even the spars and masts may, in desperate cases, be cut up for this purpose, which will fill the hold with light matter, and as soon as the water inside is level with that outside, no more will enter. If every hogshead buoy up 875 lbs., they will be a great help to buoy up the ship, (but care must be taken not to put the empty casks too low, which would overset the ship,) and she will float, although half the bottom be torn off. Mariners, for want of this knowledge, often leave their ships too soon, taking to their boats, although the ship be much the safest, and do not sink for a long time after being abandoned—not considering that, although the water gain on their pumps at first, they may be able to hold way with it when risen to a certain height in the hold, because the velocity with which it will enter, will be in proportion to the square root of the difference between the level of the water inside and outside—added to this, the fuller the ship the easier the pumps will work, because the water has to be raised to a less height; therefore, they ought not to be too soon discouraged.

Description of the Thrashing Machine, with elastic Flails; invented by JAMES WARDROP, *of Ampthill, Virginia.**

PLATE XXV.

A—the floor on which the flails are fixed.

B—the part of the floor on which the grain is laid, made of wicker-work, through which the grain falls, and is conveyed to the fan or screen below; the pivot of the fan is seen at P, and is turned by a band from the wheel or wallower.

C C C—a thin board raised round the floor to confine the wheat, and made shelving outwards, to render raking off the straw more easy.

D—the wallower or wheel.

E—Crank handle to turn the wheel.

F F—Flails.

G G G—Lifters, with ropes fixed to the flails.

I I I—Catches or teeth to raise the lifters.

K—Post on which the wallower is fixed.

L—Beam on which the lifters rest and are fixed by an iron rod passing through the lifters, and let into this beam.

M—Check-beam to stop the end of the lifters from rising.

N—Keeps in which the lifters work.

O—Beam in which the ends of the flails are mortised.

Q—Fly-ends loaded with lead, not necessary in a horse machine.

R—Showing the lifters and keeps, how fixed.

The machine, to be worked by two men, was made on a scale of a 12 foot flail, having a spring which required a power of 20 lbs. to raise it three feet high at the point:—A spring of this power, and raised three feet high, being found to get out wheat with great effect.

---

* The flail thrashing machine has been superseded by that with cylindrical beaters and a concave, variously modified. This is now so generally introduced as not to require any description. The flail machine having been originally engraved for this work, has been retained.

APPENDIX,

CONTAINING

A Description of a Merchant Flour Mill, on the most approved Construction, with the recent Improvements, with two additional Plates,

By CADWALLADER AND OLIVER EVANS, ENGINEERS;

AND

EXTRACTS

FROM SOME OF THE BEST MODERN WORKS ON THE SUBJECT OF MILLS, WITH OBSERVATIONS BY THE EDITOR.

Description of a Merchant Flour Mill, driving four Pairs of five feet Mill-Stones; arranged by CADWALLADER *and* OLIVER EVANS, *Engineers, Philadelphia.*

PLATE XXVII.

1—A hollow cast-iron shaft, circular, 15 inches in diameter, except at those points where the water and main bevel wheels are hung, where it is increased to 19 inches in diameter. The water-wheel is secured on this shaft by 3 sockets, as represented in Plate XXVIII., fig. 3, and makes 10 revolutions per minute.

2—The main driving bevel-wheel, on the water-wheel shaft, 8 feet in diameter, to the pitch line; 100 cogs, 3 inches pitch, and 8 inches on the face; revolving 10 times per minute, and driving.

3—A bevel-wheel on the upright, 4 feet in diameter to pitch line; 50 cogs, same pitch and face of cogs as above, revolving 20 times per minute.

4—The large pit spur-wheel, making 20 revolutions per minute, 9 feet ⅛th inch diameter, to pitch line; 114

cogs, 3 inches pitch, face 10 inches; this wheel gives motion to

5, 5, 5, 5—Four pinions on the spindles of the mill-stones, 18,1 inches in diameter to pitch line, 19 cogs, same face and pitch.

6, 6, 6, 6—Iron upright shafts, extending the height of the building, and coupled at each story.

7, 7, 7, 7—Are 4 pairs of five feet mill-stones, making 120 revolutions per minute. Two of them shown in elevation; and the position of the 4, shown in Plate XXVIII. as represented by the dotted lines, fig. 1.

8—A pulley on the upright shaft, which, by a band, gives motion to

8—The fan for cleaning grain, revolving 140 times per minute, wings 3 feet long, 20 inches in width.

9—A bevel wheel 2 feet diameter, cogs 2 inches pitch, face 2,5 inches, on the upright shaft, gearing into a bevel wheel, the face of which is shown, drives the bolting screen 18 revolutions per minute.

10—A bevel wheel on upright shaft, 56 cogs, 2 inches pitch, 2,5 inches face, gearing into

10—A bevel wheel on the shaft of the bolting reels, 31 cogs, same pitch and face.

10, 10—Are two of four bolting reels shown, 18 feet long, 30 inches diameter, revolving 36 times per minute.

11—A large pulley on the upright shaft, which, by a band, gives motion to the rubbing stones 11.

12—A bevel wheel, on the top of the upright shaft, gearing into

12—A bevel wheel, on the horizontal shaft, at one end of which is

13—A bevel wheel, 1 foot diameter, gearing into a bevel wheel

14—of 5 feet diameter, which reduces the motion of the hopper-boy down to 4 revolutions per minute, which sweeps a circle of 20 feet.

15—Meal elevator attending 4 pairs of stones.

16—Grain elevator.

17—Packing-room and press.

APPENDIX.

PLATE XXVIII.

Figure 1.

A bird's eye view of the mode of giving motion to 4 pairs of mill-stones.

4—The large pit spur-wheel, driving at equal distances on its periphery, the pinions

5, 5, 5, 5—attached to the spindles of the mill-stones.

7, 7, 7, 7—Mill-stones, 5 feet diameter, represented by dotted circles.

Figure 2.

An enlarged view of the couplings of the upright shaft. They are of cast-iron, with their holes truly reamed, to receive the ends of the iron upright shafts.

2—The face of a coupling, divided into 6 equal parts, radiating from the centre: three of the parts project, and three are depressed; so that when two of them are coupled, the projections of one will fill the depressions in the other, as 1, the coupling connected.

Figure 3.

A cast-iron socket for the water-wheel; it is a plate $\frac{3}{4}$ths of an inch thick; the eye for the shaft to pass through, $1\frac{1}{4}$ inch thick, and 12 inches deep: the sockets, for receiving the arms, are 14 inches long, and have projections 5 inches deep; 3 3 3, &c., are the projections; the intermediate space, between the sockets, are cut out to lessen the weight of metal, but in such a manner as to preserve the strength. It requires three of these sockets for a large water-wheel; the arms for receiving the buckets, are dressed to fit tightly in the sockets, and secured firmly by bolts, as 2 2.

Figure 4,

Is an arm for the water-wheel, as dressed; 1, the end to be bolted in the socket; 2, the end for screwing on the bucket.

The advantages of this mode of constructing water-wheels, is, that the shaft is not weakened, by having mortises cut in to receive the arm : that it is not so liable to decay, and if an arm, or bucket, be destroyed by accident, they can be dressed out, and the mill stopped, only while you unscrew the broken part, and replace it by a new one.

Figure 5.

An elevation of the flour press. 1, the barrel of flour; 2, the funnel; 3 3, the driver; 4 5, the lever; 4 3, the connecting bars, fastened by a strong pin to each side of the lever, at 4, and to the driver at 3. 6, a strong bolt, passing through the floor, and keyed below the joist: there is a hole in the upper part of the bolt, to receive a pin which the lever works on, which, when brought down by the hand, moves the pin 4, in the dotted circle; the connecting bars drawing down the driver 3 3, pressing the flour into the barrel; and as it becomes harder packed, the power of the machine increases; as the pin 4 approaches the bolt 6, the under sliding part of the lever is drawn out, to increase its length; and is assisted in rising by a weight fastened to a line passing over pulleys.

When the pin 4 is brought down within half an inch of the centre of the bolt 6, or plumb line, the power increases from 1 to 288; and with the aid of a simple wheel and axis, as 1 to 15, from 288 to 4320; or, if the wheel and axis be as 1 to 30, it will be increased to 4320; that is to say, one man will press as hard with this machine as 8640 men could do with their natural strength. It is extremely well calculated for cotton, tobacco, cider, or, in short, any thing that requires a powerful press.

Operation of the Mill :—The grain, after having been weighed, by drawing a slide, is let into the grain elevator 16, then hoisted to the top of the building, and by a spout moving on a circle, can be deposited into spouts leading to any part of the mill, when wanted for use : by drawing sliders in other spouts, converging to the grain elevator 16, it can be re-elevated, and thrown into the hopper of the rubbing stones 11; after passing through which, it descends into the bolting screen 9, and when screened, falls into the fan 8, is there cleaned, and from that descends into a very large hopper, over the centre of the four pairs of mill-stones, which are supplied regularly with grain. After being ground, the meal descends into a chest, is taken by the elevator 15, to the top of the building, there deposited under the hopper-boy, which spreads, cools, and collects it to the bolting reels, where the several qualities are separated, and the flour descends into the packing room 17, where it is packed in barrels.

By this arrangement, we dispense with all conveyers, and have only one grain, and one flour elevator, to attend two pair of stones; we also dispense with one-half the quantity of gearing usually put into mills, and, consequently, occupy much less space, leaving the rest of the building for stowing grain, &c.

All the wheels in this mill are of cast-iron, and the face of the cogs very deep; for experience justifies the assertion that depth of face in cog-wheels, when properly constructed, does not increase friction; and that the wheels will last treble the time, by a small increase of depth; we recommend the main driving wheels to be 10 inches on the face. The journals of all shafts, when great pressure is applied, should be of double the length now generally used; increase of length does not increase friction; say for water wheels, journals of from 8 to 14 inches.

Draughts of mills are furnished by the subscribers; and the cast-iron work can be obtained, at the Steam Engine Manufactory and Iron Foundry of Messrs. Rush & Muhlenberg, Bush Hill, Philadelphia.

CADWALLADER EVANS,
OLIVER EVANS.

June 15, 1826.

WATER-WHEELS.

On the Construction of Water-Wheels, and the Method of applying the Water for propelling them, so as to produce the greatest Effect.

The following article is from the pen of a practical engineer of experience and talents; his observations are, in general, in perfect accordance with those of the editor. The principles which he advocates are undoubtedly correct, and it is hoped that their publication in this work will induce some of our most intelligent mill-wrights to forsake the beaten track, and to practise the modes recommended. Let them recollect that Mr. Parkin was not a mere theorist, but a practical man, like themselves. The death of this gentleman has deprived society of the services of one of its members, from whose liberality, experience, and skill, much was expected.

[FROM THE FRANKLIN JOURNAL.]

In constructing water-wheels, especially those of great power, the introduction of iron is a most essential improvement; and if this metal, and artisans skilled in working it, could be obtained at reasonable rates, water-wheels might be made wholly of it, and would prove, ultimately, the cheapest; as, if managed with due care, and worked with pure (not salt) water, they would last for centuries; but, as the first cost would be an objection, I would recommend, in all very large wheels, that the axis be made of cast-iron; and, in order to obtain the greatest strength with the least weight, the axis or shaft ought to be cast hollow, and in the hexagon or octagon form, with a strong iron flanch, to fix each set of arms, and the cog-wheel, upon; these flanches to be firmly fixed in their places with steel keys.

On the adaptation of water-wheels to the different heights of the water falls by which they are to be worked, I will remark that falls of from 2 to 9 feet are most advantageously worked with the undershot wheel; falls of

10 feet and upwards by the bucket or breast wheel, which, up to 20 or 25 feet, ought to be made about one-sixth higher than the fall of water by which it has to be worked; and in wheels of both descriptions, the water ought to flow on the wheel from the surface of the dam. I am aware that this principle is at direct variance with the established practice, and perhaps there are few wheels in these States that can be worked, as they are now fixed, by thus applying the water; the reasons will be apparent from what follows.

In adjusting the proportions of the internal wheels by which machinery is propelled, it is necessary, in order to obtain the greatest power, to limit the speed of the skirt of the water-wheel, so that it shall not be more than from 4 to 5 feet per second; it having been ascertained by accurate experiments, that the greatest obtainable force of water, is within these limits. As a falling body, water descends at the speed of about 16 feet in the first second, and it will appear evident that if a water-wheel is required to be so driven, that the water with which it is loaded has to descend 10, 11, or 12 feet per second, at which rate wheels are generally constructed to work, a very large proportion of the power is lost, or, rather, is spent, in destroying, by unnecessary friction, the wheel upon which it is flowing.

In the common way of constructing mill work, and of applying water to wheels, it has been found indispensably necessary to have a head of from 2 to 4 feet above the aperture through which the water flows into the buckets, or against the floats of a water-wheel, in order to be able to load the wheel instantaneously, without which precaution, it could not be driven at the required speed: from this circumstance it has been erroneously inferred, that the impulse or shock which a water-wheel, so filled, receives, is greater than the power to be derived from the actual gravity of the water alone. This theory I have heard maintained among practical men; but it is, in fact, only resorting to one error to rectify another. Overshot wheels have been adopted, in numerous cases, merely for the purpose of getting the water more readily into the buckets; but confine the wheel to the proper working speed, and that difficulty will not exist.

In consequence of the excessive speed at which water-wheels are generally driven, a small accumulation of back water either suspends or materially retards their operations; but, by properly confining their speed, the resistance from back water is considerably diminished, and only amounts to about the same thing as working from a dam as many inches lower, as the wheel is immersed; and in undershot wheels worked from a low head, or situated in the tide-way, the resistance from back water may be farther obviated by placing the floats not exactly in a line from the centre of the wheel, but deviating 6 or 8 inches from it, so as to favour the water in leaving the ascending float.

In constructing water-wheels to be driven at the speed of 4 or 5 feet per second, it will be requisite to make them broader, to work the same quantity of water which is required to drive a quick-speeded wheel. Thus, if a person intending to erect a mill, has a stream sufficient to work a wheel 5 feet broad, the skirt to move 10 feet per second, it is evident that if he wishes to work all the water which such wheel takes, he must have his wheel 10 or 12 feet broad, instead of 5, otherwise the water must run to waste, as there would not be room, in a slow-moving wheel of 5 feet broad, to receive more than half of it. The principal advantages resulting from the proposed method of adapting wheels to the falls by which they are to be worked, and the method of applying water, are—

1. The lessening of friction upon the main gudgeons, (and first pair of cog-wheels) by which, with a little care, they may be kept regularly cool, and the shaft or axis be preserved much longer in use than when the gudgeons cannot be kept cool.

2. By working water upon the principle of its actual gravity alone, and applying it always at the height of the surface of the dam, its power is double what is obtained by the common method of applying it.

3. The expensive penstock required to convey the water to the wheels, generally used, will not be needed,

as one much shallower, and consequently less expensive, will be sufficient.

4. The resistance of back water is reduced as far as possible.

5. The risk of fire is less, as the friction is reduced.

W. PARKIN, *Engineer*.

September 24, 1825.

That water, whenever the fall is sufficient, ought always to be applied upon the principle of its actual gravity, appears to be a conclusion so obvious, that it is astonishing it should ever be disputed. The acknowledged difference between the effect of overshot and undershot wheels, is an evidence of the truth of the principle. The whole moving power of water is derived from its gravity; it is this which causes it to fall, and although in falling from a given height it acquires velocity, its gravitating force remains the same, and all the effect which this might have produced, has been expended upon itself, and not in moving any other body. The force with which water strikes, after it has fallen from any height, is calculated to deceive those who are not well grounded in the principles of hydrostatics; but it is admitted, both by Mr. Evans and Mr. Ellicott, that the effect upon overshot wheels is diminished, by increasing the head, and the reason given for leaving the head so great as they prescribe, is the necessity of filling the buckets with sufficient rapidity; this necessity, however, is created by giving too much velocity to the wheel.

It has been stated by Mr. Evans, and is generally believed by mill-wrights, that it is necessary to give a much greater velocity to wheels, than that which is recommended by Smeaton and others, in order to cause the mill to run steadily, and prevent its being suddenly checked by any increased resistance. This is saying that the water-wheel ought to be made to operate as a flywheel, which it will not do if its motion be slow. The objection to this it twofold. By giving to the skirt of the wheel a motion much exceeding 4 or 5 feet per second, its power is considerably reduced below the maximum, and this loss of power is perpetual; wasting a considerable portion of water, to convert the water-wheel into a fly-wheel, which water might be employed in giving greater power to the mill. When a mill, from the nature of the work which it has to perform, requires the action of a fly-wheel, the situation of the water-wheel is often the worst that could be devised for this purpose, especially where there is any considerable gearing in the mill. A fly-wheel does not add actual power, but it serves to collect power, where the resistance is unequal; and in order to its producing this effect most perfectly, it ought to be placed as near as possible to the working part of the machinery. In grist mills there is no necessity for a fly-wheel; the stones perform this office in the most effectual manner, and the same remark is applicable to every kind of mill without a crank, and in which the resistance is equal, or nearly so, during the whole time of its action.

Although we have spoken highly of the general views given by Mr. Parkin, in his communication to the Franklin Journal, he has fallen into some mistakes, which were pointed out by a writer in the same publication, Vol. IV., page 166. A part of this communication is subjoined, as it contains remarks, and a tabular view of the velocities attained, and the distances fallen through, by bodies, in fractional parts of a second, which may be of great practical utility:—

"I suppose that, at the present day, no man who professes to be capable of directing the construction of a water-wheel, or of estimating the amount of a water power, is ignorant of the fact, that water falls through a distance of about sixteen feet in the first second. But I suspect that many who assume the above qualifications, do not know the ratio of increase, either in the distance, or the velocity. I have drawn this conclusion, not only from conversations with several practical engineers, but, also, from essays published in our scientific journals. As an instance of the latter, I will select, for its convenience of reference, an article on water-wheels, published in this Journal, (Vol. I. p. 103,) which, being the production of a practical engineer, and having passed the inspection of a scientific committee, may be considered as corroborating my commencing observations. In the third paragraph of that article, is the following sentence : ' As a falling body, water descends at the *speed* of nearly 16 feet in the first second, and it will appear evident, that if a water-wheel is required to be so driven, that the water with which it is loaded has to descend 10, 11, or 12 feet per second, at which rates wheels are generally constructed to work, that a *very large proportion* of the power is lost.'

"Here, in the first place, we find *speed*, or *velocity*, confounded with the *distance* fallen in the first second; whereas, the latter is 16 feet, and the former is accelerated, from nothing, at the commencement, to 32 feet per second, at the end of the first second; so that this part of the sentence conveys, strictly, no intelligible meaning; it is, nevertheless, made a standard, by a comparison between which, and any given velocity of a water-wheel, we are to infer the loss of power sustained through excess of speed; thus, in the case of a wheel whose velocity is 10 or 12 feet per second, comparing these numbers with the mysterised number 16, the writer concludes, ' that a very large proportion of the power is lost.' The height of the fall which indicates the whole amount of the power, is not mentioned, but surely, to estimate the *proportion* between a defined part, and an undefined whole, is impossible.

* * * * * * *

"I have made a calculation of the distances, and velocities, attained by falling bodies, in various fractional parts of a second, which is here introduced for the information of those practical and theoretical engineers, who have avoided the labour of doing it for themselves.

"I have proceeded on the following established data; namely :—

"Heavy bodies fall through a distance of 16 feet, in the first second; at the end of which, they have acquired a velocity of 32 feet per second.—The velocity increases

APPENDIX.

as the times.—The distance increases as the square of the times.

Time of Descent.	Distance fallen.	Velocity attained per second.
	feet. inches.	feet. inches.
1 ⎫	0 0$\frac{1}{90}$	0 3
2	0 0$\frac{1}{21}$	0 6
3	0 0$\frac{2}{10}$	0 9
4 ⎬ 128ths of a sec.	0 0$\frac{3}{18}$	1 0
5	0 0$\frac{2}{4}$	1 3
6	0 0$\frac{3}{4}$	1 6
7 ⎭	0 0$\frac{3}{5}$	1 9
2 ⎫	0 0$\frac{3}{4}$	2
3	0 1$\frac{2}{3}$	3
4 ⎬ 32nds of a sec.	0 3	4
5	0 4$\frac{2}{3}$	5
6	0 6$\frac{1}{4}$	6
7 ⎭	0 9$\frac{1}{4}$	7
1-4th of a sec.	1 0	8
9 ⎫	1 3$\frac{1}{3}$	9
10	1 6$\frac{3}{4}$	10
11	1 10$\frac{2}{3}$	11
12 ⎬ 32nds of a sec.	2 3	12
13	2 7$\frac{7}{8}$	13
14	3 0$\frac{3}{4}$	14
15 ⎭	3 6$\frac{2}{11}$	15
1 half of a sec.	4 0	16
17 ⎫	4 6$\frac{1}{5}$	17
18	5 0$\frac{3}{5}$	18
19	5 7$\frac{2}{3}$	19
20 ⎬ 32nds of a sec.	6 3	20
21	6 10$\frac{2}{3}$	21
22	7 6$\frac{3}{4}$	22
23 ⎭	8 3$\frac{1}{3}$	23
3-4ths of a sec.	9 0	24
25 ⎫	9 9$\frac{1}{5}$	25
26	10 6$\frac{3}{4}$	26
27	11 4$\frac{2}{3}$	27
28 ⎬ 32nds of a sec.	12 3	28
29	13 1$\frac{2}{3}$	29
30	14 0$\frac{3}{4}$	30
31 ⎭	15 0$\frac{1}{5}$	31
1 ⎫	16	32
2	64	64
15 ⎬ seconds.	3600	480
30 ⎭	14400	960
1 minute.	57600	1920

"To determine what *proportion* of a given water power is lost by a given velocity of the wheel, it is only necessary to ascertain what distance the water must descend to acquire that velocity. Then, this distance compared with the whole fall, answers the question. Thus: suppose the whole fall to be 10 feet, and the velocity of the wheel 4 feet per second; this velocity is due to a fall of 3 inches, or one-fortieth part of the whole fall, which is the proportion sought. Or, suppose the velocity to be 13 feet per second, which is due to a fall of 2 feet 7⅜ inches, then the loss is rather more than one-fourth of the whole fall of 10 feet. But, it must be especially noted, that these estimates embrace the supposition, that the water issues upon the wheel in the direction of the motion of its skirt, and precisely at that distance below the surface of the dam, which answers to the velocity of the wheel. Inattention to this particular, is a very important and frequent cause of loss.   L. M."

With respect to the actual advantage of giving to overshot wheels a motion much less rapid than that usually given, the following example will probably have more effect on the mind of the mere practical workman, than any reasoning that could be offered; and, in fact, reasoning would be of little value, were it not supported by practical results.

The subjoined account is from the Technical Repository, a work published in London:—

"*On the comparative Advantages of different Water-Wheels, erected in the United States of America, by* JACOB PERKINS, *Esq.; and in this Country, by Mr.* GEORGE MANWARING, *Engineer.*

"Mr. PERKINS erected, at Newburyport, a water-wheel of 30 feet in diameter, on the plan of what is termed in America, a *pitch-back;* but, in this country, a *back-shut;* that is, one which receives the water near to its top, but not upon it, as in *overshot-wheels;* this is, indeed, the most judicious mode of laying water upon the wheel; as, in case of floods, the wheel moves in the same direction with the water, and not in the opposite one; neither is it encumbered, as in the *overshot* wheel, with a useless load of water at its top, where it does nothing but add to the weight upon the necks or pivots of the wheel-shaft, and to the consequent loss of power by the increased friction upon them; whereas, in the *pitch-back*, or *back-shut wheel*, the water is laid on at a point, where it acts by its leverage in impelling the wheel, and has yet time to become settled in the buckets, previously to its reaching the point level with the axis, where it acts with its greatest power. The wheel itself was constructed of oak, but with iron buckets; and it had a ring of teeth around it, which drove a cast-iron pinion, of three feet in diameter, which gave motion to three lying shafts, each of thirty feet in length, coupled together, so as to form a line of ninety feet; and from which, the necessary movements were communicated to the machinery for making nails.

" Mr. Perkins placed his pinion as close as possible under the pentrough, which delivered the water upon the wheel; and he thus greatly lessened the weight upon the necks or gudgeons of the wheel-shaft, by suspending it, as it were, upon the pinion; whereas, had he, as is usual, placed it on a horizontal line with the axis of the wheel, and on the opposite side of it, he would have loaded the necks with a double weight; namely, the water upon one side of the wheel, and the resistance opposed by the machinery to be driven by it, on the other. He also took care that the teeth upon the wheel, and the pinion, should always *be kept wet, or run in water*, instead of being *greased*, as is usual, and this he found sufficient to cause them to run smoothly and without the least noise. The motion of the wheel's periphery was about three feet per second, agreeably to the improved theory, so ably demonstrated by the late scientific Mr. Smeaton; and it continued to perform its work, with great satisfaction to its owners, for ten years, when it was unfortunately destroyed by fire.

" An opportunity soon presented itself of comparing the advantages of this water-wheel with another, which the proprietors were induced to erect, on the representations of a mill-wright, that the wheel was too high, and that it would be much better, were it only twenty-three feet in diameter, and received its water at the breast.

The trial, however, proved, that in driving the nail machinery, which had escaped the fire that destroyed the water-wheel, *the new wheel required twice the quantity of water to work it which actuated the former one, and only did the same work.*

" Mr. Manwaring has also had an opportunity of verifying, in this country, the advantages of a construction similar to Mr. Perkins's, in a cast-iron *back-shut* water-wheel of the same diameter as his, (namely, thirty feet,) and which also has a ring of teeth around it, driving a pinion of three feet in diameter, posited on the same side of the wheel as Mr. Perkins's, but not quite so high, it being a little above the centre of the wheel, and the teeth of the wheel and pinion are always kept wet. This wheel is employed in grinding flour, at a corn-mill in Sussex, and drives six pair of stones, besides the other necessary machinery, it moving at the rate of about three feet per second; and so great satisfaction has it given, that Mr. Manwaring is now constructing another water-wheel upon the same plan, and for the same proprietor; only, that it will be wider, and is calculated to drive eight pairs of stones.

" We are glad to have this opportunity of communicating these valuable practical facts: the same results being also obtained in two countries so widely separated as the United States and England, make them more valuable; and prove, that when persons think rightly, they will naturally think alike."

The foregoing example, although it relates to a *pitch-back* wheel, may serve our purpose as well as if it had been an overshot; there being an evident similarity between an overshot, with the water delivered on the top, with but little head, and the *pitch-back*, as constructed by Mr. Perkins; and also, between an overshot with considerable head, and the breast wheel.

The remarks made upon pitch-back wheels, are worthy the serious attention of the mill-wright. Mr. Evans very correctly compares them, in their action, to overshots: Mr. Ellicott thinks "an overshot with equal head and fall, is fully equal in power," and has dismissed them in a very few words. The reason of this is evident; the *head*, which they thought to be necessary, was not so easily managed with the pitch-back, as with the overshot; but when it is admitted, that the water should be delivered at the surface of the dam, that the velocity of the wheel should not exceed 4 or 5 feet per second, and that its capacity for containing water should be increased, the difficulty vanishes altogether. The water, when emptied from the buckets, has its impulse in the right direction to carry it down the tail-race; and in case of back water, the greater facility with which it will move is undeniable.

With respect to undershot wheels, Mr. Evans concludes that they ought to move with a velocity nearly equal to two-thirds of that of the water, and Mr. Ellicott estimates the velocity at quite two-thirds. It would be saying but little to assert that this did not agree with theory; but it does not accord with the opinions of many intelligent and experienced mill-wrights. It was asserted, upon theory, that the power of an undershot wheel would be at a maximum, when the velocity of the floats of the wheel was equal to one-third of the velocity of the water; practice, however, did not confirm the truth of this theory; and Borda has shown that the conclusion was theoretically incorrect, applying only to the supposition that the water impelled a single float-board; but that in the action upon a number of float-boards, as in a mill-wheel, the velocity of the wheel will be *one-half* the velocity of the water, when the effect is a maximum. The demonstration of this may be seen under the article Hydrodynamics, in the Edinburgh Encyclopædia. This was fully confirmed by the experiments of Smeaton, who, in speaking upon them, observes, that "in all the cases in which most work is performed in proportion to the water expended, and which approach the nearest to the circumstances of great works, when properly executed, the maximum lies much nearer to one-half, than *one-third*, one-half seeming to be the true maximum."

APPENDIX.

The succeeding observations are extracted from "Practical Essays on Mill-work and other Machinery, by Robertson Buchanan." Cast iron is very generally employed in England, not only for the wheel-work of mills, but, also, for many parts of the framing; the same practice obtains in those parts of our own country, where castings can be procured with facility, and will gain ground as its real value becomes known. Of course, the following extracts apply, in many instances, to the use of this material; but it will be found that the principles upon which they are founded, will, in general, apply to wood, as well as to iron.

" A Practical Inquiry respecting the Strength and Durability of the Teeth of Wheels used in Mill-work.

" Having treated of the forms of the teeth of wheels, I come now to consider their proportional strength, with relation to the resistance they have to overcome.

" I am aware, that owing to a great variety of circumstances, this subject is involved in much difficulty, and that it is no easy task to form any general rule with regard to the pitches and breadths of the teeth of wheels I do not pretend to more than a mere approximation towards general rules; yet, were this judiciously done, I am of opinion that it might be useful to the mill-wright, who has not had leisure or opportunity for scientific inquiries. A rule, though not absolutely perfect, is better in all cases than to have no guide whatever.

" And it is too evident to require proof, that it is essential to the beauty and utility of any machine, that the strength and bulk of its several parts be duly proportioned to the stress, or wear, to which the parts may be subject.

" Some general observations on the wheel work of mills, will serve greatly to simplify our inquiries on the subject."

"General Observations on the Wheel-Work of Mills.

"Mistaken attempts at economy have often prompted the use of wheels of too small diameter. This is an evil which ought carefully to be avoided. Knowing the pressure on the teeth, we cannot, with propriety, reduce the diameter of the wheel below a certain measure.

"Suppose, for instance, a water-wheel of 20 horses' power, moving at the pitch line with a velocity of $3\frac{1}{2}$ feet per second. It is known, that a pinion of 4 feet diameter might work into it, without impropriety; but we also know that it would be exceedingly improper to substitute a pinion of only one foot diameter, although the pressure and velocity at the pitch lines, in both cases, would be, in a certain sense, the same. In the case of the small pinion, however, a much greater stress would be thrown on the *journeys* (or *journals*) of the shaft. Not, indeed, on account of torsion or twist, but on account of transverse strain, arising as well from greater direct pressure, as from the tendency which the oblique action of the teeth, particularly when somewhat worn, would have to produce great friction, and to force the pinion from the wheel, and make it bear harder on the journals. The small pinion is also evidently liable to wear much faster, on account of the more frequent recurrence of the friction of each particular tooth.

"That these observations are not without foundation, is known to mill-wrights of experience. They have found a great saving of power, by altering corn mills, for example, from the old plan of using only one wheel and pinion, (or *trundle*,) to the method of bringing up the motion, by means of more wheels and pinions, and of larger diameters and finer pitches.

"The increase of power has often, by these means, been nearly doubled, while the tear and wear has been much lessened; although it is evident, the machinery, thus altered, was more complex.

"The due consideration of the proper communication of the original power, is of great importance for the construction of mills, on the best principles. It may easily be seen that, in many cases, a very great portion of the original power is expended, before any force is actually applied to the work intended to be performed.

"Notwithstanding the modern improvements in this department, there is still much to be done. In the usual modes of constructing mills, due attention is seldom given to scientific principles. It is certain, however, that were these principles better attended to, much power, that is unnecessarily expended, would be saved. In general, this might be, in a great measure, obtained, by bringing on the desired motions in a gradual manner, beginning with the first very slow, and gradually bringing up the desired motions, by wheels and pinions of larger diameters. This is a subject which should be well considered, before we can determine, in any particular case, what ought to be the pitch of the wheels. In the case above alluded to, where the supposition is a pinion of 4 feet diameter, or of 1 foot diameter, it is obvious, that the same pitch for both would not be prudent: that for the small pinion, ought to be much less than that which might be allowed in the case of the larger pinion. It is also equally obvious, that the breadth of the teeth, in the case of the small pinion, ought to be much greater than that in the case of the larger pinion.

"It is evident, however, that although great advantage may often be derived from a fine pitch, that there is a limit in this respect, as also with regard to the breadth. We shall endeavour to find some trace of this limit in what follows; and that we may the better do this, we shall call in the aid of propositions, which are true with respect to pieces of timber, or metal, subjected to ordinary causes of pressure. It is allowed, that they cannot here, in strictness, be *demonstrated*, as applicable to wheel-work. Yet they will, for want of better light, serve at least to prevent any material practical error, with regard to the strength of the teeth of the wheels. For it is to be remembered, that we are not so much here in search of truths of curious or profound mathematical speculation, as of that kind of evidence of which the subject admits, and which may be sufficiently satisfactory for any practical purpose.

"As cast-iron pinions are now generally used, and as the teeth of the pinion are most subject to wear, I think we are safe in the present inquiry, in considering them all as cast-iron.

"The laws to which I have alluded in this investigation, are these:—

"'*Principles of proportioning the Strength of Teeth of Wheels.*'

"'PROPOSITION I.

"'*The Strength of any Piece of Timber or Metal, whose Section is a Rectangle, is in direct Proportion to the Breadth, and as the Square of the Depth.*'*

"Hence may be inferred, that the strength of the teeth of wheels, moving at the same velocity, and under the same circumstances, is directly in proportion to their breadth, and as the square of their thickness. Thus, for example, if we double the breadth, we only double the strength; but if we double the thickness, or, in other words, double the pitch, keeping the original breadth, we increase the strength four times.

"For although when wheels are working accurately, the strain is, at the same time, divided over several teeth; yet as a very small inaccuracy, or even the interposition of any small body, such as a chip of wood, or stone, throws the whole stress upon a single tooth, in practice; therefore, and in order to simplify this case, we may consider the strength of a single tooth, as resisting the pressure of the whole work.

"But as the length of the teeth commonly varies with the pitch, this circumstance must be taken into account, and the most simple view we can take of it seems to be that of having the strain of each tooth, thrown all to the outward extremity: we have then the following proposition to guide this part of our inquiry:—

"'PROPOSITION II.

"'*If any Force be applied laterally to a Lever or*

---

* See Emerson, Prop. 67.

Beam, *the Stress upon any Plate is directly as the Force and its Distance from that Plate.'**

"'PROPOSITION III.

"'*The Pitch being the same, the Stress is inversely as the Velocity.*'†

"For example—if the pitch lines of one pair of wheels be moving at the rate of 6 feet in a second, and another pair of wheels, in every other respect under the same circumstances, be moving at the rate of 3 feet in a second, the stress on the latter, is double of that on the former."

"*Of arranging the Numbers of Wheel-Work.*

"In a machine, the velocity of the impelled point should be to that of the working point, in the ratio which is adapted to the maximum effect of the moving power on the one part, and the best working effect on the other part. Any other arrangement of the relative motions of the parts of a machine must clearly be attended with a loss of power, or the work will not be done properly. But when the best working velocity is known, and, also, that which enables the first mover to produce the greatest effect, the proper arrangement of the numbers of the teeth of the wheels and pinions, is a very simple operation.

"It will be an advantage to advertise the young mechanic of one or two essential particulars, before proceeding to the principal object.

"In the first place, when the wheels drive the pinions, the number of teeth in any one pinion should not be less than 8; but rather let there be 11 or 12 if it can be done conveniently. And in the particular form of teeth previously described, the number of teeth in a pinion should not be less than 10; but it would be better to have 13 or 14.

"Secondly—When the pinions drive the wheels, the number of teeth on a pinion may be less; but it will not,

---

* See Emerson, Prop. 69. † See Emerson, Prop. 119, Rule 8.

in any case, be desirable to have fewer than 6 teeth on a pinion; and give the preference to 8 or 9, where it can be done with convenience.

"Thirdly—The number of teeth in a wheel should be prime to the number of teeth in its pinion; that is, the number representing the teeth in the wheel should not be divisible by the number of teeth in the pinion without a remainder. And as the numbers of pinions will, in general, be first settled, it will be an advantage to take a prime number for each pinion, as 7, 11, 13, 17, 19, 23, &c., because such numbers are more seldom factors than others. But when it happens that a prime number can be directly fixed upon for the wheel, any whole number which approaches near to the required ratio will answer for the pinion; as minute accuracy is not required. A prime number for the wheel, or one which is not divisible by the number of the pinion, is esteemed the best, because the same teeth will not always come together, and the wear will be more uniform.

"Fourthly—If it be desired that a given increase or decrease of velocity should be communicated with the least quantity of wheel work, it has been shown that the number of teeth on each pinion should be to the number on its wheel, as 1 : 3,59 (Dr. Young's Nat. Phil. Vol. II. Art. 366.) But, on account of the space required for several wheels, and the expense of them, it will often be necessary to have 5 or 6 times the number of teeth on the wheel, that there is on the pinion. The ratio of 1 : 6 should, however, not be exceeded, unless there be some other important reason for a higher ratio."

"*Practical Observations with regard to the making of Patterns of Cast-Iron Wheels.*

"Having determined the pitch of the wheel strong enough for the purpose to which it is to be applied, the thickness of the tooth serves to regulate the proportionate strength of the other parts.

"A very respectable mill-wright informs me, that he

has for a considerable time adopted the following rule for determining the length of the teeth of wheels, the practical efficacy of which he has found quite satisfactory:—

"Rule—*Make the length of the teeth equal to the pitch, deducting freedom,* (by the freedom is meant the distance at the top of one tooth and the root of another, measured at the line of centres,) in other words, the distance from root to root of the teeth, at the line of teeth, when the wheels are in action, exactly equal to the pitch.

"For example—he makes the teeth of two inches pitch, one inch and thirteen-sixteenths in length, which is allowing three-sixteenths of freedom.

"Another respectable mill-wright, who has had much experience, particularly in mills moved by horses, has, for a considerable time past, made the teeth of his wheels only one-half of the pitch in length, and works them as deep as possible, without the point touching the bottoms. Before he fell on this expedient, he found the teeth exceedingly liable to be broken from any sudden motion of the horses.

"Indeed, upon reflection, it will be found that there is no occasion for more freedom, than that the point of the tooth of the one wheel, shall just clear the ring of the other; more than this must only serve to weaken the teeth. The mode of gearing, however, above alluded to, is more necessary in horse mills than where the moving power is steady and regular.

"Hutton (on clock-work) recommends making the distance of the pitch line three-fourths of what we call the thickness of the tooth. Thus, suppose the rule applied to a two inch pitch, and that the tooth and space were exactly equal, then the tooth would project three-fourths of an inch beyond the pitch line, and its root would be as far within the pitch line, as to receive freely the tooth intended to act on it: suppose it also three-fourths, then the tooth would be one and a half inch long, besides the freedom, which making, as above, three-sixteenths, the tooth would be in all one and eleven-sixteenths inch long.

"But it is to be remarked, that the mill-wright, in making his pattern for a cast-iron wheel, has to attend to a circumstance arising from the nature of that material. The pattern must not only be of such a form as to be sufficiently strong, calculating by the bulk of the parts, but also proportioned, so that when the fluid metal is poured in the mould, it may cool in every part nearly at the same time.

"When due attention is not paid to this circumstance, as the metal is cooling, if it contract faster in one part than in another, it will be apt to break somewhere, just as a drinking glass is broken by suddenly cooling or heating in any particular part of it. In all patterns for cast-iron, about one-eighth of an inch to the foot should be allowed for the contraction of the metal in cooling.

"Attention must also be paid to taper the several parts, so that they may rise freely without injuring the mould, when the founder is drawing them out of the sand. A little observation of the operations of a common foundry, will better instruct in this part of the subject than many words. We may observe, however, that about one-sixteenth of an inch, in a depth of 6 inches, is commonly a sufficient taper.

"Attending to those circumstances, I beg leave to offer the following proportions as having been found to answer in practice.

"Make the thickness of the ring equal to the thickness of the tooth near its root. When the ring is made thinner than the root of the tooth, the ring commonly gives way to a strain, which would not break the tooth.

"Make the arm, at the part where it proceeds from the ring, of the same breadth and thickness as the ring, and, at the junction, let it be so formed as to take off any acute angle which would be apt to break off in the sand.

"The arms should become larger as they approach the centre of the wheel, (see Emerson, Prop. 119, Rule 8,) and the eye should be sufficiently strong to resist the driving of the wedges, by means of which it is to be fixed on the shaft. This cannot be brought easily to calculation.

"On the other hand, care must be taken not to make the eye so thick as to endanger unequal cooling.

"It should be somewhat broader than the breadth of the teeth, in order that it may be the firmer on the shaft: this breadth must be greater in proportion as the wheel is large.

"When the ring is about an inch thick, it is common to make the eye about an inch and a quarter in thickness, and about one-fifth broader than the ring, when the wheel is about four feet in diameter.

"Small wheels have generally but four arms, but it being improper to have a great space of the ring unsupported, the number of arms should be increased in large wheels.

"In order to strengthen the arms with little increase of metal, it is not unusual to make them feathered, which is done by adding a thin plate to the metal at right angles to the arm.

"The same rules apply to bevelled wheels; of the practical mode of laying down the working drawings of which we have already spoken. But it is proper to observe that the eye of a bevelled wheel should be placed more on that side which is farthest from the centre of the ideal cone, of which the wheel forms a part.

"When wheels are beyond a certain size, it becomes necessary to have patterns sometimes made for them, cast in parts, which are afterwards united by means of bolts.

"A very good mode to prevent the bad effects of unequal contraction, is to have the arms curved; the curved parts are commonly of the same radius as the wheel, and spring from the half length of the arms."

"*Of Malleable or Wrought-Iron Gudgeons.*

"Professor Robinson states,[*] that the cohesive force of a square inch of cast-iron is from 40,000 to 60,000 lbs. wrought iron from 60,000 to 90,000 lbs.

"In the year 1795, I had occasion to substitute cast-

---

[*] Encyclopædia Britannica, article, Strength of Materials, 40.

iron gudgeons for those of wrought iron, and made some experiments on those metals, from which I drew the following inference: *that gudgeons of the same size, of cast and of wrought iron, in practice, are capable, at a medium, of sustaining weights without flexure, in the proportion of* 9 *to* 14.

"Taking it for granted that this proportion is near the truth, we may find the diameter which any wrought iron gudgeon ought to have when its lateral pressure is given, in the following manner:—

"1. Find the diameter which a cast-iron gudgeon should have to sustain the given pressure; then say, as 14 is to the cube of the diameter of the cast-iron gudgeon, so is 9 to the cube of the diameter of the wrought iron gudgeon.

"2. The root of this last number gives the diameter required of the wrought iron gudgeon.

EXAMPLE.

"Suppose the lateral pressure to be 125 hundred weights, the cube root of which is 5, the diameter in inches of the cast-iron gudgeon: then say,

$$\begin{array}{ll} \text{As} & 14 \\ \text{Is to} & 125 \\ \text{So is} & 9 \\ \text{To} & 80{,}357 \end{array}$$

"The cube root of which is 4,30887."

"*Of the Bearings of Shafts.*

"The bearings on which gudgeons and journals rest and revolve, are sometimes termed *Pillows*, and frequently *Brasses*, from being often made of that substance.

"It has become general to fix pillows in blocks of cast-iron. Hence the term *Pillow Block*, and sometimes, corruptly, *Plumber Block*.

"At the cotton works of Deanston, near Down, a water-wheel has run nearly 30 years on pillows of cast-iron, with little sensible wear on the gudgeons, nor were they ever found liable to heat.

"The outer skin of cast-iron, particularly when cast in metallic moulds, is remarkably hard, and it is reasonable to suppose that it would make a durable pillow, as we have seen is the case in the above instance."

" On the Framing of Mill-Work.

" Mill-work, from its motion, occasions a tremor on all the parts of its framing, which subjects it to much more speedy decay than the mere pressure upon carpentry.

" Besides this general tremor, it is often subjected to violent, sudden thrusts, from the bad actions of the wheels, or from reciprocating motions.

" It ought, therefore, to be not only sufficiently *strong* and *stiff*, but sufficiently *heavy*, to give solidity and steadiness.

" Where the framing of the machinery is not firm and well bound, a vibratory motion in its parts, of course, takes place; which vibratory motion expends a considerable portion of the power applied. This loss of power is very difficult of investigation. It is certain, however, that whatever motion of a vibratory nature is communicated to the framing and objects in contact with it, (abstracted from the elasticity of the parts,) must be lost to the effect the machine would produce, were the parts sufficiently strong and well bound together; and it is to be observed, that firm and well-bound framing is much preferable to heavy framing, not so well connected in its parts. It is as certain, that though the framing in either case may be constructed so as to be equally strong, yet the heavy framing, from its vibration, will expend more of the original power than that which is less heavy, but firmly connected.

" Besides *strength, stiffness,* and *solidity,* the framing of mill-work requires to be constructed so as *to be easy of repair;* and so contrived, that *any particular part may be repaired or renewed* with the least possible derangement to the other parts of the framing.

"There is another circumstance in this species of framing which demands great attention. *The shafts often require to be restored to their true situations,* from which they may have deviated by the wearing of the parts. Now the framing ought to be so constructed as easily to admit of this *restoration of the shafts,* as also of any other shifting of them which may in practice become necessary.

"But though the framing which supports the parts of mills and machines should be firm, it is an advantage that the part on which any axis rests should have a small degree of elastic tremor, when the machine is in motion. Such tremor has considerable power in diminishing the friction. It may farther be observed, that framing to support machinery should be as independent of the building as possible, because the tremor it always communicates is exceedingly injurious."

On Reaction Wheels.

These wheels were slightly noticed at page 176; and a description of Barker's mill is to be found in nearly every work upon hydraulics, together with the improvement made in it by Rumsey. Within a few years past, wheels which operate upon the principle of the rotary trunk, in Barker's mill, have been extensively brought into use. We are not informed by whom they were invented; Mr. Evans alludes to them in the first edition of this work, published in 1795; but it does not appear certain that he had then seen them; it is manifest, at all events, that they were not publicly known. His words are, "One of these is said to do well where there is much back water; it being small, and of a true circular form, the water does not resist it much. I shall say but little of these, supposing the proprietors mean to treat of them."

Their great merit, certainly, is their simplicity; and where there is a plentiful supply of water, they may, in many cases, be preferable to any other. Those interested in them aver that they are but little, if at all, inferior in economy to overshot mills; this, however, we are, by no means, prepared to admit. In back water they will undoubtedly operate better than any other, as there will not be any sensible loss from their wading, but only from the diminution of the effective head. In an eight feet fall, for example, should there be four feet of back water, the remaining four feet will produce nearly, or quite, its full effect.

Many patents have been obtained for modifications of, and variations in, this wheel; and from the specification of one of them, as published in the Journal of the Franklin Institute, at Philadelphia, we will give such extracts as will suffice to exhibit their nature and mode of action. In doing this, we shall omit the claims of the patentee, as this is a point with which we, in this place, have nothing to do.

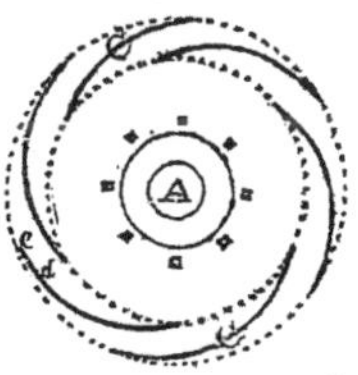

"Fig. 1, a bird's eye view of the wheel, the end to which the shaft is to be attached, at the perforation, A, being downwards, and the open end, or rim, upwards. To show the floats, the upper rim, which covers them, is not represented. The lines C C exhibit the form of the floats, or buckets, and the manner in which they are arranged. The diameter of this wheel, and the width of the floats between the two heads, and the depth of aperture between the floats, will, of course, be varied according to the quantity and head of water which can be obtained, and the purpose to which it is to be applied. The curved floats, it will be seen, are made to lap over each other; and, in practice, it has been found that the proportion in which they do so is a point of considerable importance. The proportion between the aperture and lap, which was found to be the best, is as three to two; that is, for every inch of aperture, measuring from float to float, at the point where the water escapes, the floats should pass each other one and a half inch. It will be manifest that a slight deviation from this proportion, in either way, will not be attended by any sensible loss of power. Any considerable deviation, however, is found

Figure 1.

to be injurious. The mechanic should be careful so to construct his wheel that the part of the aperture seen at *e* should be less than that seen at *d*.

"Upon the inner edge of the rim there is a projecting fillet, or flanch, which may be seen in the section D, of this wheel, at the lower part of Fig. 3, with this difference, that said fillets or flanches are to be made flat, as they are to work against, and not within, each other.

"Wheels so constructed may be applied either on a horizontal or vertical shaft, and either singly or in pairs, according to circumstances.

"Fig. 2 represents the double reacting wheel, placed on a horizontal shaft, in which manner they are to be used, whenever it is desirable to obtain motion from such a shaft. S is the horizontal shaft, A the penstock, and B the cistern; the heads, or sides of the cistern, are formed in whole, or in part, of cast-iron plates, securely bolted together. D D are two water-wheels, one of which is placed on each side of the cistern B, their open ends standing against the side plates of the cistern, which are perforated, having openings in them equal in size to those on the heads of the wheels, and being concentric with them. The fillet, or flanch, upon the rim of each wheel, is made flat, and is fitted to run as closely to a similar fillet or flanch on the cistern head as may be, without actually bearing against it, so as to prevent too much waste of water, and yet to avoid friction by touching it.

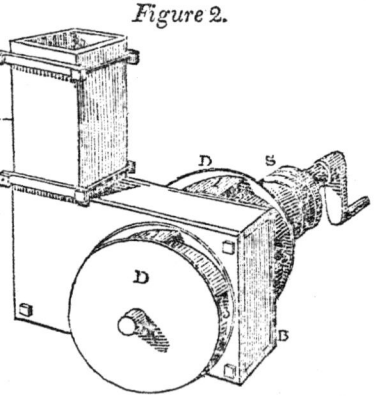

Figure 2.

"The size of the orifices in the wheel and cistern plates is a point of essential importance, and should greatly exceed what has been heretofore thought necessary. Their area should be such as to permit the whole column of water to act unobstructedly on the wheel, whatever may be the height of the head. It is found that for a head of four feet, the area of the orifice should never be permitted to fall short of three times the number of square inches which can be delivered by all the openings of the floats. The penstock, or gate way, should also be sufficiently large to admit freely the same proportionate quantity of water through every part of its section; say about three times the area of the orifices of the cistern heads and wheels.

"For a greater head these openings must be proportionally increased, or the whole intention will be defeated, as it has been from want of attention to this principle, that numerous failures have occurred in the attempt to drive mills by reaction wheels. Whenever it is practicable, the limit which has been given should be exceeded, but never can be diminished without loss.

"Instead of using a trunk or penstock, smaller than the horizontal section of the cistern B, extend the sides front and back of said cistern, upwards in one continued line, whenever the same can be done; the cistern and penstock then form one trunk, of equal section throughout.

"When greater power is requisite, place other reacting wheels, or pairs of wheels, upon the same shaft, so that each may operate in the same way.

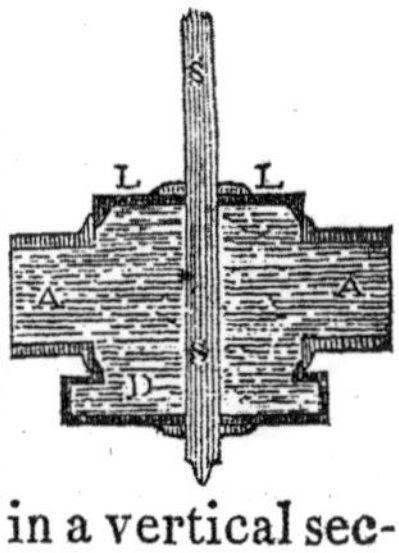

"Fig. 3 represents one of the reacting wheels, placed upon a vertical shaft, with the cistern by which it is supplied with water; to this is also attached what is denominated *the lighter*, which is intended to relieve the lower gudgeon and step from the pressure of the column of water, and also, when desired, the weight of the wheel, and whatever is attached thereto. The whole being shown in a vertical section through the axis of the wheel.

Figure 3.

"A A is the cistern of water, the construction of which, with its penstock, may be seen at B A, fig. 4.

"D the wheel, the flanch on its upper side passing within the edge of that on the lower plate of the cistern.

"L L the *lighter* for relieving the gudgeon and step of the shaft and wheel from the downward pressure.

"The lighter is a circular plate of iron, concentric with the wheel, and attached to the same shaft. Upon its lower side is a flanch, or projecting rim, fitting into an orifice in the upper plate of the cistern, in the same manner in which that of the wheel fits into the lower plate; allowing, therefore, of a vertical motion of the shaft to a certain extent, without binding upon the plates of the cistern.

"From the equal pressure of fluids in all directions, the lighter, (when equal in its area to that of the orifice of the wheel,) will be pressed upwards with the same degree of force with which the latter, (the wheel,) is pressed downwards; and if made larger, it will be pressed upwards with a greater force; and may be so proportioned as to take off the weight both of the machinery and of the water, from the gudgeon and its step.

"When a single wheel is placed upon a horizontal shaft, the lighter will take the place of the second wheel, and so also in the case of any odd number of wheels, either on a vertical or a horizontal shaft.

"Fig. 4 represents the double reacting wheel on a vertical shaft. A being the penstock—B the cistern—D D the wheels, revolving within the plates of the cistern in the same manner as the wheel and lighter in Fig. 3.

Figure 4.

"The upper wheel in this arrangement answers all the purposes of the lighter in the former, the orifice of which may be enlarged, if desired, with the same views."

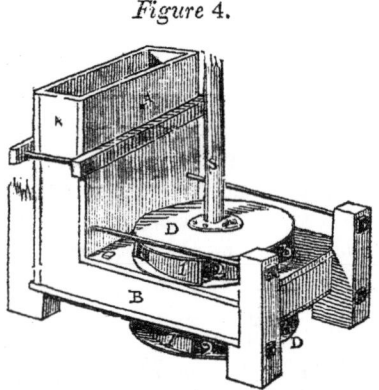

The foregoing is a description of the reaction wheel, as patented by Mr. Calvin Wing, and is given in the language of his specification; it exhibits, therefore, *his* views upon the subject. The buckets are sometimes so made as not to lap, the inner end of one terminating in a line with the outer end of another. Some persons construct them so that the buckets are adjustable, thus allowing the apertures to be enlarged or diminished, according to the quantity of water employed, or of machinery to be driven. There are, in fact, not fewer, we believe, than eight or ten patents for different modifications of this wheel, and from the interest which it has excited, it may be considered as in a fair way to have its relative merits fully tested.

APPENDIX. 391

Explanation of the Technical Terms, &c., used in this Work.

Aperture—The opening by which water issues.

Area—Plain surface, superficial contents.

Algebraic signs used are $+$ for more, or addition. $-$ Less, subtracted. $\times$ Multiplication. $\div$ Division. $=$ Equality. $\sqrt{}$ The square root of; 86^2 for 86 squared; 88^3 for 88 cubed.

Biquadrate—A number squared, and the square multiplied into itself—the biquadrate of 2 is 16.

Corollary—Inference.

Cuboch—A name for the unit or integer of a power, being the effect produced by one cubic foot of water in one foot perpendicular descent.

Cubic foot of water—What a vessel one foot square and one foot deep will hold.

Cube of a number—The product of a number multiplied by itself twice.

Cube root of a number—Say of 8;—the number which multiplied into itself twice will produce 8; namely, 2. Or, it is that number by which if you divide a number twice, the quotient will be equal to itself.

Decimal point—Set at the left hand of a figure, shows the whole number to be divided into tens, as ,5 for $\frac{5}{10}$ths; ,57 for $\frac{57}{100}$ths; ,557 for $\frac{557}{1000}$ths parts.

Equilibrio, Equilibrium—Equipoise or balance of weight.

Elastic—Springy.

Friction—The act of rubbing together.

Gravity—That tendency all matter has to fall downwards.

Hydrostatics—The science which treats of the weight of fluids.

Hydraulics—The science which treats of the motion of fluids, as in pumps, water-works, &c.

Impulse—Force communicated by a stroke, or other power.

Impetus—Violent effort of a body inclining to move.

Momentum—The force of a body in motion.

Maximum—Greatest possible.

Nonelastic—Without spring.

Octuple—Eight times told.

Paradox—Contrary to received opinion; an apparent contradiction.

Percussion—Striking together, impact.
Problem—A question proposed.
Quadruple—Four times, fourfold.
Radius—Half the diameter of a circle.
Right angle—A line square, or perpendicular to another.
Squared—Multiplied into itself; 2 squared is 4.
Theory—Speculative plan existing only in the mind.
Tangent—A line perpendicular to, or square with, a radius, and touching the periphery of a circle.
Theorem—Position laid down as an acknowledged truth. A rule.
Velocity—Swiftness of motion.
Virtual or effective descent of water—(See Article 61.

SCALE FROM WHICH THE FIGURES ARE DRAWN IN THE PLATES FROM II. TO XI.

PLATE II. Fig. 11, 12, 8 feet to an inch; fig. 19, 10 feet to an inch.
    III. Fig. 19, 20, 23, 26, 10 feet to an inch.
    IV. Fig. 28, 29, 30, 31, 32, 33, 10 feet to an inch.
    VI. Fig. 1, 10 feet to an inch; fig. 2, 3, 8, 9, 10, 11, 2 feet to an inch.
    VII. Fig. 12, 13, 14, 15, 2 feet to an inch; fig. 16, 10 feet to an inch.
    X. Fig. 1, 2, 18 feet to an inch; fig. H, I, in fig. 1, 4 feet to an inch.
    XI. Fig. 1, 2, 3, 2 feet to an inch; fig. 6, 8, 1 foot to an inch.

THE END.

Plate I

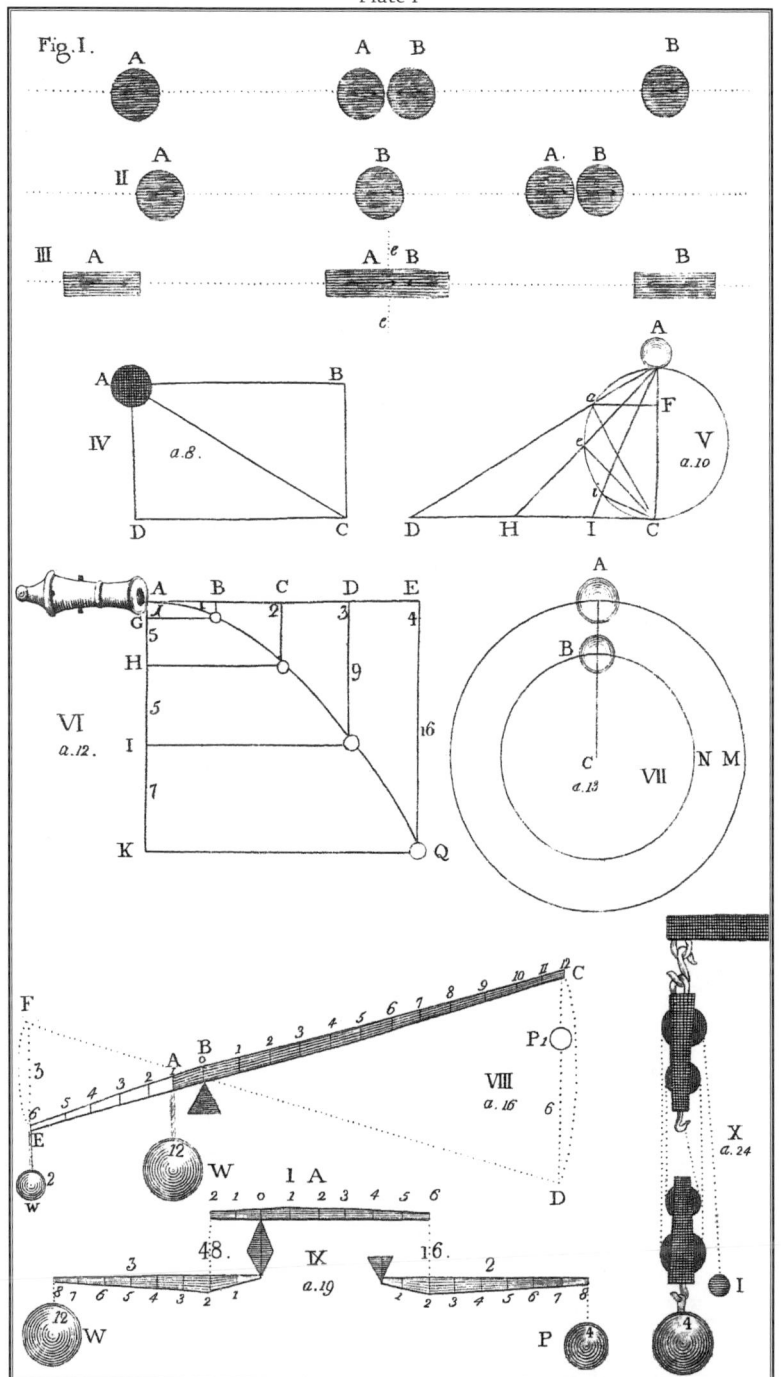

Plate II

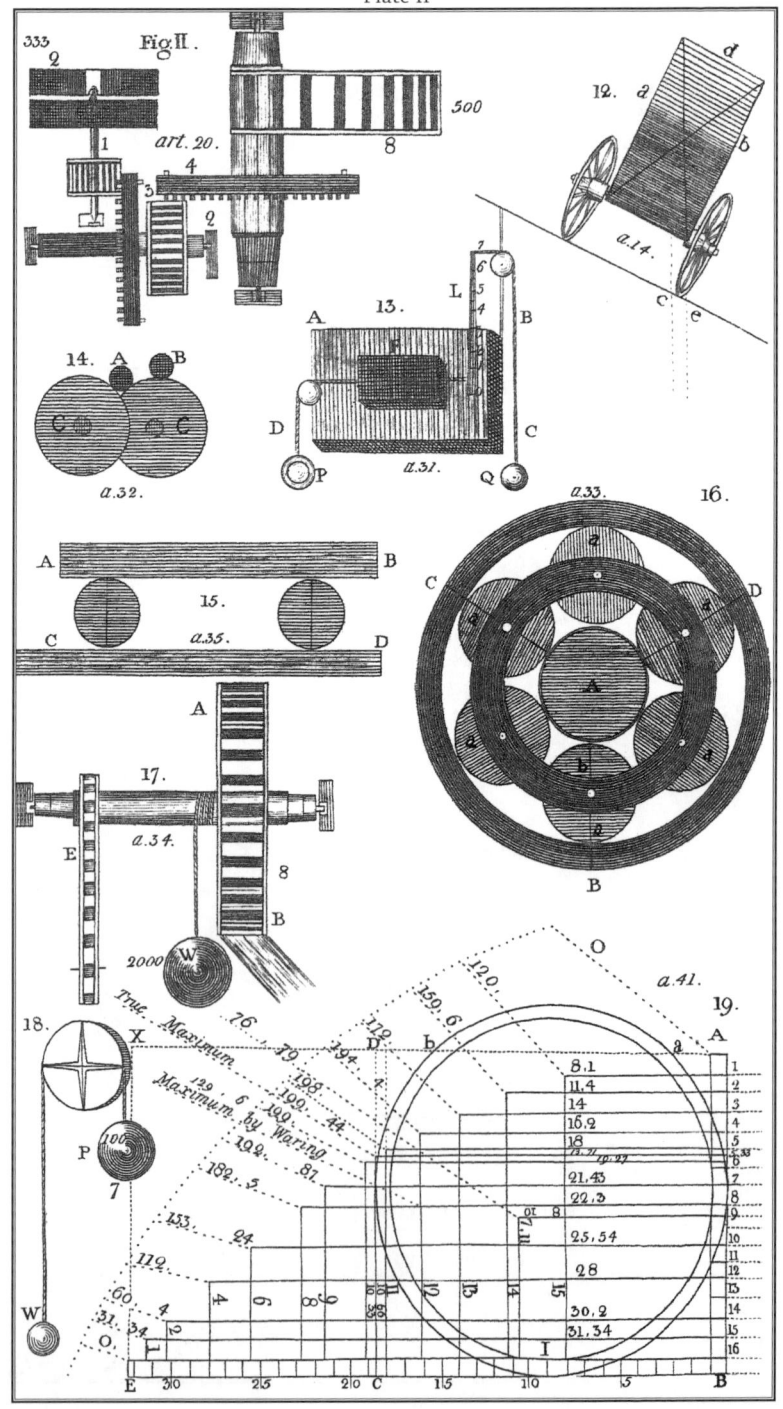

Plate III

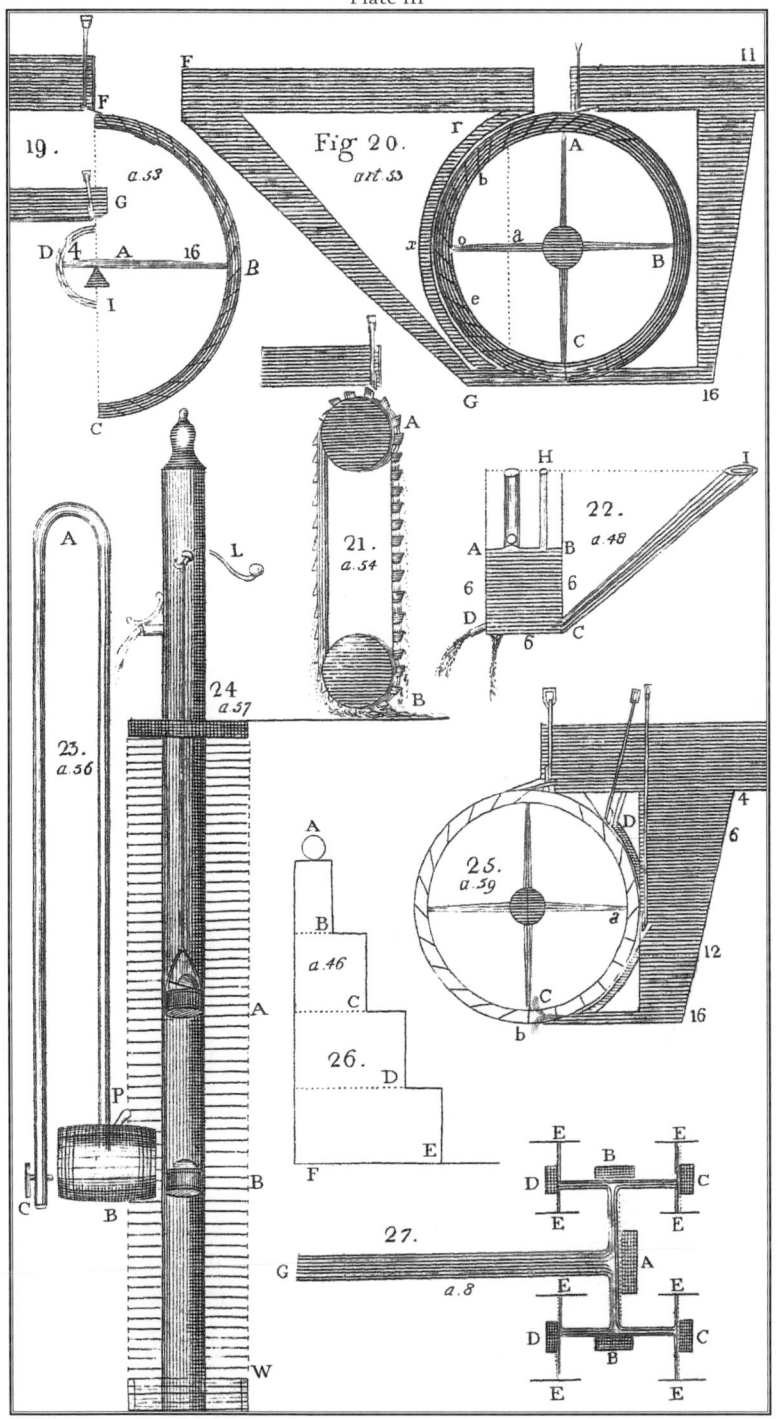

Plate IV

Plate V

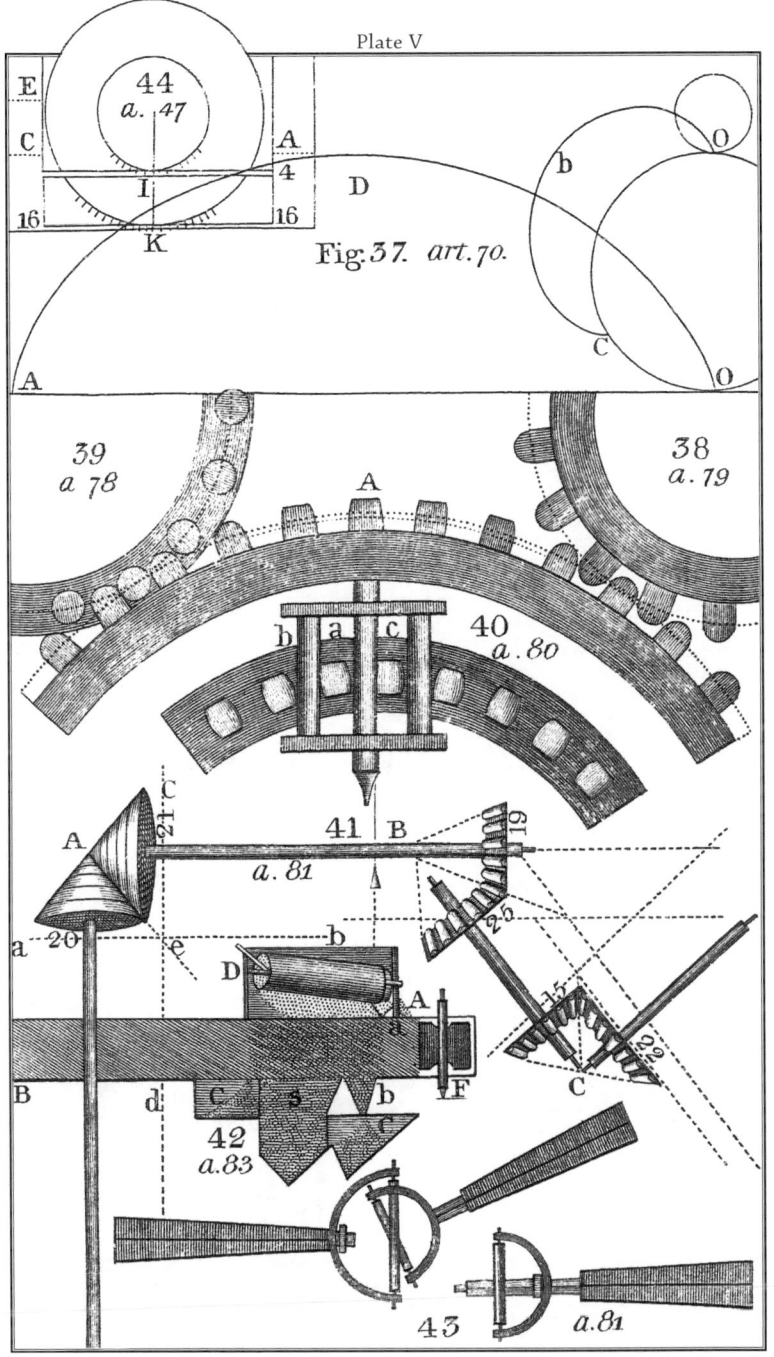

Fig. 37. art. 70.

Young & Delleker Sc.

Plate VI

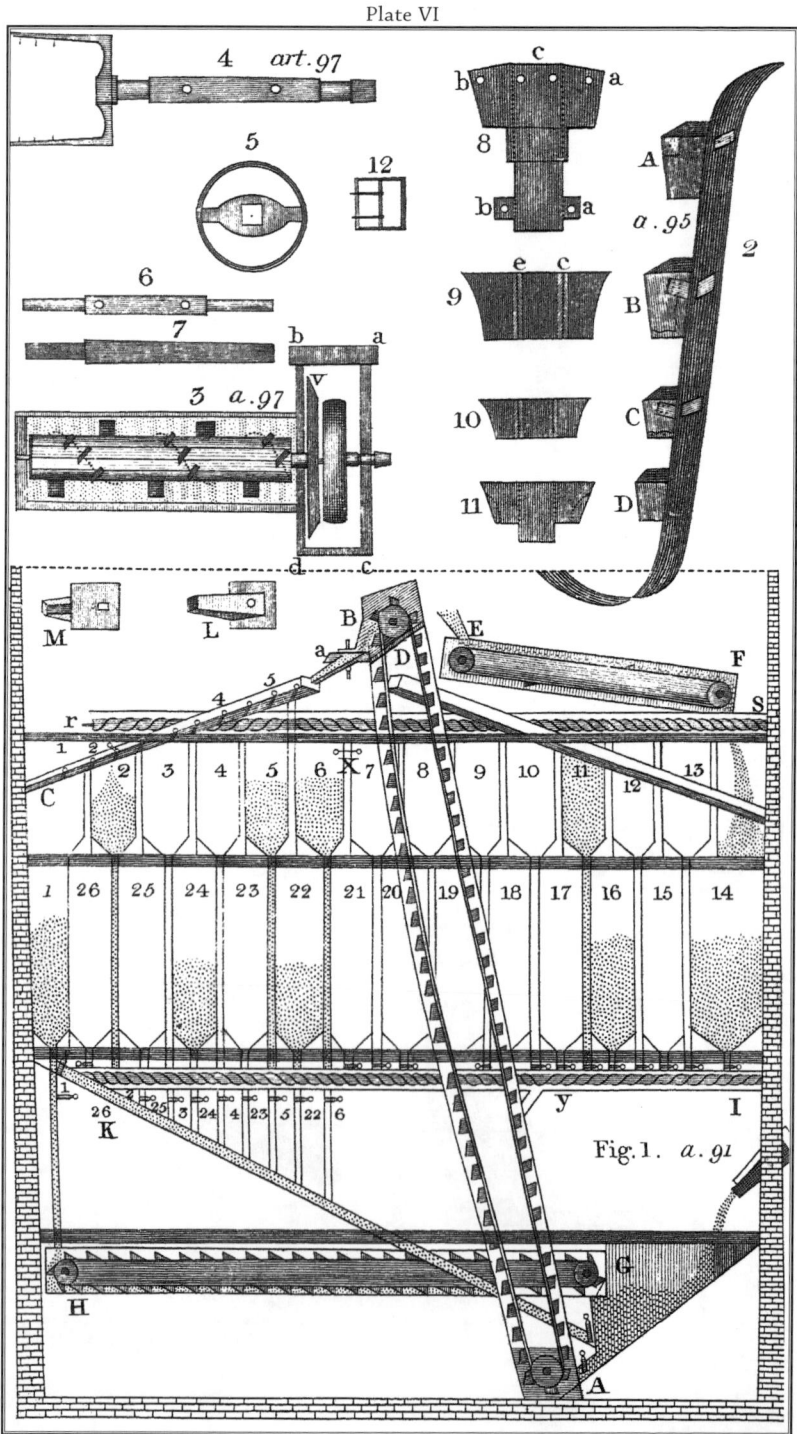

Fig. 1. a. 91

Young & Delleker Sc.

Plate VII

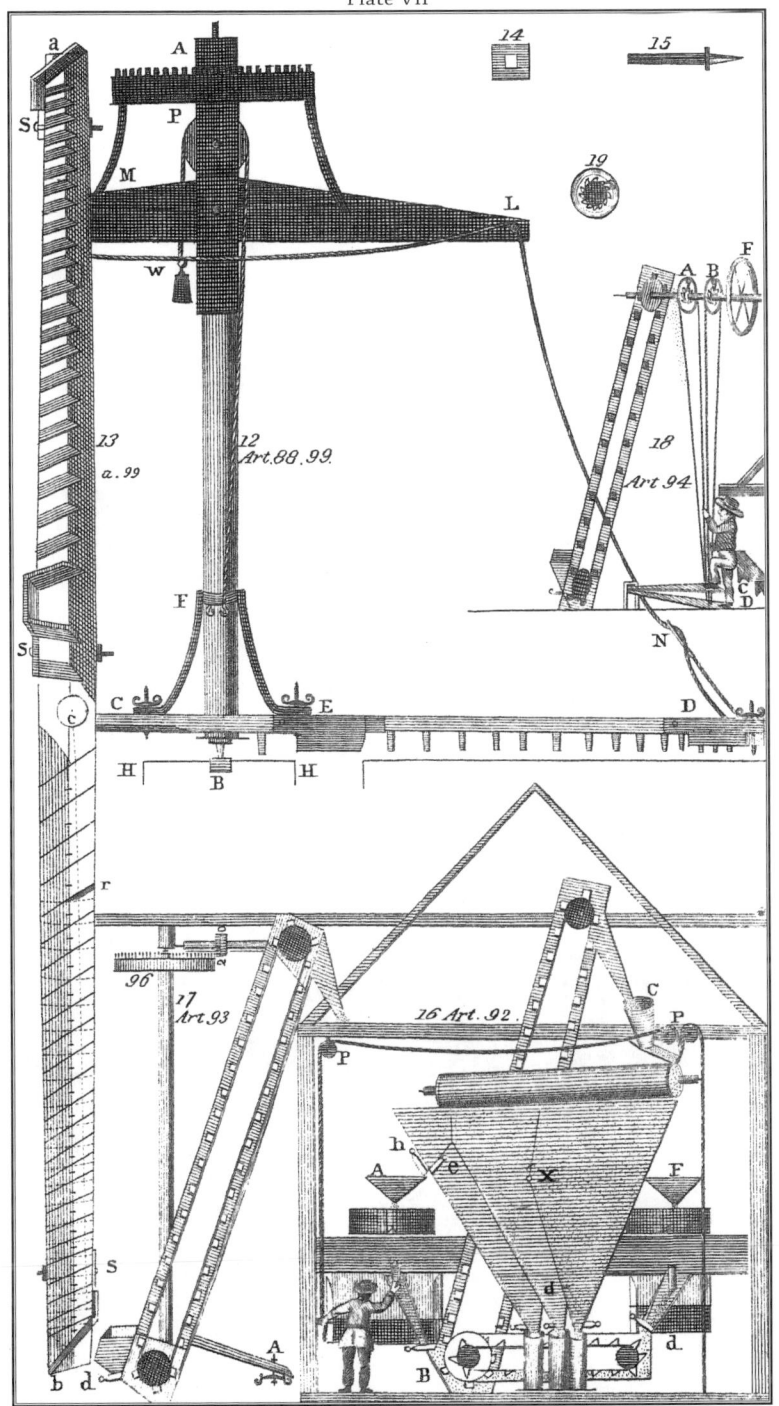

Plate VIII   Art.e 89.

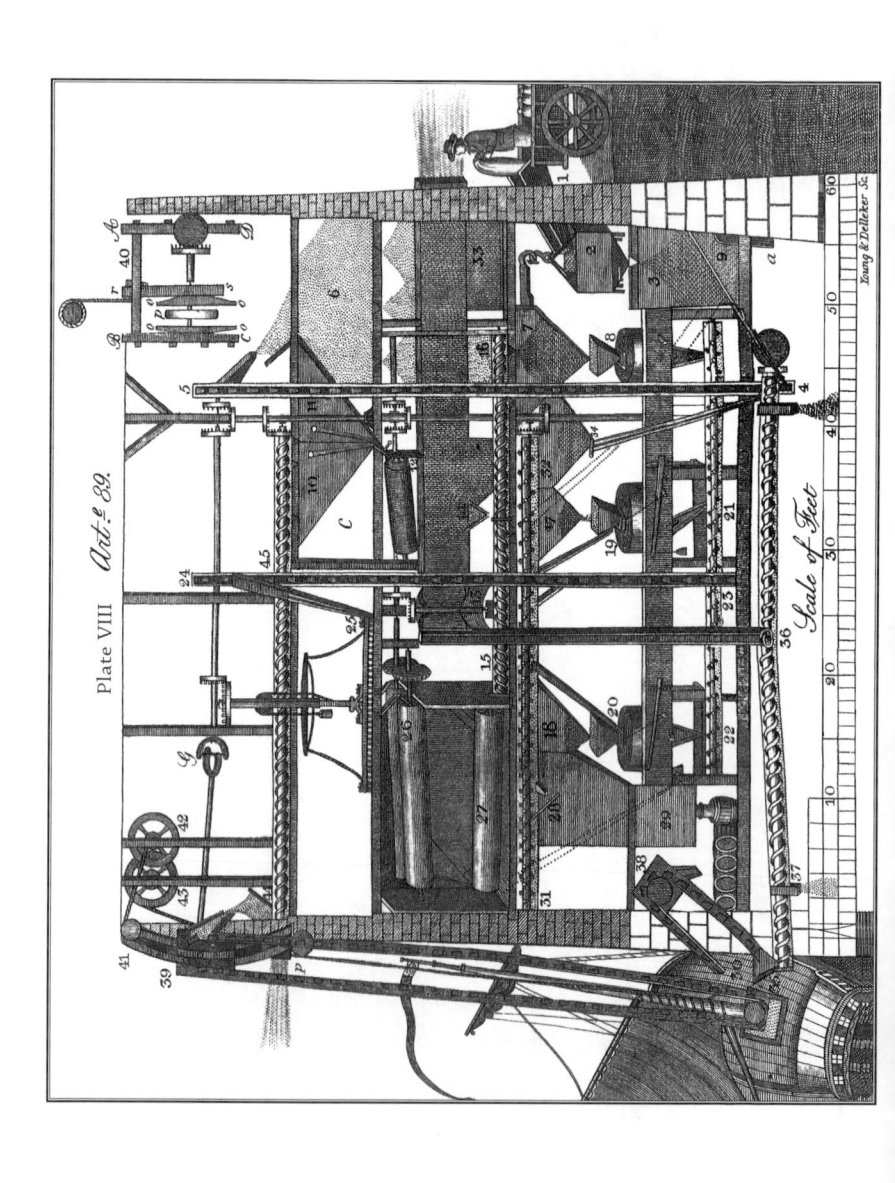

Plate IX

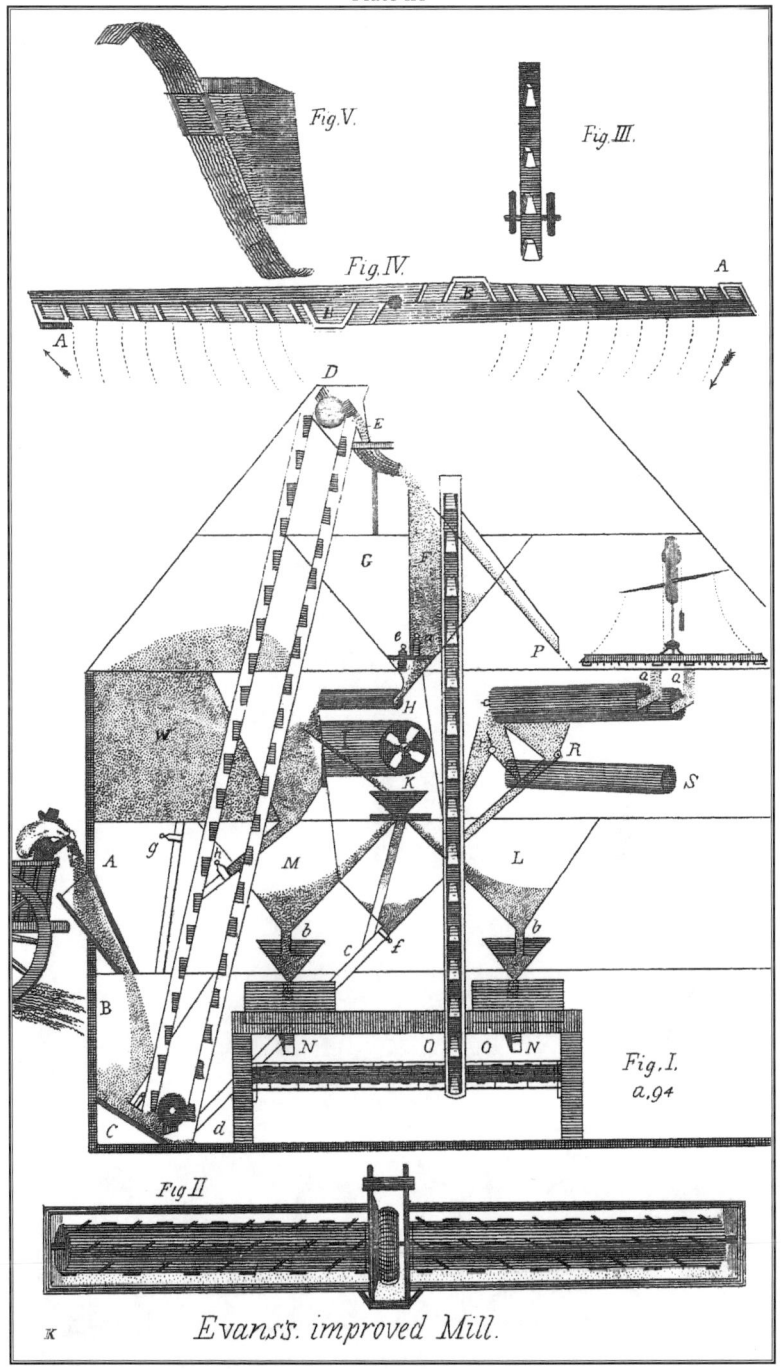

Fig. V.

Fig. III.

Fig. IV.

Fig. I.
a, 94

Fig. II.

Evans's improved Mill.

Plate X

Plate XI

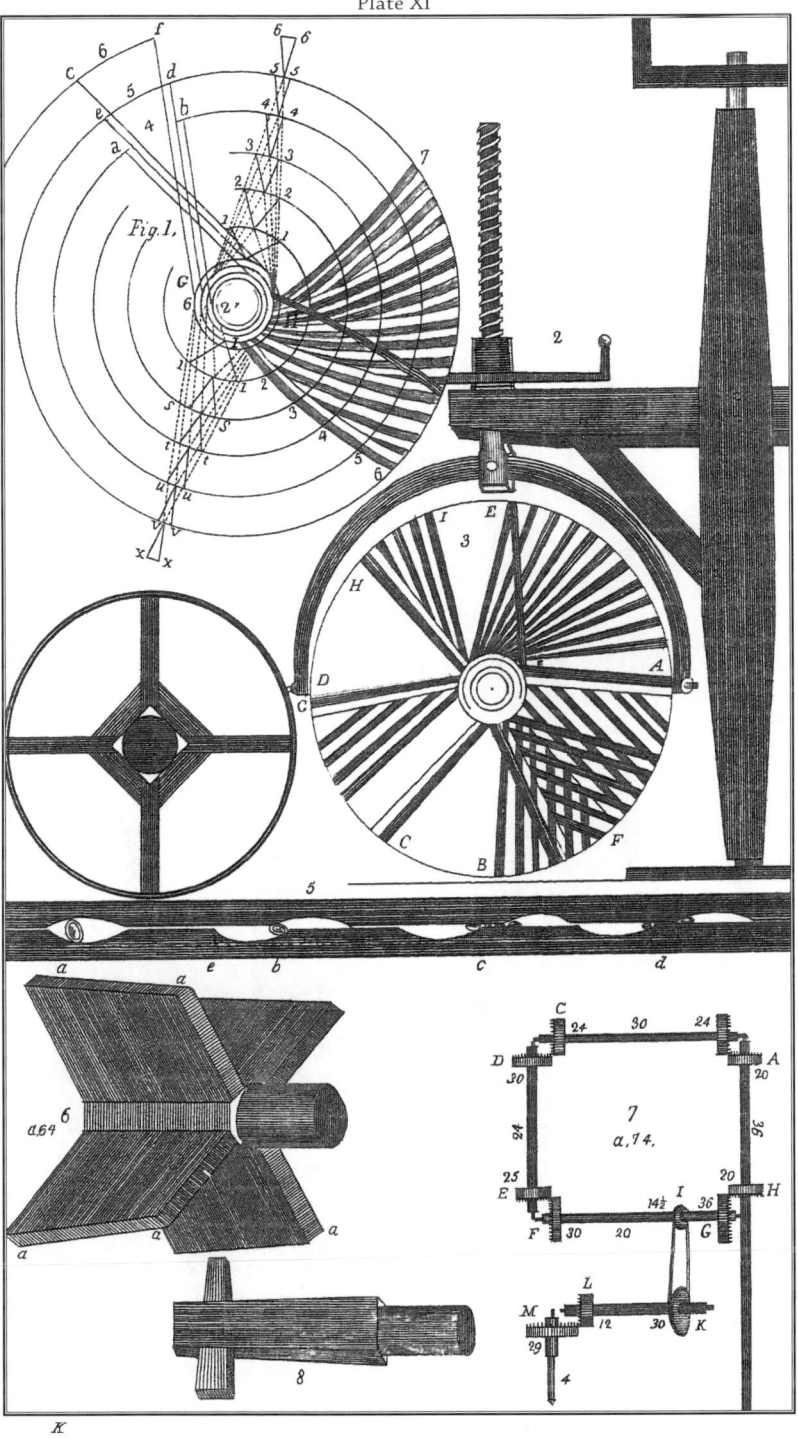

Plate XII

Plate XIII

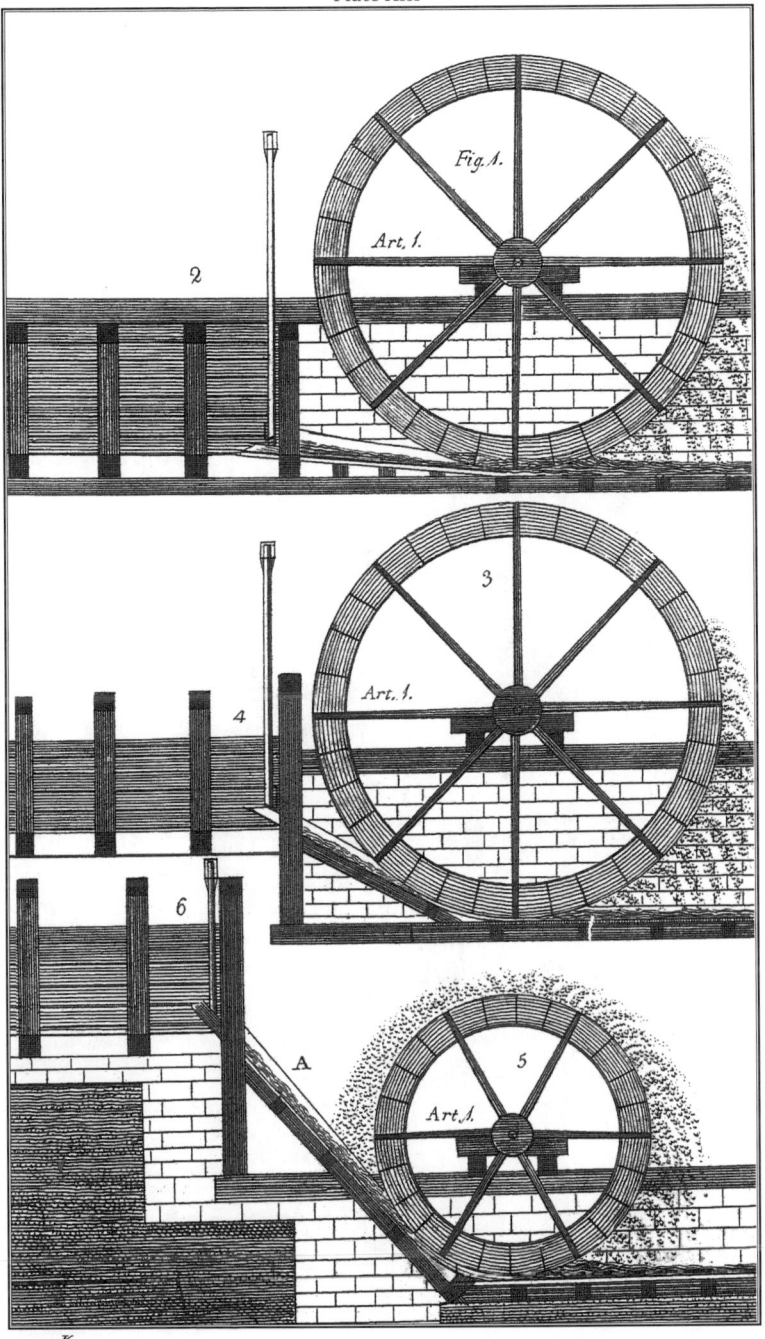

Plate XIV

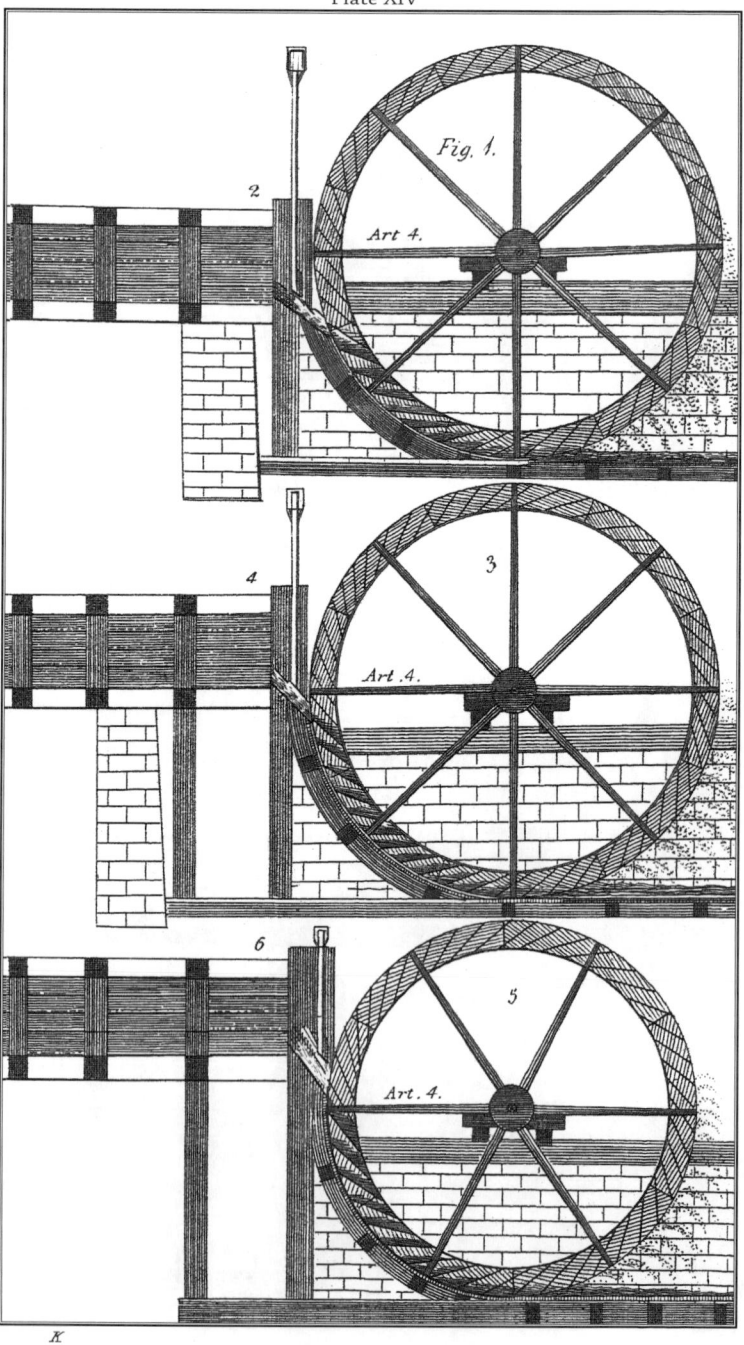

Plate XV

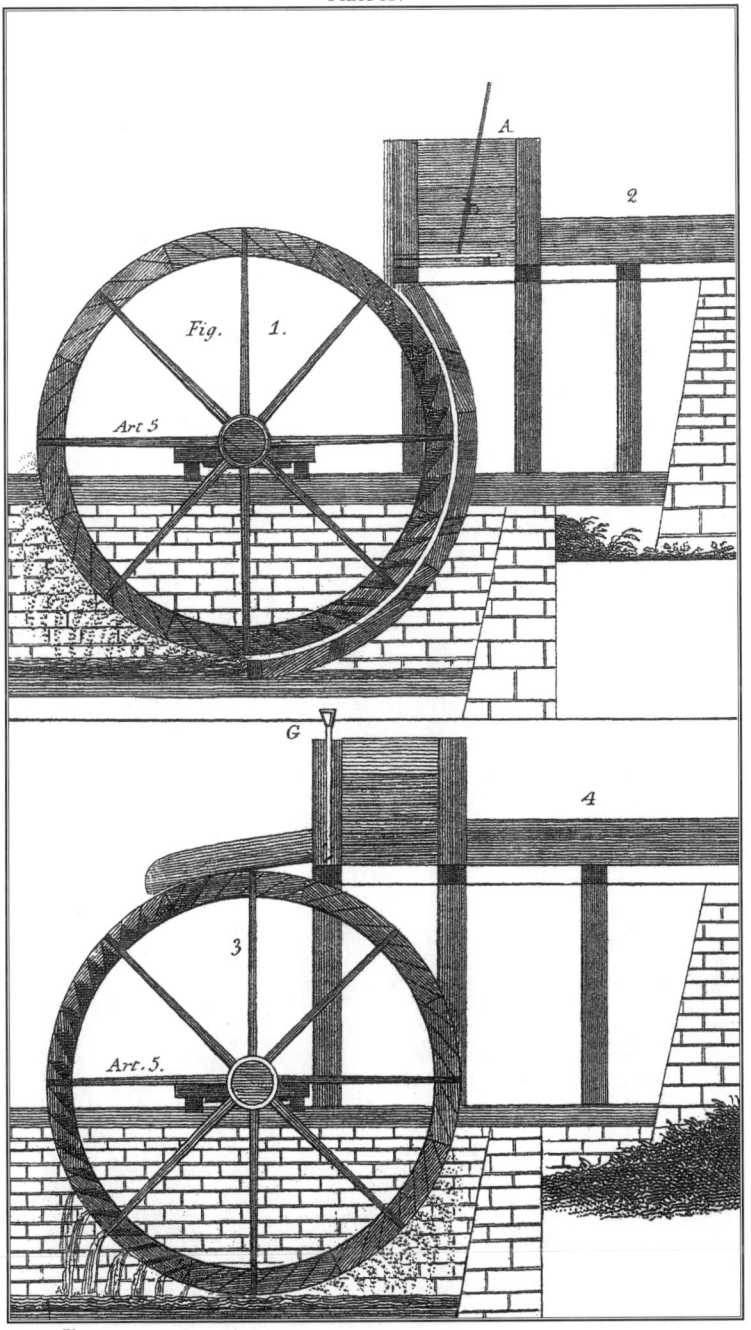

Plate XVI

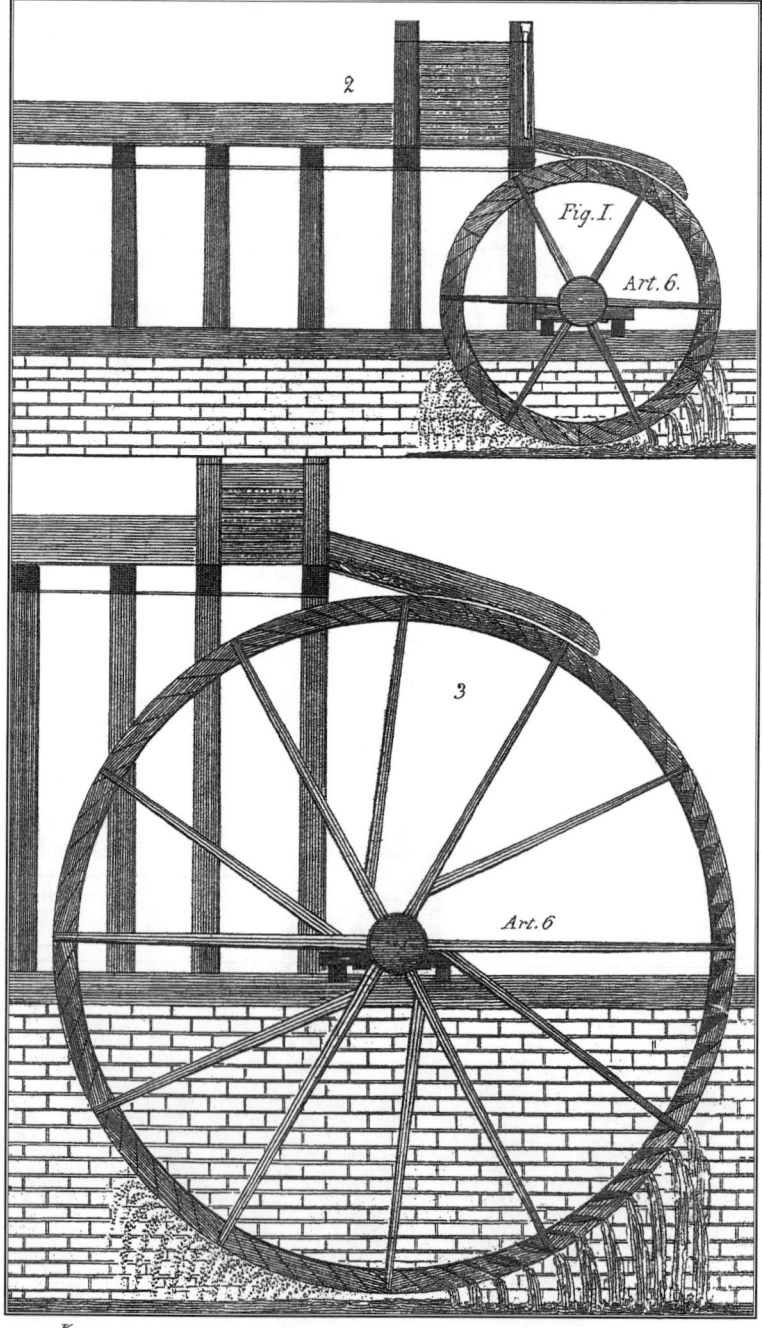

Plate XVII

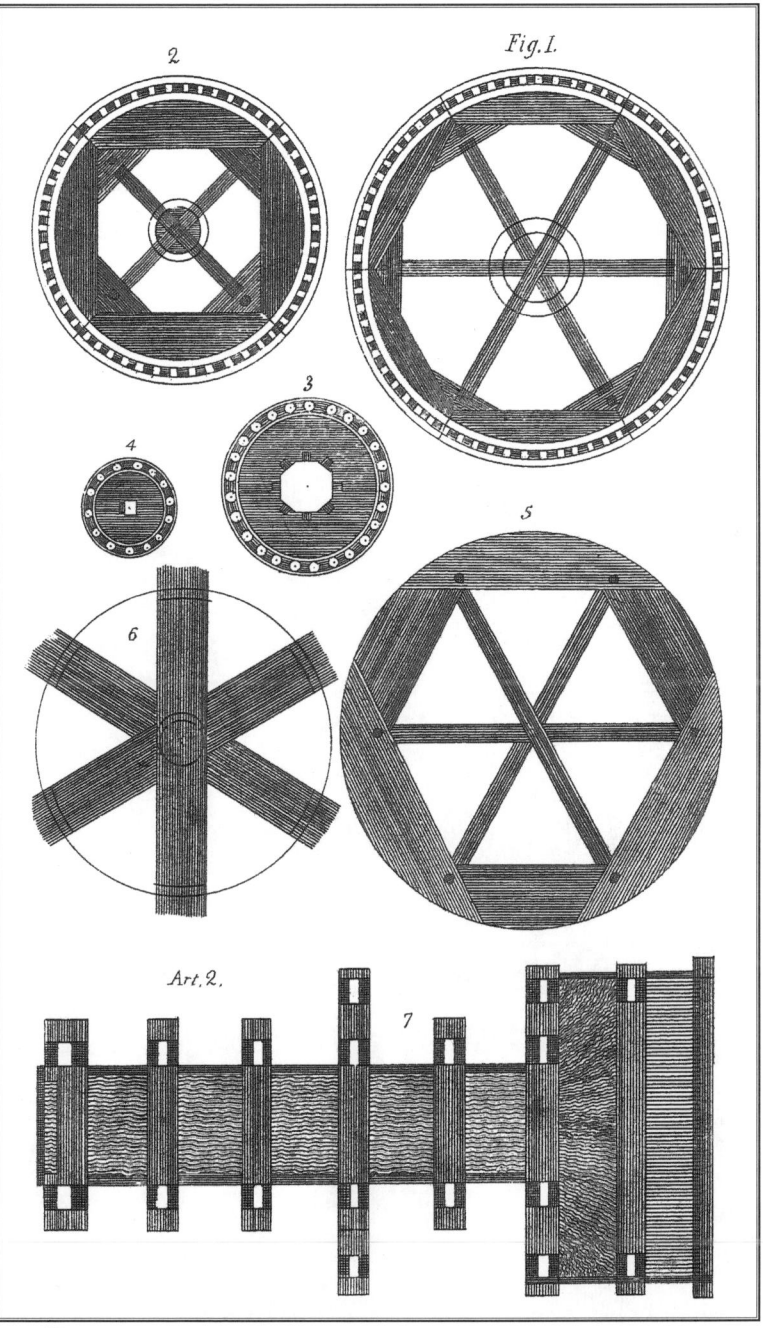

Plate XVIII

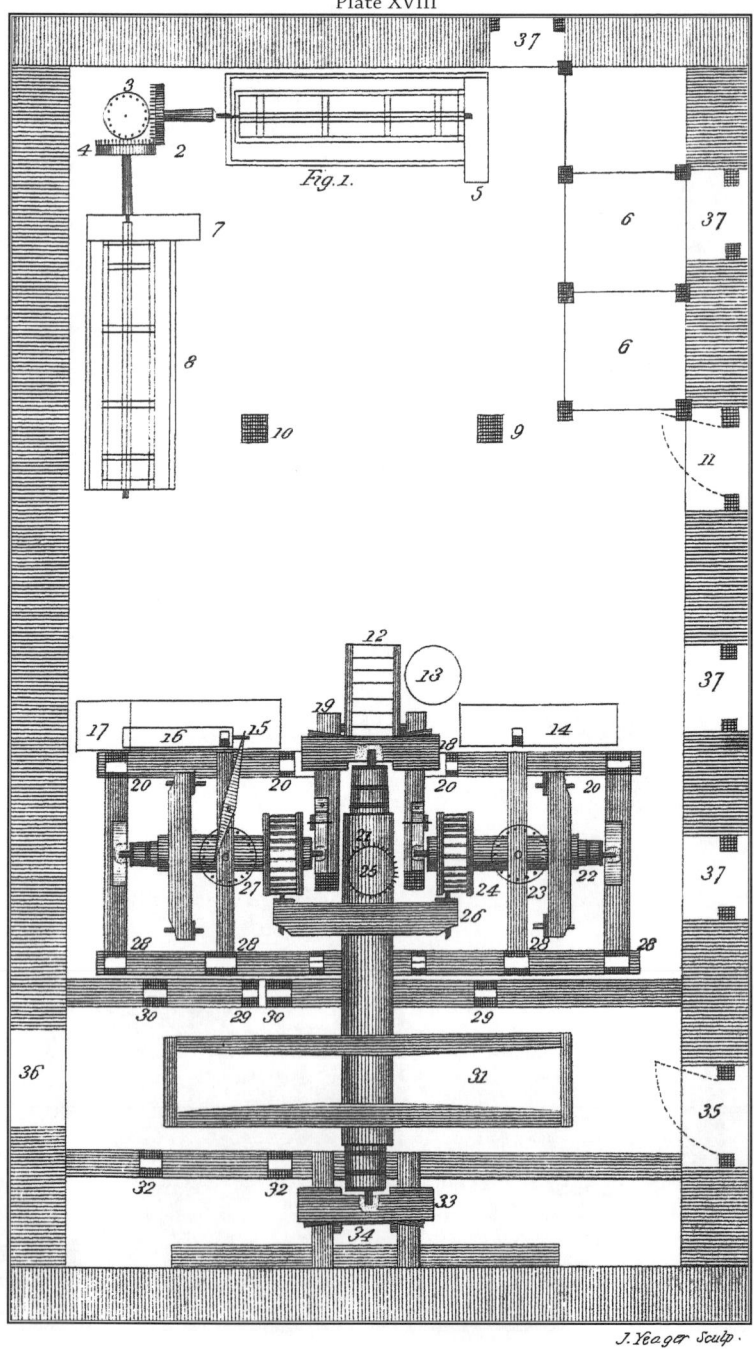

Fig. 1.

J. Yeager Sculp.

Plate XIX

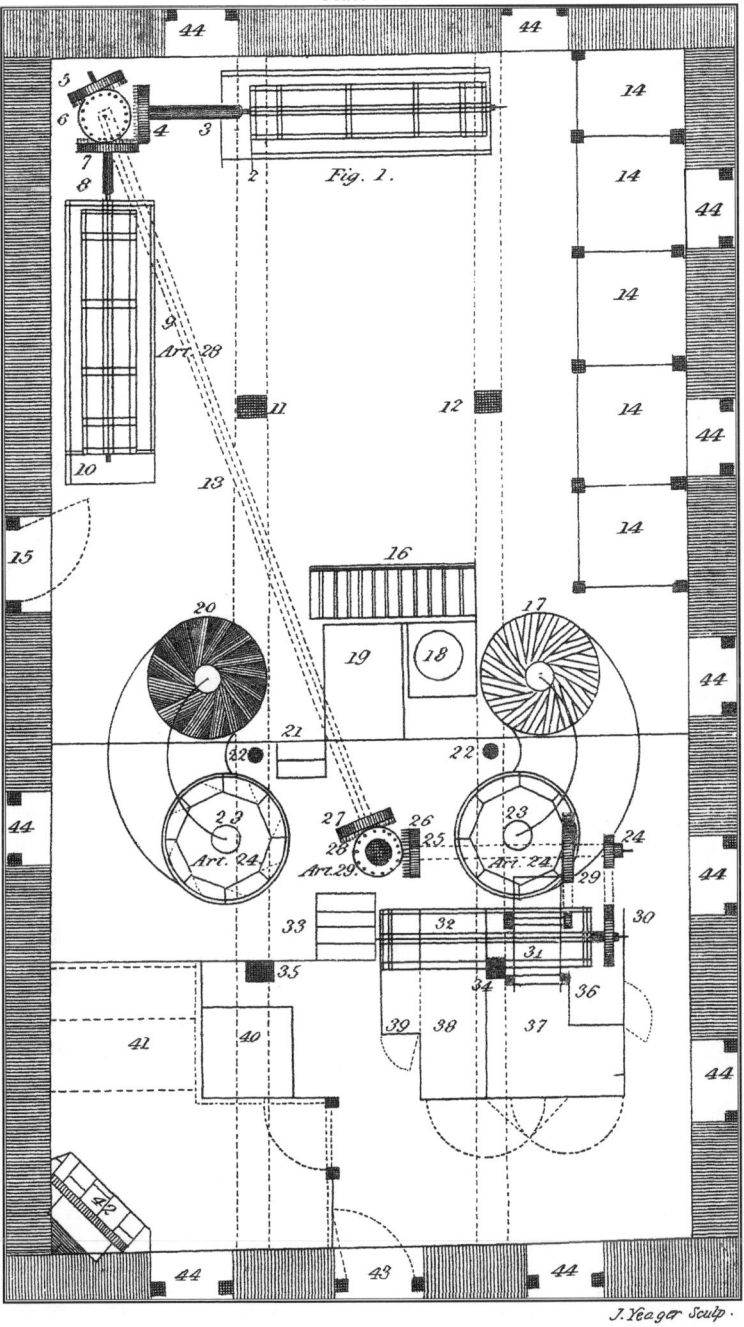

Fig. 1.

Plate XX

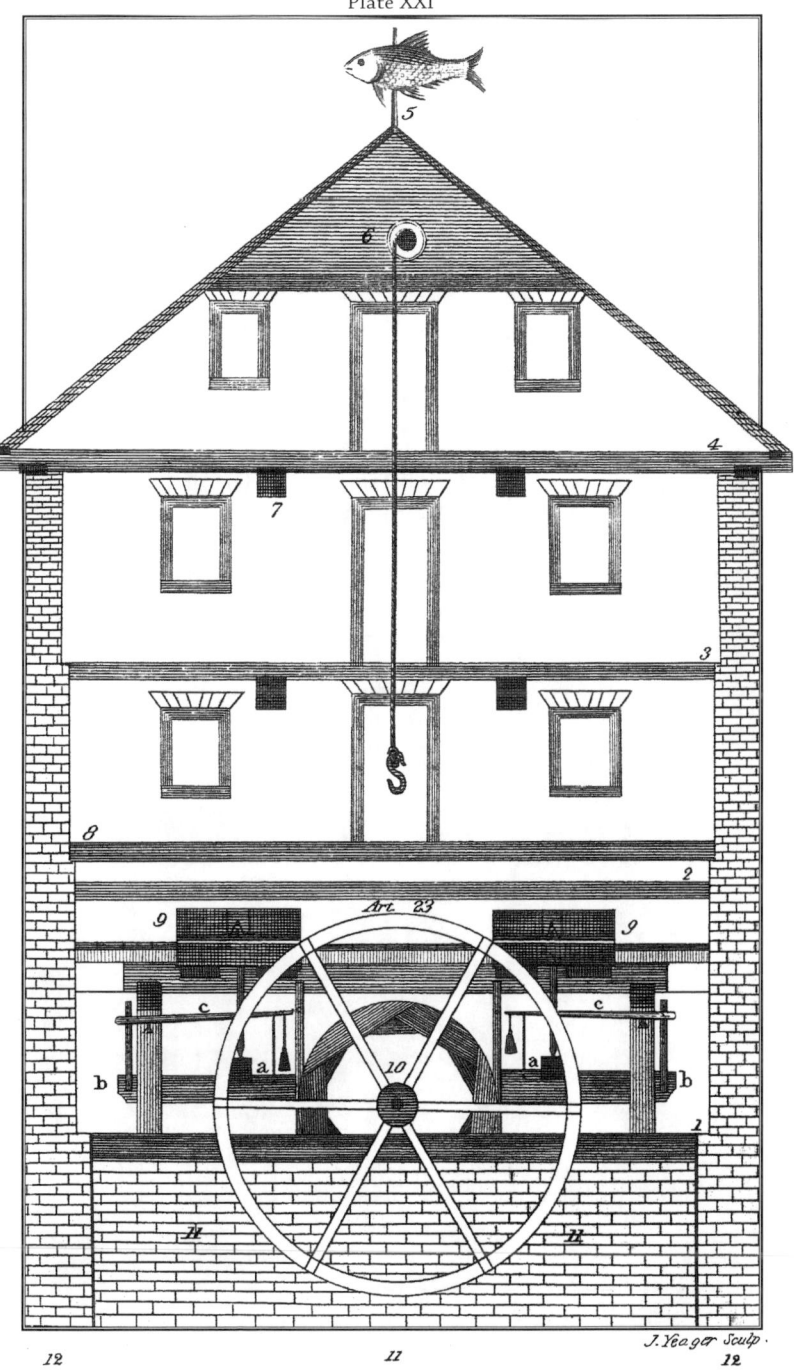

Plate XXI

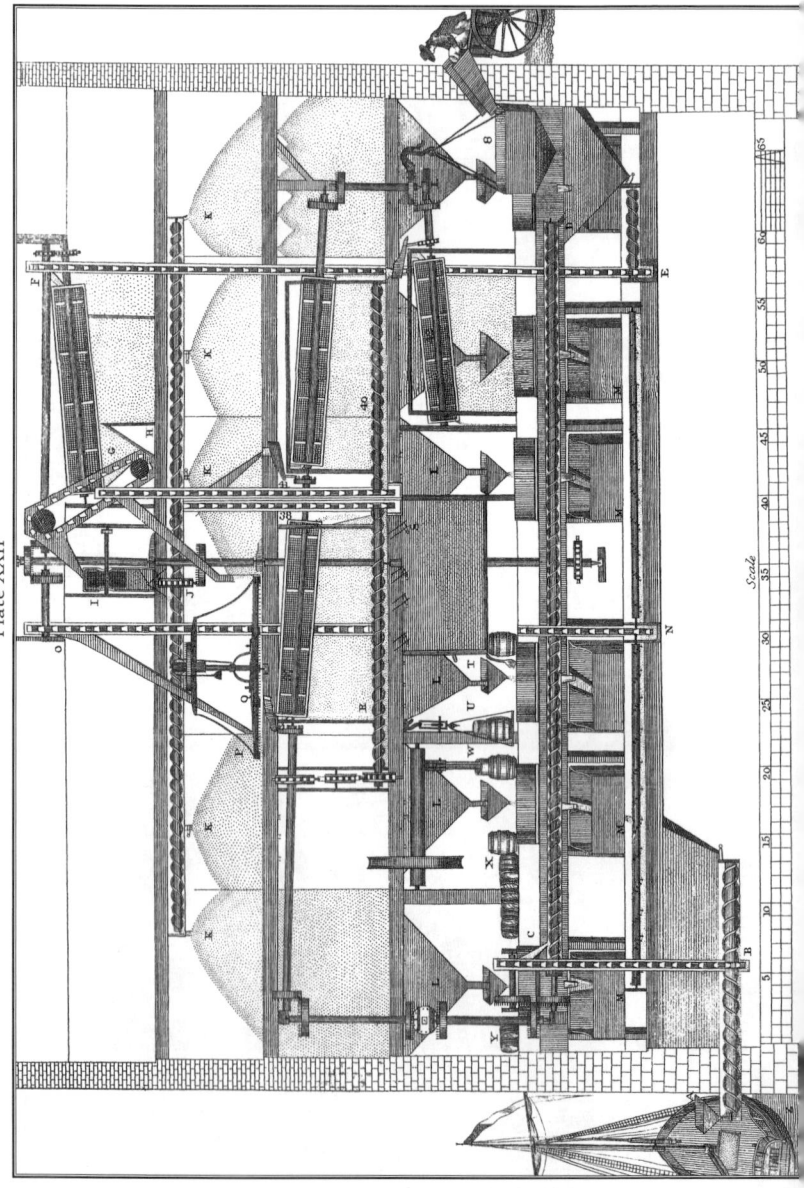

Plate XXII

Fig. 0.

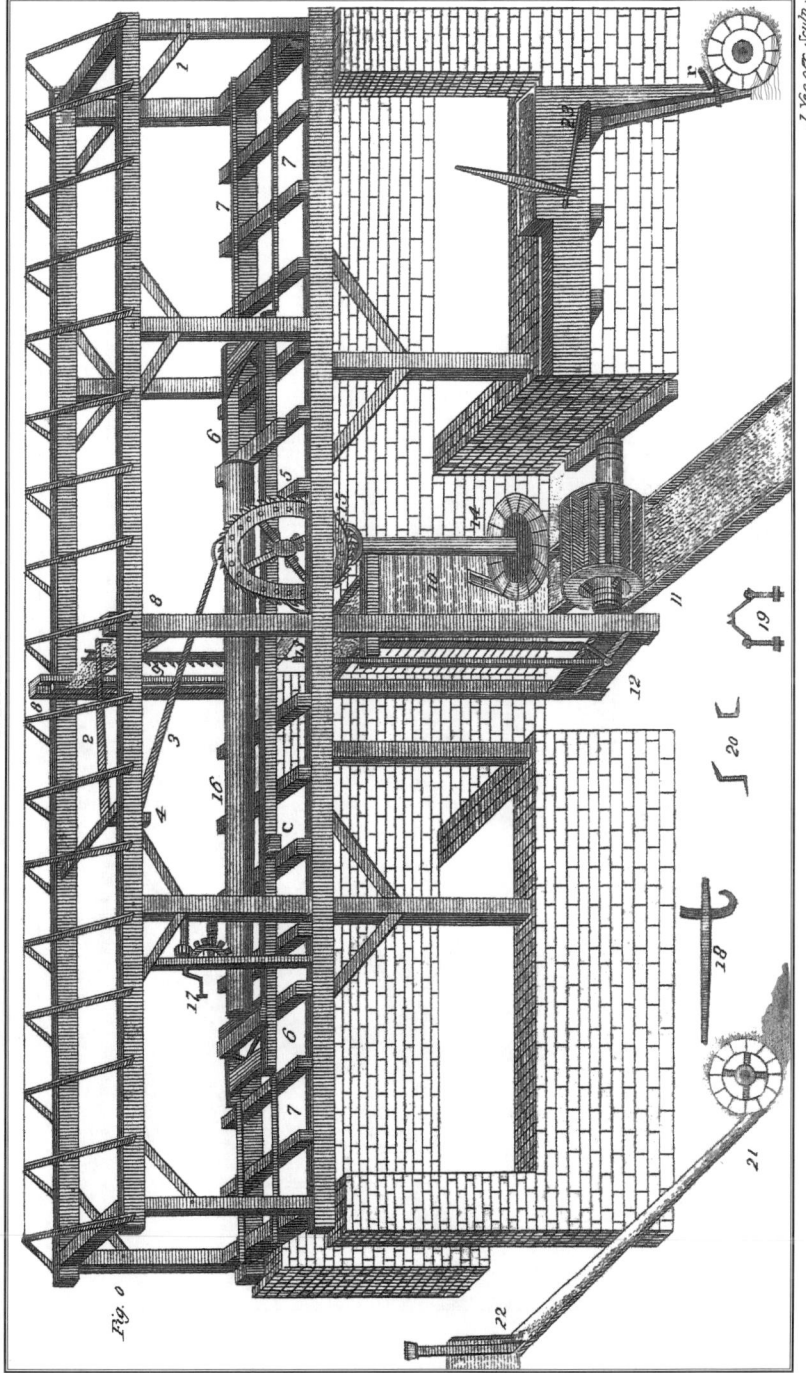

Plate XXIV

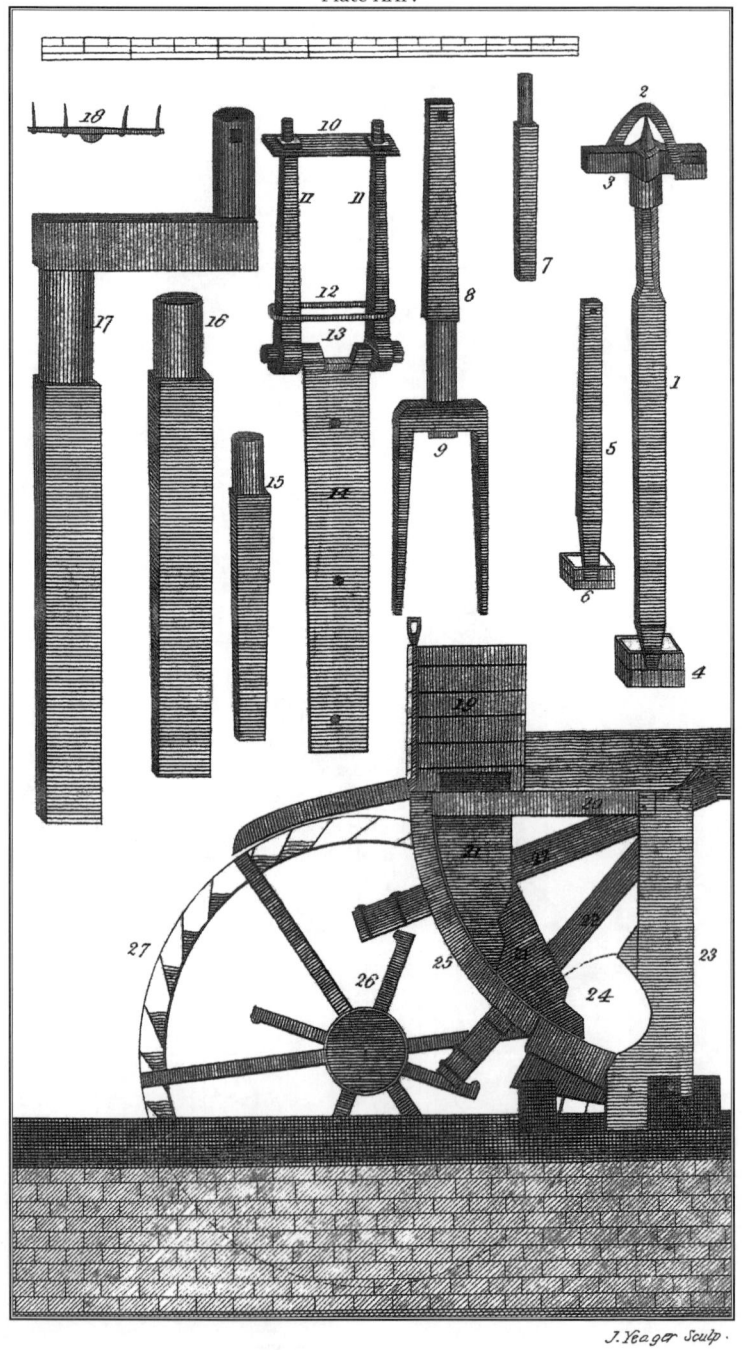

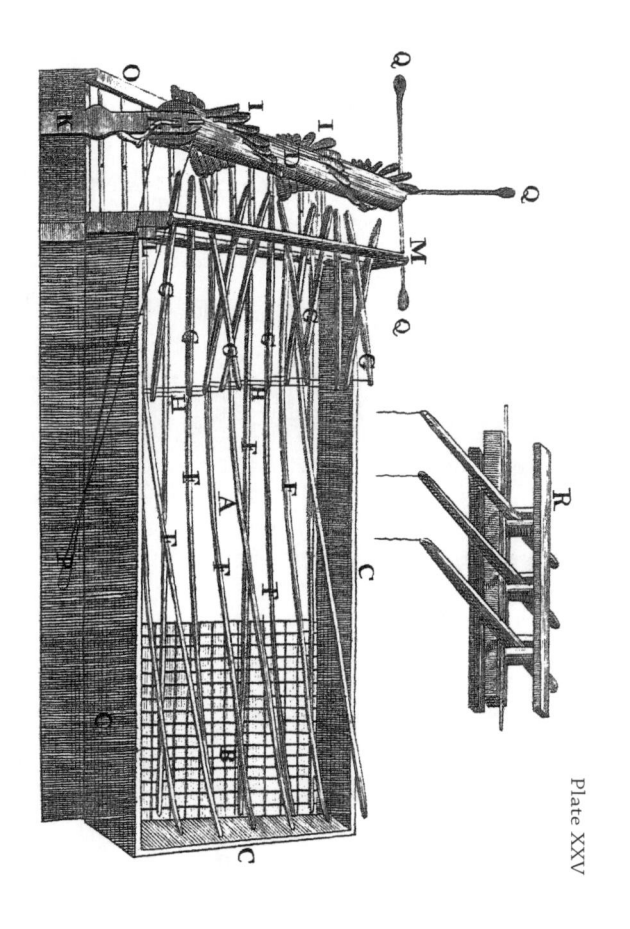

Plate XXV

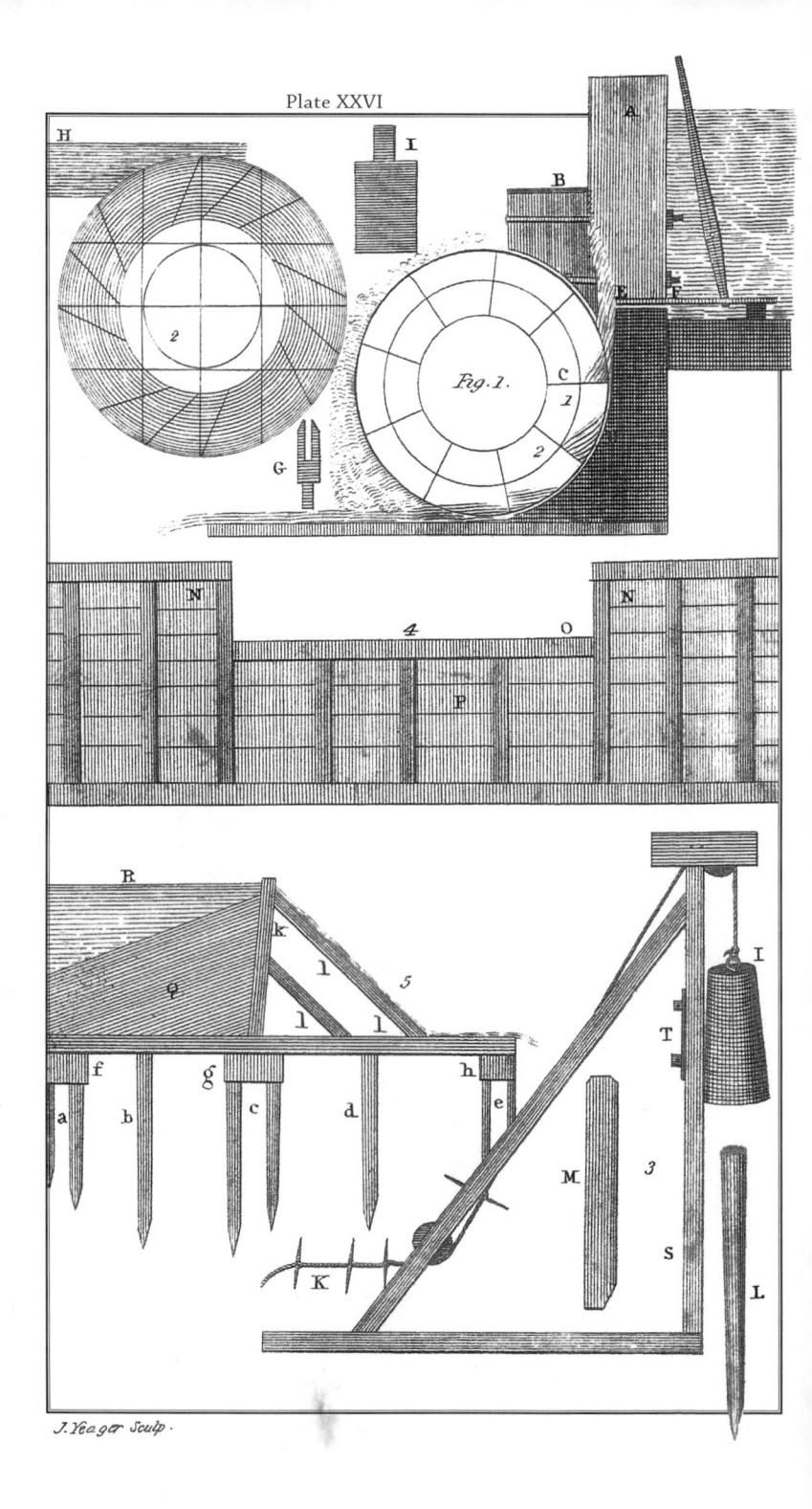

Plate XXVI

Plate XXVII

FLOUR MILL.

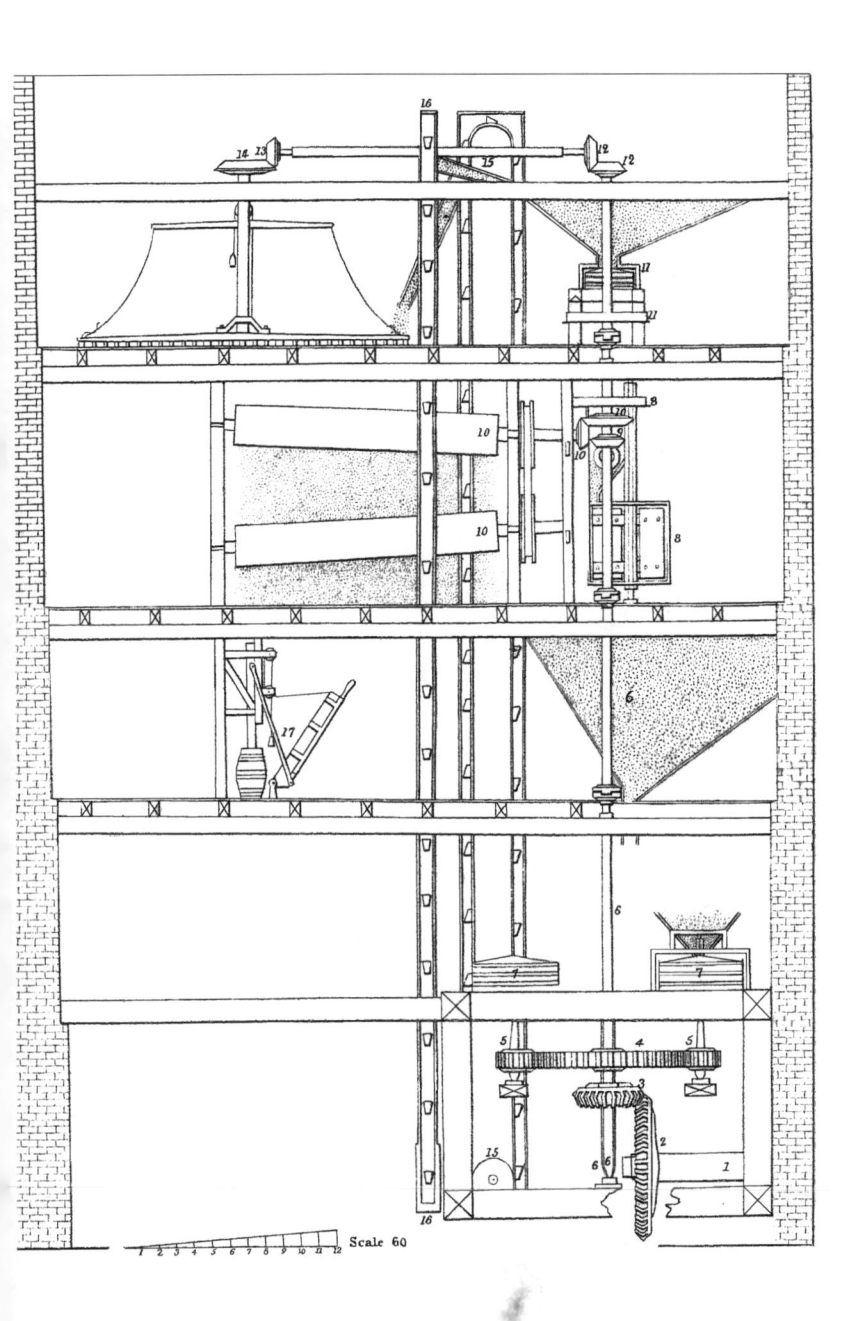

Plate XXVIII

FLOUR MILL.

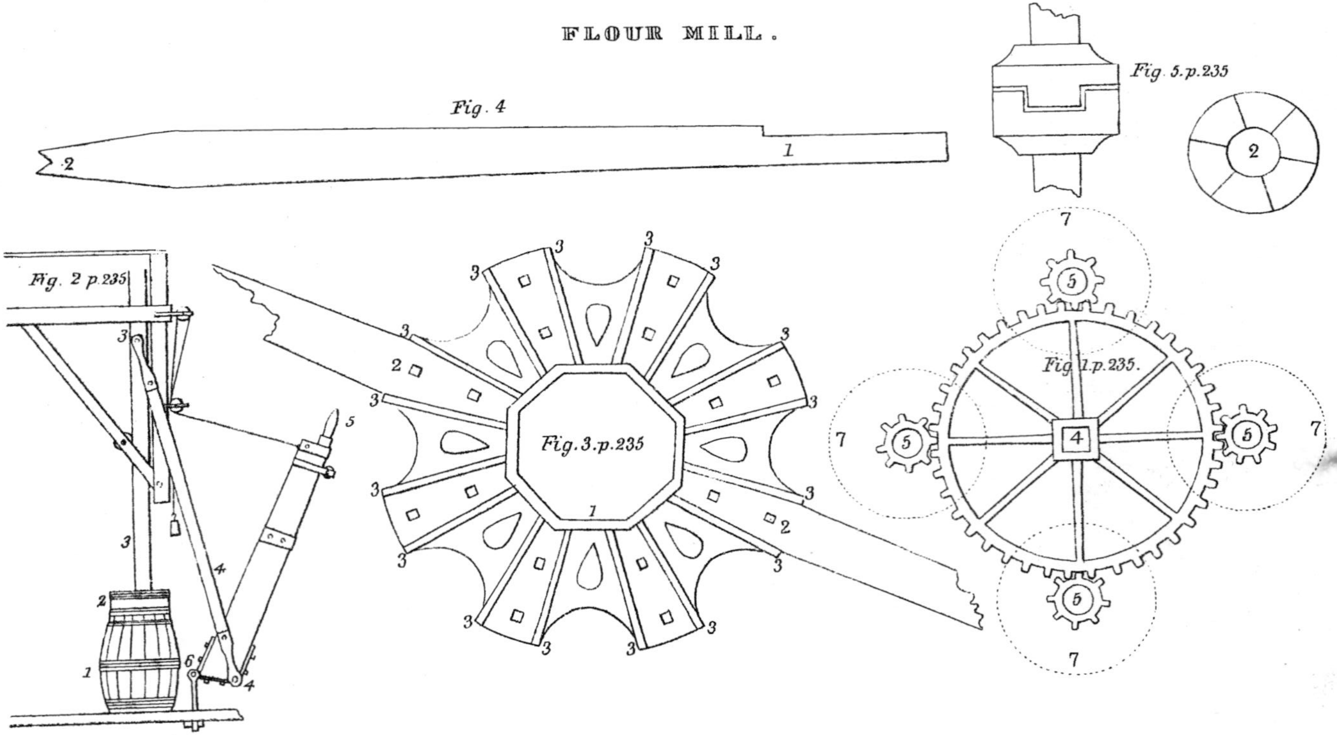